Advancements in Mechanical Power Transmission and Its Elements

Advancements in Mechanical Power Transmission and Its Elements

Guest Editors

Gang Li
Weidong Zhu
Yawen Wang
Jing Wei

MDPI

Basel • Beijing • Wuhan • Barcelona • Belgrade • Novi Sad • Cluj • Manchester

Guest Editors

Gang Li
Michael W. Hall School of Mechanical Engineering
Mississippi State University
Starkville
USA

Weidong Zhu
Department of Mechanical Engineering
University of Maryland Baltimore County
Baltimore
USA

Yawen Wang
Department of Mechanical and Aerospace Engineering
University of Texas at Arlington
Arlington
USA

Jing Wei
College of Mechanical and Vehicle Engineering
Chongqing University
Chongqing
China

Editorial Office
MDPI AG
Grosspeteranlage 5
4052 Basel, Switzerland

This is a reprint of the Special Issue, published open access by the journal *Machines* (ISSN 2075-1702), freely accessible at: https://www.mdpi.com/journal/machines/special_issues/3OW18A1VD8.

For citation purposes, cite each article independently as indicated on the article page online and as indicated below:

Lastname, A.A.; Lastname, B.B. Article Title. *Journal Name* **Year**, *Volume Number*, Page Range.

ISBN 978-3-7258-4259-9 (Hbk)
ISBN 978-3-7258-4260-5 (PDF)
https://doi.org/10.3390/books978-3-7258-4260-5

Cover image courtesy of Gang Li

Contents

Preface . . . . **vii**

Yuhui Liu, Liping Chen and Gang Li
Multi-Point Control for Face-Milled Spiral Bevel Gears with a Predesigned Fourth-Order Motion Curve
Reprinted from: *Machines* **2024**, *12*, 34, https://doi.org/10.3390/machines12010034 . . . . **1**

Jianmin Wen, Haoyu Yao, Hui Li and Bindi You
Research on the Hobbing Processing Method of Marine Beveloid Gear
Reprinted from: *Machines* **2024**, *12*, 35, https://doi.org/10.3390/machines12010035 . . . . **16**

Yuhui Liu, Liping Chen, Xian Mao and Duansen Shangguan
A Semi-Analytical Loaded Contact Model and Load Tooth Contact Analysis Approach of Ease-Off Spiral Bevel Gears
Reprinted from: *Machines* **2024**, *12*, 623, https://doi.org/10.3390/machines12090623 . . . . **38**

Fulin Liao, Xingyuan Zheng, Jianliang Huang and Weidong Zhu
Analysisof Nonlinear Dynamics of a Gear Transmission System Considering Effects of the Extended Tooth Contact
Reprinted from: *Machines* **2025**, *13*, 155, https://doi.org/10.3390/machines13020155 . . . . **57**

Daniel Vietze, Josef Pellkofer and Karsten Stahl
Study on the Potential of New Load-Carrying Capacity Descriptions for the Service Life Calculations of Gears
Reprinted from: *Machines* **2024**, *12*, 304, https://doi.org/10.3390/machines12050304 . . . . **83**

Yan He, Linzheng Ye and Yao Liu
Feature Extraction of a Planetary Gearbox Based on the KPCA Dual-Kernel Function Optimized by the Swarm Intelligent Fusion Algorithm
Reprinted from: *Machines* **2024**, *12*, 82, https://doi.org/10.3390/machines12010082 . . . . **99**

Lorenz Constien, Martin Weber, Josef Pellkofer and Karsten Stahl
Efficiency of Bevel and Hypoid Gears—Test Rig Development and Experimental Investigations
Reprinted from: *Machines* **2024**, *12*, 647, https://doi.org/10.3390/machines12090647 . . . . **119**

Di Cui, Yongjian Yu, Yujun Xue, Pengge Guo, Hongbiao Han and Haichao Cai
Effect of Roundness Error of the Grooves on the Inner Ring Runout of Angular Contact Ball Bearings
Reprinted from: *Machines* **2024**, *12*, 532, https://doi.org/10.3390/machines12080532 . . . . **134**

Taisu Liu, Yuan Liu, Peitong Liu and Xiaofei Du
Method of Dynamic Modeling and Robust Optimization for Chain Transmission Mechanism with Time-Varying Load Uncertainty
Reprinted from: *Machines* **2024**, *13*, 166, https://doi.org/10.3390/machines13020166 . . . . **154**

Yook Wah Liew, Owen Matthews, Dzung Viet Dao and Huaizhong Li
Power Transmission Mechanism and Tribological Performance of Modern Bicycle Drivetrains—A Review
Reprinted from: *Machines* **2025**, *13*, 66, https://doi.org/10.3390/machines13010066 . . . . **179**

Yongming Yang
Research on Active–Passive Training Control Strategies for Upper Limb Rehabilitation Robot
Reprinted from: *Machines* **2024**, *12*, 784, https://doi.org/10.3390/machines12110784 . . . . **202**

Preface

It is with great pleasure that we present this reprint, "Advancements in Mechanical Power Transmission and Its Elements," which compiles significant research contributions in the field of mechanical power transmission systems. In an era where efficiency, reliability, and performance optimization are paramount across industries, understanding the complex interactions between gears, bearings, chains, and other drivetrain components has become increasingly critical.

This reprint emerged from our recognition of the rapid technological advancements occurring in both theoretical modeling and experimental investigation of mechanical transmission systems. The growing integration of time-varying contact analysis, dynamics simulation, and robotics applications has created new opportunities and challenges that deserve focused scholarly attention. The aim is to provide a comprehensive resource that bridges fundamental mechanical engineering principles with cutting-edge research methodologies. The articles collected in this reprint explore innovative approaches to reducing vibration, minimizing power losses, enhancing load capacity, and improving the overall performance of power transmission elements. From advanced gear tooth contact analysis and smart bearing designs to AI-assisted drivetrain monitoring systems, the scope of the reprint encompasses both theoretical frameworks and practical applications that will shape the future of mechanical engineering.

We extend our sincere gratitude to all contributing authors whose dedication and expertise have made this reprint possible. Special thanks go to our reviewers for their meticulous evaluation and valuable feedback, and to the editorial team at MDPI for their continuous support throughout the publication process.

This reprint is addressed to researchers, engineers, industry professionals, and graduate students interested in the latest developments in mechanical power transmission. We hope it serves as both an informative reference point and an inspiration for future innovations in this dynamic field.

Gang Li, Weidong Zhu, Yawen Wang, and Jing Wei
Guest Editors

machines

MDPI

Article

Multi-Point Control for Face-Milled Spiral Bevel Gears with a Predesigned Fourth-Order Motion Curve

Yuhui Liu [1], Liping Chen [1] and Gang Li [2,*]

[1] School of Mechanical Science and Engineering, Huazhong University of Science and Technology, Wuhan 430074, China; yuhuiliu@hust.edu.cn (Y.L.); chenlp@hust.edu.cn (L.C.)
[2] Department of Mechanical Engineering, Mississippi State University, Starkville, MS 39762, USA
* Correspondence: gli@me.msstate.edu; Tel.: +1-662-325-5362

Abstract: This paper presents an ultimate motion methodology of a face-milling spiral bevel gear pair to synthesize the mating tooth surfaces with a predesigned fourth-order motion curve. The methodology is to control some contact points along the contact path in the process of tooth contact analysis via application of an extended local synthesis which permits some transmission errors rather than zero at the concerned contact point. The modified offset motion correction is selected to demonstrate the proposed methodology. Applied torque corresponding to an elastic approach of 0.00635 mm at the mean contact point is calculated and the loaded tooth contact analysis is performed. Numerical results show that the extended local synthesis can effectively control the transmission errors on the predesigned fourth-order motion curve at arbitrarily predesigned contact points along the contact path of the spiral bevel gear pair. The tooth contact pattern for the actual tooth pair is scattered into three segments since the rotational motion of the driven gear at any instant angular position is dependent on the tooth pair with the least transmission error among the three adjacent tooth pairs. The actual tooth contact patterns of the spiral bevel gear pair become continuous when meshing tooth surfaces are elastically deformed.

Keywords: fourth-order motion curve; loaded tooth contact analysis; transmission error; tooth contact analysis; spiral bevel gears

1. Introduction

Spiral bevel gears are widely used in aviation, helicopters, navigation systems, automobiles, and renewable energy devices. Similar to other types of gears, operation performances, e.g., vibration, efficiency, and reliability [1], of spiral bevel gear pairs mainly depend on their contact patterns and motion curves (MCs). Much research work has been dedicated to the synthesis and manufacturing of face-milling spiral bevel gears [2–4]. Litvin [5] presented a local synthesis method to control its contact patterns actively with a parabolic MC. Litvin [6] contrasted the pinion tooth surfaces produced by head cutters with circular and straight blades, respectively. After obtaining the stabilized contact pattern with a limited magnitude of MC, they proceeded to investigate tooth root bending stress and the contact pressure by using loaded tooth contact analysis (LTCA). Fan [7] discussed distinctive characteristics of the face-milling and the face-hobbing processes of spiral bevel gears and discounted mechanisms of tooth contact analysis (TCA) that are suitable for free-form numerical controlled machines and LTCA, considering tooth deformations and the housing deflection. Liu [8] presented a semi-analytical LTCA method for spiral bevel gears which uses analytical formulas and finite element analysis correction and involves Tredgold's approximation, an optimization model, and a contact judgment strategy. Simon [9] conducted a numerical analysis to investigate the effects of pinion misalignments and tooth spacing errors on contact patterns and motion relationships of the spiral bevel gear. Sheveleva [10] introduced the concept of intermediate tangent grids for the pinion and the gear in a spiral bevel gear pair. Meshing performances can be observed by measuring the distance between the gear and the pinion as the gear pair rotated around the respective axes. A face-milling method for hypoid gears with a fourth-order transmission error model was developed to reduce error sensitivity due to misalignments [11]. Mu [12] proposed a tooth surface modification method to prevent tooth edge contact for high-contact-ratio (HCR) spiral bevel gears under misalignments or heavy loads based on cutter blade profile corrections. Alves [13] proposed a method to calculate bending displacements of a spiral bevel gear pair combining a finite element model with interpolation techniques to improve the calculation efficiency. This helps for a more efficient analysis of the bending behavior of spiral bevel gears under various operating conditions. Tsai [14] focused on designing the machine-tool settings for a four-axis milling machine tool for spiral bevel gears to improve contact performances using a predesigned motion curve. Ma [15] developed a nonlinear dynamic model to optimize operation performances of four-point contact ball bearings in a three-point contact state considering lubrication traction and dynamic characteristics of the bearing assemblies. Xiang [16] conducted an analysis of geometric errors of a spiral bevel gear pair in motion axes and their impacts on gear meshing performances. They focused on six-axis CNC grinding machines and used forward and inverse kinematics modeling techniques based on screw theory to create models for predicting and compensating for volumetric errors, ultimately improving the accuracy and quality of gear grinding processes. Alvarez [17] developed the surface topology model for predicting the roughness of spiral bevel gear surfaces, which can help manufacturers better control product quality and reduce trial and error costs, and verified the accuracy of the surface topology model to achieve the size and surface quality of gear production. Mu [18] presented an approach for designing high-contact-ratio spiral bevel gears with a focus on minimizing higher-order transmission errors. This method improved meshing performances and overall functionality of spiral bevel gears by using a concept of function-oriented design to optimize the gear design process. The authors of [19] proposed a mathematical model for a logarithmic spiral bevel gear drive. This type of gear drive offers several advantages over conventional spiral bevel gears, including higher contact strength and the ability to separate their shaft angle. These characteristics make logarithmic spiral bevel gear drives highly promising for a wide range of applications, providing improved performance and functionality compared to traditional spiral bevel gears. Mu [20] proposed a higher-order tooth surface modification method aimed at reducing loaded transmission errors (LTEs) and meshing impacts for improving dynamic characteristics of an HCR spiral bevel gear pair. Chen [21] presented an integration of various factors to achieve a global tooth surface control method of spiral bevel gears to solve an ease-off surface equation and optimize producing parameters. Yang [22] introduced a taper design method for face-milled hypoid and spiral bevel gears within the completing process. This approach ensures that there are proportional modifications in both the tooth space width and tooth thickness in relation to the cone distance.

Much of the research work is concerned with the parabolic MC which can absorb tooth meshing impacts at the changeover points to some extent between adjacent tooth pairs caused by linear MC due to misalignments. However, the discontinuity of the first derivative to the parabolic MC at the changeover points still causes some tooth impact in the transition of mesh. Stadtfeld and Gaiser [23] noticed this and proposed a combined motion curve (CMC), composed of a parabolic and a fourth-order polynomial, named as ultimate motion graph (UMG), which narrows the gap of the first derivative discontinuity at the changeover point. Stadtfeld demonstrated via experimentation that the ground hypoid gears produced on a free-form CNC hypoid generator with CMC excelled the ground hypoid gears with a parabolic MC to a great extent. Fan [24] introduced a CNC six-axis generator developed by the Gleason Works. He mentioned the concept of UMG which could be implemented via instantaneous machine settings including modified roll that were expressed by the higher-order polynomial functions up to the sixth-order with respect to the cradle rotation angle. So far, the UMG enables the tooth flank crowning with the greatest flexibility. Li et al. [25] developed a gear form-grinding method with a predesigned fourth-order transmission error. Wang and Fong [26] developed a method to form a fourth-order MC which was a simplified form of CMC and used the radial motion correction to control the initial bias of the contact path and the MC, respectively. They found that the radial motion correction could be also used on the CNC machine to increase its adjustability [27]. They also used the same method along with a variant cradle rotation angle to generate the tooth surfaces of a hypoid gear pair with the fourth-order polynomial MC [28]. Since they chose several points on the predesigned contact path and the corresponding number of points in the vicinity of the predesigned fourth-order polynomial MC, the two machine settings mentioned above at these points were calculated one after another and fitted with sixth-order polynomials. In addition, they thought that the MC and the contact path were decoupled by taking the cradle rotation angle and the instantaneous radial setting as controlling parameters. Since the fluctuation degree of MC is relatively small, different MCs could correspond to almost the same contact path. In other words, the change in the contact path of the spiral bevel gear pair occurs at a slower rate than adjustments to machine settings or roll motion.

In this paper, a predesigned fourth-order CMC of the spiral bevel gear pair is developed as a design input to improve its meshing performances and determine the instantaneous offset corrections. The work presents a general method to accurately control transmission errors of the spiral bevel gear pair with an arbitrarily predesigned CMC. Numerical TCA and LTCA results show that the proposed method is very efficient in analyzing meshing performances of the spiral bevel gear pair.

2. Predesigned CMC of a Spiral Bevel Gear Pair

2.1. Basic MC

It has already been recognized by researchers that the main source of vibration and noise is MC. Reducing MC and controlling the shape of an MC function curve can improve dynamic performance and reduce vibration of a gear system. Conventionally, a general MC function is defined as

$$\Delta\phi_2 = (\phi_2 - \phi_{20}) - a_1(\phi_1 - \phi_{10}) \tag{1}$$

where ϕ_1 and ϕ_2 are actual rotation angles of the pinion and the gear, respectively; a_1 is the speed ratio of the spiral bevel gear, i.e., Z_1/Z_2, in which Z_1 and Z_2 are the tooth number of the pinion and the gear, respectively; and ϕ_{10} and ϕ_{20} are theoretical rotation angles of the pinion and the gear, respectively.

When a spiral bevel gear pair is in meshing, the difference between the gear's actual rotation angle and its mean rotation angle is defined as the MC of the gear pair, and the second-order and the fourth-order polynomial MCs can be represented as

$$\Delta\phi_2{}^{(2)} = (\phi_2 - \phi_{20}) - a_1(\phi_1 - \phi_{10}) = a_2(\phi_1 - \phi_{10})^2 \tag{2}$$

$$\Delta\phi_2{}^{(4)} = (\phi_2 - \phi_{20}) - a_1(\phi_1 - \phi_{10}) = a_2(\phi_1 - \phi_{10})^2 + a_3(\phi_1 - \phi_{10})^3 + a_4(\phi_1 - \phi_{10})^4 \tag{3}$$

respectively, where $a_i(i = 2, 3, 4)$ is the order number of the MC determined by kinematic requirements, and superscripts (2) and (4) represent the second-order and the fourth-order MC functions of the spiral bevel gear pair, respectively.

2.2. Formation of CMC

For a specific spiral bevel gear pair, we proposed a method to determine its CMC. The CMC formation process of the spiral bevel gear pair is as follows:

Step 1: According to the pinion tooth number and kinematic errors of the spiral bevel gear pair, 10 points $A_i(i = 1, 2, \cdots, 10)$, denoted by " $\circ$ " as in Figure 1a, are chosen to control the shape of the fourth-order MC. Using the least squares method, the values of $a_i(i = 2, 3, 4)$ can be determined.

Step 2: Combining two intersections of the above fourth-order polynomial MC with the abscissa, which are symmetrical about the coordinate origin and another point on the ordinate, i.e., $B_i (i = 1, 2, 3)$, denoted by " • " as in Figure 1b, the second-order polynomial MC can be represented as

$$\Delta\phi_2{}^{(2)} = (\phi_2 - \phi_{20}) - a_1(\phi_1 - \phi_{10}) = -\frac{b_2}{b_1{}^2}(\phi_1 - \phi_{10})^2 + b_2 \tag{4}$$

where b_1, or b_3, and b_2 are the coordinates of $B_i (i = 1, 2, 3)$, i.e., $B_1(-b_1, 0)$, $B_2(0, b_2)$ and $B_3(b_3, 0)$, respectively.

Step 3: The upper part of the fourth-order polynomial MC and the lower part of the second-order polynomial MC, as in the thick line in Figure 1, comprises a CMC, which can be represented as

$$\Delta\phi_2 = \begin{cases} -\frac{b_2}{b_1{}^2}(\phi_1 - \phi_{10})^2 + b_2 & (\phi_1 - \phi_{10}) < -b_1 \text{ or } (\phi_1 - \phi_{10}) > b_3 \\ a_2(\phi_1 - \phi_{10})^2 + a_3(\phi_1 - \phi_{10})^3 + a_4(\phi_1 - \phi_{10})^4 & -b_1 \le (\phi_1 - \phi_{10}) \le b_3 \end{cases} \tag{5}$$

Step 4: Moving the combined curve in the thick line downward until the maximum point coincides with the abscissa, the final combined function of transmission errors (FTEs), i.e., the CMC, as in Figure 1b, is formed, in which the minimum point M denoted by " • " represents the mean contact point. The CMC function of the spiral bevel gear pair can be represented as

$$\Delta\phi_2 = \begin{cases} -\frac{b_2}{b_1{}^2}(\phi_1 - \phi_{10})^2 + b_2 - \Delta\phi_{2\,\max} & (\phi_1 - \phi_{10}) < -b_1 \text{ or } (\phi_1 - \phi_{10}) > b_3 \\ a_2(\phi_1 - \phi_{10})^2 + a_3(\phi_1 - \phi_{10})^3 + a_4(\phi_1 - \phi_{10})^4 - \Delta\phi_{2\,\max} & -b_1 \le (\phi_1 - \phi_{10}) \le b_3 \end{cases} \tag{6}$$

where $\Delta\phi_{2\,\max}$ is the ordinate value of the maximum point of Equation (5).

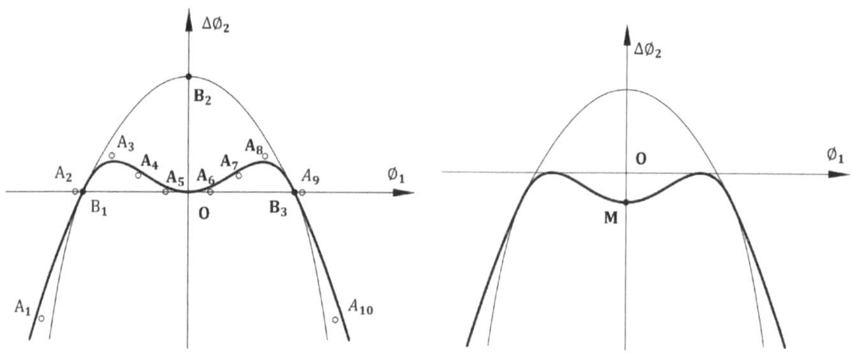

Figure 1. Formation of CMC of the spiral bevel gear pair. (**a**) CMC up to the fourth order in process; (**b**) CMC in the end.

3. Derivation of Tooth Surfaces of the Work Gear

3.1. Coordinate Systems of the Face-Milling Machine Tool

A common mathematical model for the generation of tooth surfaces of the pinion and the gear is used in Figure 2. There are six coordinate systems. A coordinate system $S_i(X_i, Y_i, Z_i)$ is rigidly attached to the work gear. A movable coordinate system $S_{ci}(X_{ci}, Y_{ci}, Z_{ci})$ is used to describe the angular position of the cradle. A coordinate system $S_{mi}(X_{mi}, Y_{mi}, Z_{mi})$ is rigidly connected to the cutting machine tool. The cradle and the work gear perform related rotations around the Z_{mi}-axis and X_{bi}-axis, respectively. A coordinate system $S_{hi}(X_{hi}, Y_{hi}, Z_{hi})$ is rigidly attached to the head cutter of the work gear. Coordinate systems $S_{ai}(X_{ai}, Y_{ai}, Z_{ai})$ and $S_{bi}(X_{bi}, Y_{bi}, Z_{bi})$ are used for assisting the installment of the work gear. Angles ψ_{ci} and ψ_i are rotation angles of the cradle and the work gear, respectively. There are six potential auxiliary motions that can be used to modify the tooth flank: the sliding base X_{Bi}, the radial setting S_{ri}, the blank offset E_{mi}, the installment angle q_i, the machine center to back X_{Di}, and the machine root angle γ_{mi}.

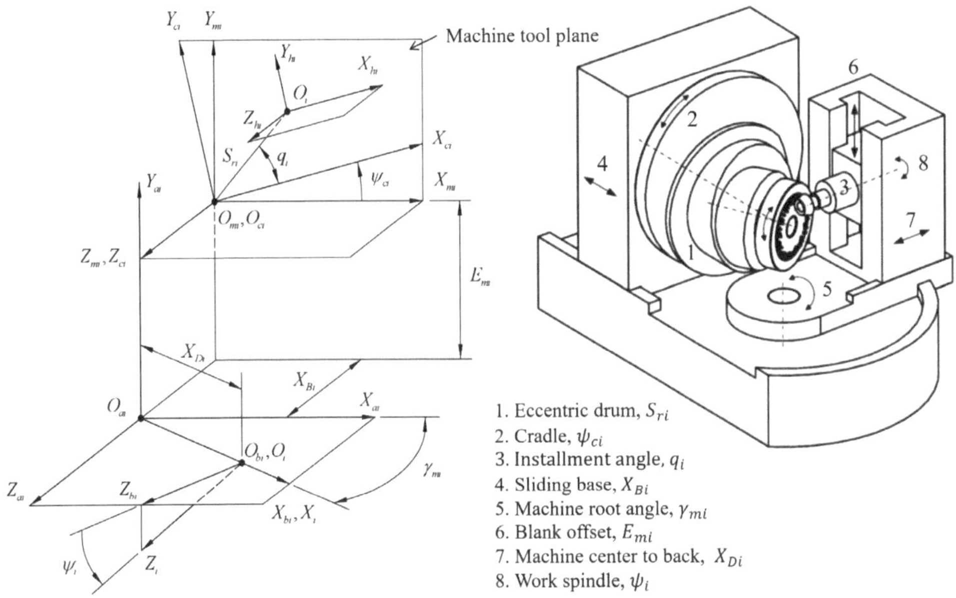

Figure 2. Coordinate systems for tooth surface generation of the gear.

3.2. Coordinate Systems of the Head Cutter

The head cutter blade usually composes a straight edge and a circular arc in Figure 3. The position vectors of the head cutter in $S_{hi}(X_{hi}, Y_{hi}, Z_{hi})$ can be represented as

$$r_{hi}{}^{(a)}(u_{hi}, \theta_{hi}) = \begin{bmatrix} x_{hi}{}^{(a)} & y_{hi}{}^{(a)} & z_{hi}{}^{(a)} \end{bmatrix}^T = \begin{bmatrix} (R_{hi} \pm u_{hi} \sin \alpha_{hi}) \cos \theta_{hi} \\ (R_{hi} \pm u_{hi} \sin \alpha_{hi}) \sin \theta_{hi} \\ -u_{hi} \cos \alpha_{hi} \end{bmatrix} \tag{7}$$

$$r_{hi}{}^{(b)}(u_{hi}, \theta_{hi}) = \begin{bmatrix} x_{hi}{}^{(b)} & y_{hi}{}^{(b)} & z_{hi}{}^{(b)} \end{bmatrix}^T = \begin{bmatrix} (T_{hi} \pm \rho_{hi} \sin \lambda_{hi}) \cos \theta_{hi} \\ (T_{hi} \pm \rho_{hi} \sin \lambda_{hi}) \sin \theta_{hi} \\ -\rho_{hi}(1 - \cos \lambda_{hi}) \end{bmatrix} \tag{8}$$

where α_{hi} and R_{hi} are the modified blade angle and head cutter point radius, respectively; u_{hi} and θ_{hi} are tooth surface parameters, where u_{hi} is defined in the range of $u_{hi} \geq \rho_{hi}(1 \mp \sin \alpha_{hi})/\cos \alpha_{hi}$; T_{hi} and ρ_{hi} are the head cutter fillet center radius and the fillet radius, respectively; λ_{hi} and θ_{hi} are tooth fillet parameters, where λ_{hi} is meaningful when it is in the range $0 \leq \lambda_{hi} \leq \frac{\pi}{2} - \alpha_{hi}$. R_{hi} and T_{hi} are related by $T_{hi} = R_{hi} \mp \rho_{hi}(1 - \sin \alpha_{hi})/\cos \alpha_{hi}$. The upper sign in " $\pm$ " and " $\mp$ " should be regarded for the concave side or outer blade, and the lower sign for the convex side or inner blade. Superscripts (a) and (b), (c) represent the straight edge and the fillet arc of the head cutter blade, respectively.

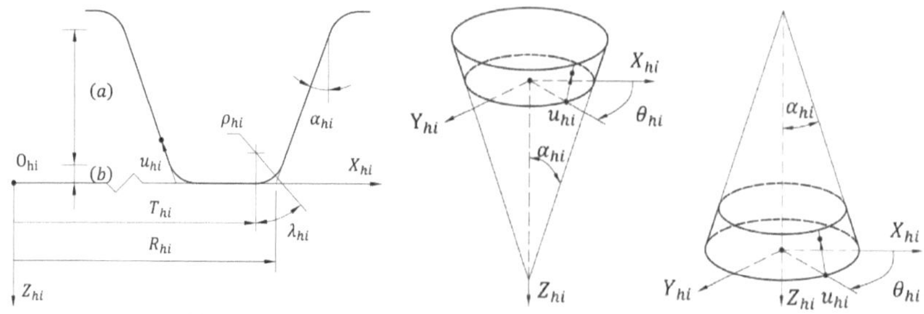

(a) **(b)**

Figure 3. Head cutter blade and generating surface. (**a**) Head cutter composition; (**b**) concave side or outer blade.

The unit normal vectors of the straight edge and fillet arc can be derived from Equations (7) and (8), respectively, which can be represented as

$$n_{hi}{}^{(a)}(\theta_{hi}) = \frac{\frac{\partial r_{hi}{}^{(a)}}{\partial u_{hi}} \times \frac{\partial r_{hi}{}^{(a)}}{\partial \theta_{hi}}}{\left|\frac{\partial r_{hi}{}^{(a)}}{\partial u_{hi}} \times \frac{\partial r_{hi}{}^{(a)}}{\partial \theta_{hi}}\right|} = \begin{bmatrix} \cos\alpha_{hi}\cos\theta_{hi} \\ \cos\alpha_{hi}\sin\theta_{hi} \\ \pm\sin\alpha_{hi} \end{bmatrix} \tag{9}$$

$$n_{hi}{}^{(b)}(\theta_{hi}) = \frac{\frac{\partial r_{hi}{}^{(b)}}{\partial u_{hi}} \times \frac{\partial r_{hi}{}^{(b)}}{\partial \theta_{hi}}}{\left|\frac{\partial r_{hi}{}^{(b)}}{\partial u_{hi}} \times \frac{\partial r_{hi}{}^{(b)}}{\partial \theta_{hi}}\right|} = \begin{bmatrix} \sin\lambda_{hi}\cos\theta_{hi} \\ \sin\lambda_{hi}\sin\theta_{hi} \\ \pm\cos\lambda_{hi} \end{bmatrix} \tag{10}$$

The straight edge of the blade forms a cone, and the fillet arc forms a revolution surface. They are used to form the tooth surface and fillet part, respectively.

3.3. Generation of Tooth Surfaces of the Gear

Transforming Equations (7)–(10) of the head cutter blade from coordinate system $S_{hi}(X_{hi}, Y_{hi}, Z_{hi})$ to the coordinate system $S_i(X_i, Y_i, Z_i)$, a family of generating surfaces are generated. The work gear tooth surface is the envelope to the family of generating surfaces, which can be represented as

$$\begin{cases} r_i{}^{(a)}(u_{hi}, \theta_{hi}, \psi_{ci}) = \mathbf{M}_{i\_hi} r_{hi}{}^{(a)}(u_{hi}, \theta_{hi}) \\ n_i{}^{(a)}(\theta_{hi}, \psi_{ci}) = \mathbf{L}_{i\_hi} n_{hi}{}^{(a)}(\theta_{hi}) \\ f_i{}^{(a)}(u_{hi}, \theta_{hi}, \psi_{ci}) = 0 \end{cases} \tag{11}$$

$$\begin{cases} r_i{}^{(b)}(\lambda_{hi}, \theta_{hi}, \psi_{ci}) = \mathbf{M}_{i\_hi} r_{hi}{}^{(b)}(\lambda_{hi}, \theta_{hi}) \\ n_i{}^{(b)}(\theta_{hi}, \psi_{ci}) = \mathbf{L}_{i\_hi} n_{hi}{}^{(b)}(\theta_{hi}) \\ f_i{}^{(b)}(\lambda_{hi}, \theta_{hi}, \psi_{ci}) = 0 \end{cases} \tag{12}$$

where $\mathbf{M}_{i\_hi} = \mathbf{M}_{i\_bi}\mathbf{M}_{bi\_ai}\mathbf{M}_{ai\_mi}\mathbf{M}_{mi\_ci}\mathbf{M}_{ci\_hi}$ is the transformation matrix from the head cutter axis to the work gear axis. $\mathbf{L}_{i\_hi}$ is the upper-left 3×3 submatrix of $\mathbf{M}_{i\_hi}$. $f_i{}^{(a)}(u_{hi}, \theta_{hi}, \psi_{ci})$, and $f_i{}^{(b)}(\lambda_{hi}, \theta_{hi}, \psi_{ci})$ are the equations of meshing for the tooth surface and the fillet part of the head cutter, respectively. The equations of meshing of the straight edge and the fillet arc of the head cutter and the work gear can be represented as

$$f_i{}^{(a)}(u_{hi}, \theta_{hi}, \psi_{ci}) = n_{hi}{}^{(a)}(\theta_{hi}, \psi_{ci}) \cdot v_{tr}{}^{(hi,i)} \tag{13}$$

$$f_i{}^{(b)}(\lambda_{hi}, \theta_{hi}, \psi_{ci}) = n_{hi}{}^{(b)}(\theta_{hi}, \psi_{ci}) \cdot v_{tr}{}^{(hi,i)} \tag{14}$$

respectively, where $v_{tr}{}^{(hi,i)}$ is the relative transitional velocity between the head cutter and the work gear for both the straight edge and the fillet arc of the head cutter. When only the blank offset is chosen as the design variable, the relative transitional velocity of the head cutter and the work gear can be represented as

$$v_{tr}{}^{(hi,i)} = \frac{dr_{hi}}{dt} - \frac{dr_i}{dt} = \begin{Bmatrix} v_{trx} \\ v_{try} \\ v_{trz} \end{Bmatrix}^{(hi,i)}$$

$$= \begin{Bmatrix} \left\{ \begin{aligned} & E_{mi}\frac{d\psi_i}{dt}\sin\gamma_{mi} + \left(\frac{d\psi_{ci}}{dt} - \frac{d\psi_i}{dt}\sin\gamma_{mi}\right)[S_{ri}\sin(q_i+\psi_{ci}) + R_{hi}\sin(q_i+\psi_{ci}+\theta_{hi})] \\ & - u_{hi}\left(\frac{d\psi_{ci}}{dt} - \frac{d\psi_i}{dt}\sin\gamma_{mi}\right)\sin\alpha_{hi}\sin(q_i+\psi_{ci}+\theta_{hi}) \end{aligned} \right\} \\ \left\{ \begin{aligned} & \frac{dE_{mi}}{dt} - X_{Bi}\frac{d\psi_i}{dt}\cos\gamma_{mi} + \left(\frac{d\psi_{ci}}{dt} - \frac{d\psi_i}{dt}\sin\gamma_{mi}\right)[S_{ri}\cos(q_i+\psi_{ci}) + R_{hi}\cos(q_i+\psi_{ci}+\theta_{hi})] \\ & - u_{hi}\left[\left(\frac{d\psi_{ci}}{dt} - \frac{d\psi_i}{dt}\sin\gamma_{mi}\right)\sin\alpha_{hi}\cos(q_i+\psi_{ci}+\theta_{hi}) - \frac{d\psi_i}{dt}\cos\alpha_{hi}\cos\gamma_{mi}\right] \end{aligned} \right\} \\ \left\{ \begin{aligned} & - E_{mi}\frac{d\psi_i}{dt}\cos\gamma_{mi} + [S_{ri}\sin(q_i+\psi_{ci}) + R_{hi}\sin(q_i+\psi_{ci}+\theta_{hi})]\frac{d\psi_i}{dt}\cos\gamma_{mi} \\ & + u_{hi}\sin\alpha_{hi}\sin(q_i+\psi_{ci}+\theta_{hi})\frac{d\psi_i}{dt}\cos\gamma_{mi} \end{aligned} \right\} \end{Bmatrix} \quad (15)$$

4. Extended Local Synthesis and Multi-Point Control Approach

4.1. Coordinate Systems for TCA

When a spiral bevel gear pair is in meshing, five coordinate systems are used to describe the meshing state, as shown in Figure 4. A coordinate system $S_f\left(X_f, Y_f, Z_f\right)$ is rigidly attached to the frame. Coordinate systems $S_1(X_1, Y_1, Z_1)$ and $S_2(X_2, Y_2, Z_2)$ are rigidly connected to the pinion and the gear, respectively. Coordinate systems $S_{d1}(X_{d1}, Y_{d1}, Z_{d1})$ and $S_{d2}(X_{d2}, Y_{d2}, Z_{d2})$ are auxiliary reference coordinate systems for the pinion and the gear, respectively. Four misalignments are defined: ΔE_1 and ΔE_2 are axial errors of the pinion and the gear, respectively; ΔE is the shortest distance between the axes of the gear and the pinion when pinion–gear axes are crossed rather than intersected; and $\Delta\gamma$ is the error of the shaft angle γ. When the gear set is aligned, all four misalignments are counted as zeros.

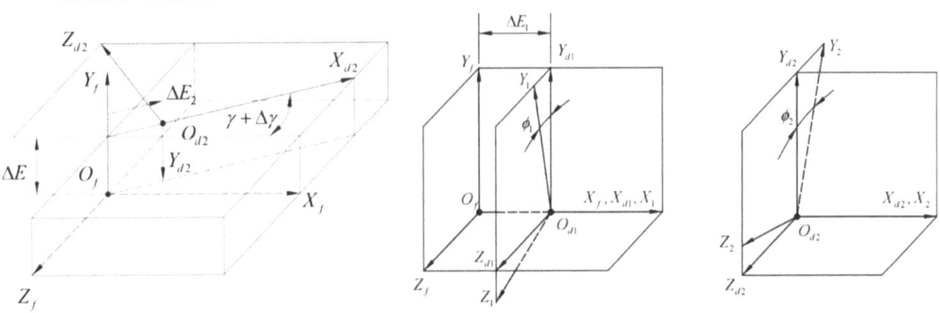

Figure 4. Coordinate systems for TCA of the spiral bevel gear pair.

4.2. Equations for Tooth Meshing and Contact

The pinion tooth surface Σ_1 and the gear tooth surface Σ_2 are in continuous tangency when the following equations stand in the coordinate system $S_f\left(X_f, Y_f, Z_f\right)$ of meshing

$$\begin{aligned} & r_{f1}{}^{(a)}(u_{h1}, \theta_{h1}, \psi_{c1}, \phi_1) - r_{f2}{}^{(a)}(u_{h2}, \theta_{h2}, \psi_{c2}, \phi_2) = 0 \\ & n_{f1}{}^{(a)}(\theta_{h1}, \psi_{c1}, \phi_1) - n_{f2}{}^{(a)}(\theta_{h2}, \psi_{c2}, \phi_2) = 0 \\ & f_i{}^{(a)}(u_{hi}, \theta_{hi}, \psi_{ci}) = 0 \ (i = 1, 2) \end{aligned} \quad (16)$$

4.3. Extended Local Synthesis

Local synthesis controls the meshing performance at the mean contact point, where the transmission error is zero, meaning that the ratio of the rotational speed of the gear pair is equal to the inverse ratio of the tooth numbers. In order to realize the predesigned CMC, the meshing performance of any concerned point on the CMC must be controlled, not just the mean contact point. Generally, the transmission error at these points is not equal to zero. This requires that the local synthesis needs to be extended to suit this situation, and thereafter, the local synthesis becomes global.

4.3.1. Multi-Point Control Approach

The predesigned CMC is taken as the design objective, of which the following inequality must sustain for the meshing point at the instant angular position of the mating gear pair

$$\frac{\Delta\phi_2 - \Delta\phi_{2t}}{\Delta\phi_2} < \varepsilon \tag{17}$$

where $\Delta\phi_{2t}$ is the actual value of CMC, and ε is an arbitrarily convergent limit, i.e., 10^{-9}. As we can see, only one variable is needed to satisfy Equation (17). We choose the blank offset of the pinion E_{v1} as the design variable. From the above analysis, we can see that there are eight equations and nine variables in Equations (16) and (17). The independent design variables are $u_{h1}, \theta_{h1}, \psi_{c1}, \phi_1, u_{h2}, \theta_{h2}, \psi_{c2}, \phi_2$ and E_{m1}. We can solve Equations (16) and (17) simultaneously by assigning an instant value of ϕ_1 for every meshing position. The process of solution is as follows.

Step 1: Based on the local synthesis, calculate the machine-tool settings of the pinion at the mean contact point M. Three design requirements are given at the mean contact before the calculation is performed. They are the direction of the tangent to the contact path, the semi-major axis of the contact ellipse, and the magnitude of the FTE. Among the three design requirements, the third one is chosen around the maximum value of the fourth-order polynomial MC, as shown in Figure 1a. The instant blank offset is related to it. The actual machine-tool settings will be formed afterward according to the predesigned fourth-order polynomial FTE. Since an error exists at the mean contact point for the CMC, as shown in Figure 1b, the machine-tool settings of the pinion should be regulated to some extent by observing Equation (17).

Step 2: Solve the set of Equations (16) and (17). Departing from the mean contact point, the pinion is rotated about its own axis, and the tooth surface of the gear must be in contact with that of the pinion. In other words, Equation (16) must be satisfied. At the same time, the predesigned CMC must also be fulfilled, so Equation (17) can be satisfied as well. The calculation is performed by using an iterative strategy.

Step 3: Perform the tooth contact analysis. After the machine-tool settings are calculated, the TCA gives the results of FTEs, the contact path, the instant ellipses, and the relative velocity of a meshing point over the tooth surface of the pinion and gear. The instant blank offset is interpolated linearly between the above serial instant blank offsets obtained in Step 2.

4.3.2. Loaded Tooth Contact Analysis (LTCA)

LTCA is used to analyze transmission errors and tooth contact patterns of the spiral bevel gear pair under load, considering tooth deformations and the load sharing among neighboring tooth pairs. It is carried out by obeying the force equilibrium condition, the deformation compatibility condition, and the non-embedding condition.

$$\begin{cases} \mathbf{C} \cdot \mathbf{F} + \mathbf{w} = (\Delta\theta \cdot \mathbf{r})_n + \mathbf{d} \\ (\mathbf{F}^T \cdot \mathbf{r})_t = T \qquad (i = 1, 2, \cdots, n) \\ F_i > 0, d_i = 0; F_i = 0, d_i > 0 \end{cases} \tag{18}$$

where n is the total number of the discrete potential contact points on the pinion and gear tooth surfaces at the instant meshing position; C is the flexibility matrix which is reciprocal of the stiffness matrix with the unit of N/m; w and d represent the initial tooth surface gap and the deformed tooth surface gap between the mating tooth surfaces, respectively; r is the distance from the contacting points to the gear shaft; F is the normal force at the contacting points; $\Delta\theta$ is the variation in the gear rotational angle under torque T; and subscripts n and t denote the normal section of the tooth and the rotational plane of the gear, respectively.

According to the experience of the Gleason Works, the amount of the tooth surface approach is approximately 0.00635 mm on a gear rolling test machine. Taking the amount as the target, the torque applied on the gear is calculated via loaded tooth contact analysis when the mating tooth pair contacts at the mean contact point.

5. Numerical Studies

An aviation spiral bevel gear set is investigated to confirm the proposed method. The design parameters of the gear drive are listed in Table 1. The predesigned CMC of the spiral bevel gear pair based on the proposed method is shown in Figure 5. Figure 5a shows a CMC for a single meshing tooth pair with the maximum transmission error of 3″. Figure 5b shows three entwined CMC from left to right representing the proceeding tooth pair, the actual tooth pair, and the following tooth pair, respectively. Design specifications of the spiral bevel gear drive are listed in Table 2. Polynomial coefficients of the CMC function of the spiral bevel gear pair are listed in Table 3. The installation parameters of the gear head cutter are listed in Table 4. They are assumed to be known according to an SB card provided by the Gleason Works. The machine-tool settings of the head cutter of the pinion are listed in Table 5.

Table 1. Blank data of the spiral bevel gear pair.

Items	Pinion	Gear
Mean spiral angle β_1, β_2	30°	30°
Shaft angle γ	90°	90°
Number of teeth Z_1, Z_2	23	65
Hand of spiral	RH	LH
Whole depth h (mm)	7.34	7.34
Pitch angle γ_1, γ_2	19°29′	70°31′
Face angle γ_{f1}, γ_{f2}	21°58′30″	71°36′40″
Root angle γ_{m1}, γ_{m2}	18°23′20″	68°1′30″
Mean cone distance A_m	115.9511	115.9511
Face width b (mm)	37	37
Module m (mm)	3.9	3.9
Clearance c_1, c_2	0.71	0.71
Addendum h_{a1}, h_{a2} (mm)	5.1	1.54
Dedendum h_{f1}, h_{f2} (mm)	2.24	5.8

Table 2. Design specifications of the spiral bevel gear pair.

Items	Pinion Concave and Gear Convex
Magnitude of function of transmission error $\Delta\phi_{2a}$	31″
Tangent to the path of contact on the gear surface η	6°
Semi-major axis of the contact ellipse a (mm)	3.7

The variation in the instant blank offset with respect to the value at the mean contact point for the concave side of the pinion tooth surface is shown in Figure 6. The blank offset of the head cutter of the pinion changes with the CMC of the spiral bevel gear pair in Figure 5. The results of tooth contact analysis have confirmed that the predesigned CMC is fulfilled with sufficient accuracy as shown in Figure 7, in which the dotted points in the middle line represent the controlled points for the actual tooth pair, while the three continuous lines represent the CMC of TCA. Based on the numerical results of TCA of the spiral bevel gear pair, the maximum transmission error is $-3.1527''$. The difference between the maximum transmission error and the predesigned CMC is $0.1527''$. As the gear is driven by the tooth pair with the least transmission error among three neighboring tooth pairs, the contact pattern of the actual tooth pair is scattered into three segments over the tooth surfaces, as in Figure 8, when the spiral bevel gears are considered as rigid bodies. The corresponding CMC of LTCA of the spiral bevel gear pair is shown in Figure 9. This is specific to the CMC and should not be taken as bridge contact, which is usually unacceptable. Based on the numerical results of LTCA of the spiral bevel gear pair, the maximum loaded transmission error is $-22.0763''$. Actually, when the loads are applied, contact patterns on tooth surfaces of the spiral bevel gear pair become continuous, and the maximum transmission error increases by $-18.9236''$ due to the elastic deformation of the meshing tooth surfaces as shown in Figure 10.

Table 3. Polynomial coefficients of the CMC function of the spiral bevel gear pair.

	a_2	a_3	a_4	b_1	b_2	b_3
Polynomial coefficients	−0.00489	−0.00075	0.00023	0.00037	0.00016	0.00008

Table 4. Installation parameters of the head cutter of the gear.

Items	Inner Blade
Blade angle α_2	22°30′
Cutter point radius R_{h2} (mm)	113.155
Point width W_{h2} (mm)	2.29
Radial setting S_{r2} (mm)	115.078
Cutter radius R_{c2} (mm)	114.3
Installment angle q_2	59°20′7″
Machine center to back X_{D2} (mm)	0.00
Sliding base X_{B2} (mm)	0.051
Blank offset E_{M2} (mm)	0.00
Fillet radius ρ_{h2} (mm)	1.31
Modified blade angle a	23°50′
Ratio of cutting m_{2c}	1.06

Table 5. Machine-tool settings of the head cutter of the pinion.

Items	Outer Blade
Cutter point radius R_{h1} (mm)	124.879
Blade angle α_1	22°30′
Radial setting S_{r1} (mm)	91.116
Installment angle q_1	56°58′56″
Sliding base X_{B1} (mm)	−0.059
Machine center to back X_{D1} (mm)	1.231
Blank offset E_{M1} (mm)	34.197
Ratio of cutting m_{1c}	2.499
Modified blade angle a	21°10′
Fillet radius ρ_{h1} (mm)	1.31

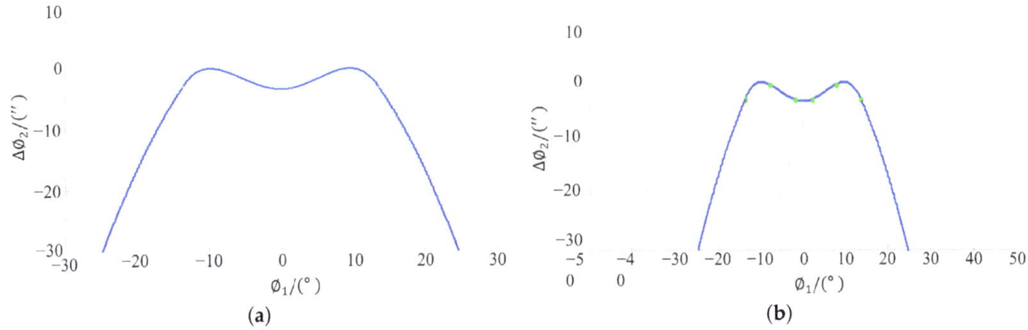

Figure 5. Predesigned CMC of the spiral bevel gear pair. (**a**) CMC for single tooth pair; (**b**) CMC for multi-tooth pairs.

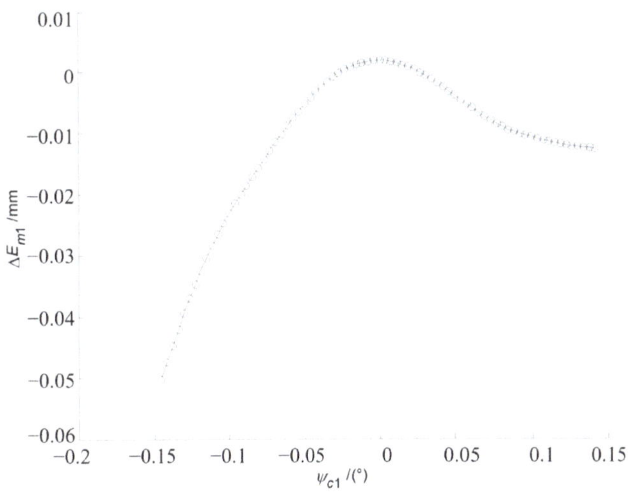

Figure 6. Variation in blank offset of the head cutter for pinion concave side.

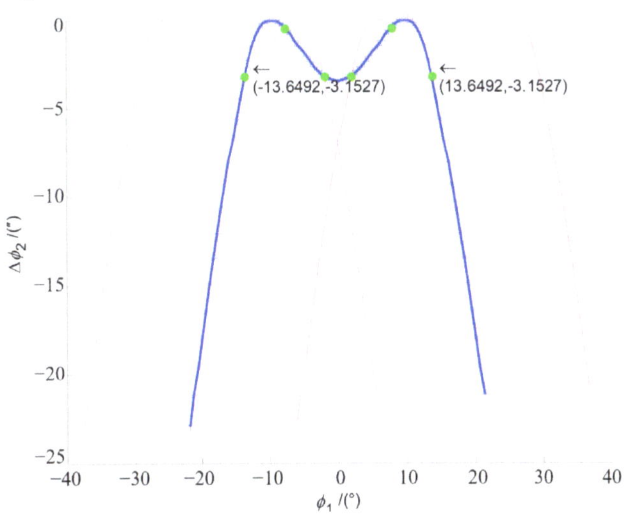

Figure 7. Numerical results of the CMC of TCA of the spiral bevel gear pair.

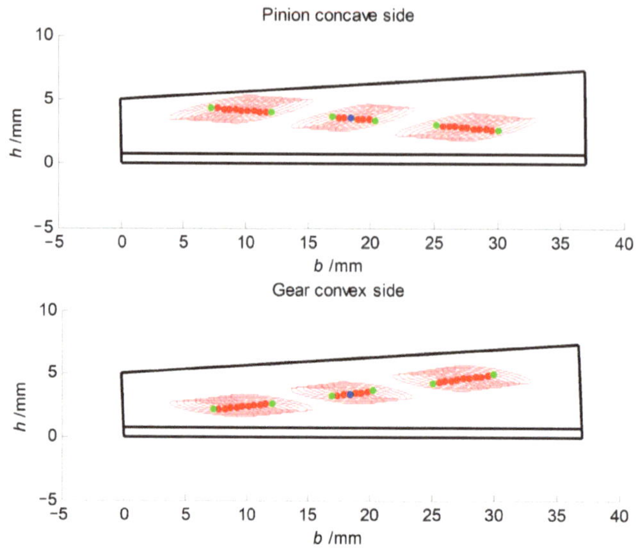

Figure 8. Contact patterns of TCA of the spiral bevel gear pair.

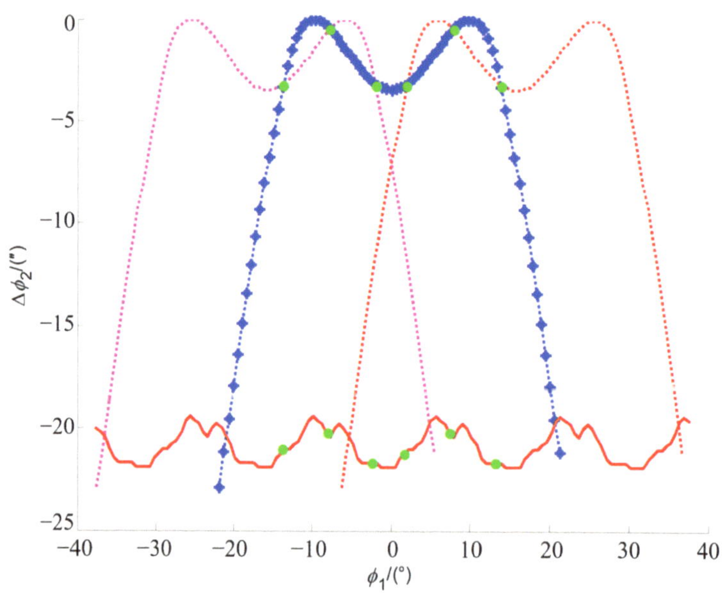

Figure 9. Numerical results of the CMC of LTCA of the spiral bevel gear pair.

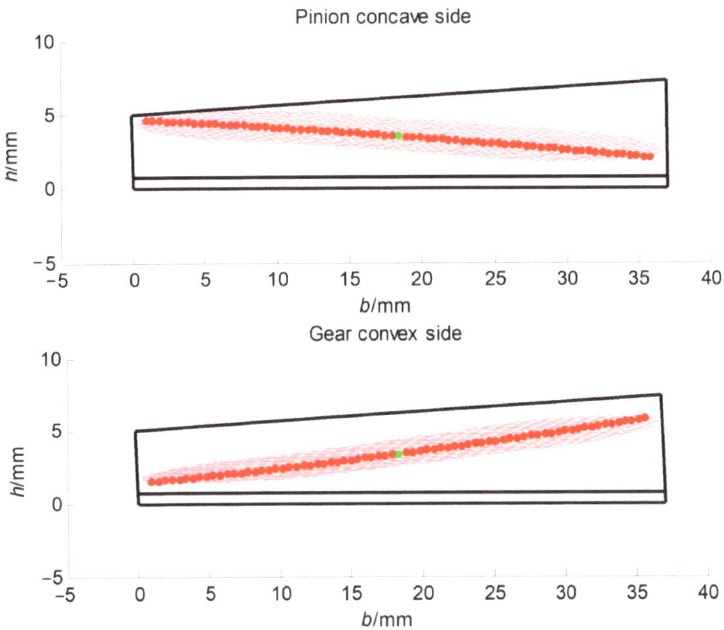

Figure 10. Contact patterns of LTCA of the spiral bevel gear pair.

6. Conclusions

Theoretically, the proposed predesigned face-milling method based on the predesigned fourth-order CMC function is not limited to gear tooth profiles of spiral bevel gears and is applicable to hypoid gears. Local gear tooth modifications of spiral bevel gears can be calculated based on the predesigned fourth-order CMC function. Especially, spiral bevel gears with completed time-varying meshing stiffness and tooth profiles are difficult to adjust the amount of modification. Based on TCA and LTCA results of contact patterns and transmission errors of the face-milled spiral bevel gear pair, the proposed predesigned fourth-order CMC function can improve its meshing performances. Some conclusions can be given as follows:

1. A method for the development of the combined MC up to the fourth-order CMC of the spiral bevel gear pair has been proposed;
2. A mathematical model is established and an extended local synthesis is used to obtain the instant blank offset for the combined motion curve. The basic idea is to introduce a closed-loop strategy for tooth contact analysis. In this way, the meshing performance of the kinematic errors over the whole tooth surface of a is controlled;
3. The proposed method was validated by a numerical study of TCA of a spiral bevel gear pair. The maximum transmission error of the TCA results is $-3.1527''$. The difference between the maximum transmission error and the predesigned CMC is only $0.1531''$, which is 4.86% of the maximum transmission error.
4. When the tooth surfaces of the spiral bevel gear pair are elastically deformed during meshing, the actual tooth contact patterns become continuous. When the loads are applied, the elastic deformation of the meshing tooth surfaces results in an increase in the maximum transmission error of $-18.9236''$.

Author Contributions: Conceptualization, Y.L. and G.L.; methodology, Y.L.; software, Y.L.; validation, Y.L. and G.L.; formal analysis, Y.L.; investigation, L.C.; resources, L.C.; data curation, Y.L.; writing—original draft preparation, Y.L.; writing—review and editing, G.L.; visualization, L.C.; supervision, L.C.; project administration, L.C. All authors have read and agreed to the published version of the manuscript.

Funding: This research was funded by Huazhong University of Science and Technology, Intelligent design and CNC basic key technology research and development Grant No. 2021AAB001.

Institutional Review Board Statement: Not applicable.

Informed Consent Statement: Not applicable.

Data Availability Statement: Data are contained within the article.

Conflicts of Interest: The authors declare no conflicts of interest.

References

1. Huang, D.; Wang, Z.; Li, G.; Zhu, W. Conjugate approach for hypoid gears frictional loss comparison between different roughness patterns under mixed elastohydrodynamic lubrication regime. *Tribol. Int.* **2019**, *140*, 105884. [CrossRef]
2. Zhou, Y.; Chen, Z.C.; Tang, J. A new method of designing the tooth surfaces of spiral bevel gears with ruled surface for their accurate five-axis flank milling. *J. Manuf. Sci. Eng.* **2017**, *139*, 061004. [CrossRef]
3. Zhang, W.; Tan, R.; Guo, X.; Chen, B.; Shu, R.; Zheng, F. Analytical synthesis of the kinematic geometry of spiral bevel gears of pure-rolling contact. *Mech. Mach. Theory* **2020**, *153*, 103992. [CrossRef]
4. Vivet, M.; Tamarozzi, T.; Desmet, W.; Mundo, D. On the modelling of gear alignment errors in the tooth contact analysis of spiral bevel gears. *Mech. Mach. Theory* **2021**, *155*, 104065. [CrossRef]
5. Litvin, F.L.; Zhang, Y. *Local Synthesis and Tooth Contact Analysis of Face-Milled Spiral Bevel Gears*; Technical Report; NASA: Washington, DC, USA, 1991.
6. Litvin, F.L.; Fuentes, A.; Mullins, B.R.; Woods, R. *Computerized Design, Generation, Simulation of Meshing and Contact, and Stress Analysis of Formate Cut Spiral Bevel Gear Drives*; NASA National Technical Information Service: Washington, DC, USA, 2003.
7. Zhang, Y.; Yan, H. New methodology for determining basic machine settings of spiral bevel and hypoid gears manufactured by duplex helical method. *Mech. Mach. Theory* **2016**, *100*, 283–295. [CrossRef]
8. Liu, Z.; Li, F.; Xu, Z.; He, Q. Semi-analytical loaded tooth contact analysis method for spiral bevel gears. *Int. J. Mech. Sci.* **2023**, *253*, 108329. [CrossRef]
9. Simon, V. Influence of tooth errors and misalignments on tooth contact in spiral bevel gears. *Mech. Mach. Theory* **2008**, *43*, 1253–1267. [CrossRef]
10. Sheveleva, G.I.; Volkov, A.E.; Medvedev, V.I. Algorithms for analysis of meshing and contact of spiral bevel gears. *Mech. Mach. Theory* **2007**, *42*, 198–215. [CrossRef]
11. Li, G.; Zhu, W. An Active Ease-Off Topography Modification Approach for Hypoid Pinions Based on a Modified Error Sensitivity Analysis Method. *ASME J. Mech. Des.* **2019**, *141*, 093302. [CrossRef]
12. Mu, Y.; Li, W.; Fang, Z. Tooth surface modification method of face-milling spiral bevel gears with high contact ratio based on cutter blade profile correction. *Int. J. Adv. Manuf. Technol.* **2020**, *106*, 3229–3237. [CrossRef]
13. Alves, J.T.; Guingand, M.; de Vaujany, J.P. Set of functions for the calculation of bending displacements for spiral bevel gear teeth. *Mech. Mach. Theory* **2010**, *45*, 349–363. [CrossRef]
14. Tsai, Y.C.; Hsu, W.Y. The study on the design of spiral bevel gear sets with circular-arc contact paths and tooth profiles. *Mech. Mach. Theory* **2008**, *43*, 1158–1174. [CrossRef]
15. Ma, S.; He, G.; Yan, K.; Li, W.; Zhu, Y.; Hong, J. Structural optimization of ball bearings with three-point contact at high-speed. *Int. J. Mech. Sci.* **2022**, *229*, 107494. [CrossRef]
16. Xiang, S.; Li, H.; Deng, M.; Yang, J. Geometric error analysis and compensation for multi-axis spiral bevel gears milling machine. *Mech. Mach. Theory* **2018**, *121*, 59–74. [CrossRef]
17. Álvarez, Á.; Calleja, A.; Arizmendi, M.; González, H.; Lopez de Lacalle, L.N. Spiral bevel gears face roughness prediction produced by CNC end milling centers. *Materials* **2018**, *11*, 1301. [CrossRef] [PubMed]
18. Mu, Y.; Li, W.; Fang, Z.; Zhang, X. A novel tooth surface modification method for spiral bevel gears with higher-order transmission error. *Mech. Mach. Theory* **2018**, *126*, 49–60. [CrossRef]
19. An, L.; Zhang, L.; Qin, S.; Lan, G.; Chen, B. Mathematical design and computerized analysis of spiral bevel gears based on geometric elements. *Mech. Mach. Theory* **2021**, *156*, 104131. [CrossRef]
20. Mu, Y.; He, X. Design and dynamic performance analysis of high-contact-ratio spiral bevel gear based on the higher-order tooth surface modification. *Mech. Mach. Theory* **2021**, *161*, 104312. [CrossRef]
21. Chen, P.; Wang, S.; Li, F.; Zou, H. A direct preset method for solving ease-off tooth surface of spiral bevel gear. *Mech. Mach. Theory* **2023**, *179*, 105123. [CrossRef]

22. Yang, Y.; Mao, S.; Cao, W.; Huang, Y. A novel taper design method for face-milled spiral bevel and hypoid gears by completing process method. *Int. J. Precis. Eng. Manuf.* **2022**, *23*, 1–13. [CrossRef]
23. Stadtfeld, H.J.; Gaiser, U. The Ultimate Motion Graph. *J. Mech. Des.* **1999**, *122*, 317–322. [CrossRef]
24. Fan, Q. Enhanced Algorithms of Contact Simulation for Hypoid Gear Drives Produced by Face-Milling and Face-Hobbing Processes. *ASME J. Mech. Des.* **2006**, *129*, 31–37. [CrossRef]
25. Li, G.; Wang, Z.; Zhu, W.; Kubo, A. A function-oriented active form-grinding method for cylindrical gears based on error sensitivity. *Int. J. Adv. Manuf. Technol.* **2017**, *92*, 3019–3031. [CrossRef]
26. Wang, P.Y.; Fong, Z.H. Mathematical model of face-milling spiral bevel gear with modified radial motion (MRM) correction. *Math. Comput. Model.* **2005**, *41*, 1307–1323. [CrossRef]
27. Wang, P.Y.; Fong, Z.H. Adjustability improvement of face-milling spiral bevel gears by modified radial motion (MRM) method. *Mech. Mach. Theory* **2005**, *40*, 69–89. [CrossRef]
28. Wang, P.Y.; Fong, Z.H. Fourth-Order Kinematic Synthesis for Face-Milling Spiral Bevel Gears With Modified Radial Motion (MRM) Correction. *ASME J. Mech. Des.* **2005**, *128*, 457–467. [CrossRef]

machines

MDPI

Article

Research on the Hobbing Processing Method of Marine Beveloid Gear

Jianmin Wen, Haoyu Yao, Hui Li and Bindi You *

School of Ocean Engineering, Harbin Institute of Technology, Weihai 264209, China; wenjm@hit.edu.cn (J.W.); 22s030030@stu.hit.edu.cn (H.Y.); 21s030114@stu.hit.edu.cn (H.L.)
* Correspondence: youbindi@hitwh.edu.cn

Abstract: Due to the particular structure of the beveloid gear, it cannot be directly hobbed by an ordinary gear hobbing machine. The existing processing method is complex and has a high cost. Therefore, the mass production and industrialization of beveloid gears are limited. To improve the machining efficiency and accuracy of processing beveloid gears, we proposed a hobbing method via the modification of ordinary hobbing machines. At first, we completed the derivation and calculation of the relevant processing parameters of the beveloid gear based on the study of the structural characteristics of the beveloid gear and the principle of hobbing machining. Then, we proposed and designed a beveloid gear hobbing method, and the modification of the ordinary hobbing machine was completed by using a hanging wheel mechanism in synchronous belt type. Finally, we completed the actual hobbing of the beveloid gear, and the feasibility of the proposed method was verified. After that, we analyzed the machining error of the trial-produced beveloid gear; the results showed that the accuracy of the trial-produced beveloid gear met the 6-level standard, which also verified the accuracy of the proposed method.

Keywords: beveloid gears; hobbing machining; modification; hanging wheel mechanism; error analysis

1. Introduction

Waterway transportation plays a key role in cargo transportation due to its advantages of large volume and low cost [1,2]. Under the trend of economic globalization, global trade volume continues to rise, and higher requirements are put forward for water transportation [3–6]. Ships are the primary component of transportation for waterway transportation, and the intelligent development and application of ships have become the future development direction and research hotspot [7–10]. In the ship power system, gears are widely used in transmission components that transmit power and torque, and the performance of gears has a significant impact on the speed, safety, durability, and fuel efficiency of ships [11,12]. As a new type of gear, the beveloid gear has been applied to marine high-speed gearboxes, marine small inclination transmission systems, and marine RV reducers that with special transmission requirements [13,14]. In addition, beveloid gears are also used in aerospace, automotive, robotics, and other fields [15,16]. The machining method of beveloid gear affects its transmission accuracy and dynamic performance directly. The research on the machining method of beveloid gears can improve the processing accuracy and efficiency of beveloid gears. It can also improve the theory and design system of beveloid gears and then promote the industrialization process of beveloid gears and the intelligent development of ships.

Beam A S. proposed the concept of beveloid gears in 1954 [17]. Compared with traditional gears, the addendum modification factor of beveloid gears varies linearly along the tooth-width direction. Adjusting the beveloid gears' axial position, compensating for axial error, eliminating turning errors, and providing smooth transmission can be accomplished, which ensures normal meshing when wear or mesh clearance occurs. Furthermore, beveloid gears can meet a higher transmission ratio and a more compact structure. They also have better adaptation to complex operating conditions and can reduce system vibration [18,19].

In general, the processing of gears is divided into the forming method and the generating method in principle. Forming methods mainly include the milling method and the grinding method, which is suitable for single-piece or small-batch production of small- and medium-sized gears [20–23]. Generating methods mainly include the hobbing method, the shaping method, the shaving method, and the honing method, which is suitable for the production of large quantities of gears [24–27]. The development of computerized numerical control (CNC) machining technology provides new processing methods for gears, especially gears with special structures and high-precision requirements [28–30].

Due to the special structure of beveloid gears, it is impossible to process them with traditional machining methods directly. Therefore, researchers have studied the machining method of beveloid gears based on the characteristics of beveloid gears.

In terms of the hobbing of beveloid gears, Shen Y proposed a method to modify the gear hobbing machine via electronic hanging wheels and microprocessors, which can complete the hobbing of involute beveloid gears [31]. Zhang Q controlled the horizontal motion of the workbench by installing a numerical control system on the ordinary gear hobbing machine so that the hobbing of the beveloid gear was completed [32]. However, the studies of Shen Y and Zhang Q lack the logical and systematic calculation of the machining parameters, and the coupling design of electronic control and mechanical systems increases the difficulty of parameter setting and adjustment. Therefore, He J and Wu X studied the hobbing principle of involute beveloid gears and deduced the calculation formula of machining parameters [33]. Huang J completed the analysis, calculation, and optimization of the machining parameters based on the geometry of the beveloid gear, which provided a theoretical reference for the machining of the beveloid gear [34]. The shaping method is mainly suitable for internal meshing beveloid gear pairs. Wu J studied the oblique inserting process of internal meshing beveloid gear pairs [35]. Hu R carried out the design and research of the shaper cutter [36]. The studies of Wu J and Hu R filled the research gap in the processing of internally meshing beveloid gear pairs. In terms of the grinding of beveloid gears, M. Brumm studied the grinding method of beveloid gears and analyzed the influence of gear grinding on the dynamic properties of beveloid gears, which promoted the research of gear grinding machining [37]. Wen J. et al. analyzed the tooth modification of beveloid gears theoretically based on the spatial meshing theory and proposed a method of modifying the non-involute beveloid gear by using a specially modified large-plane grinding wheel gear grinding machine [38]. Huang J proposed a method of modifying the beveloid gear by using a general gear grinding machine, which provided an idea for the modification of the non-involute beveloid gear [34]. Jiang P proposed an efficient precision machining method for beveloid gears based on conical worm grinding wheels [39]. Zhang Y completed all designs of the involute beveloid gear, including the processing principle, parameter calculation, and gear grinding machine modification [40]. The studies of Jiang P and Zhang Y provided a theoretical basis for the realization of efficient precision grinding of beveloid gears. In addition, Liao Y. et al. proposed a numerical control machining method for beveloid gears based on the machining principle of beveloid gears, which provided a theoretical reference for CNC machining beveloid gears [41]. Wang Z. et al. proposed a method for honing the teeth of beveloid gears, which improved the machining accuracy of beveloid gears effectively [42].

Compared with the above processing methods, considering factors such as machining accuracy, machining efficiency, processing cost, etc., the hobbing method is more suitable for the machining of external helical involute beveloid gears. The existing studies mainly carry out theoretical research on the hobbing method of beveloid gears. However, the verification of actual machining and systematic calculation of the corresponding machining parameters are still a gap. In addition, the process of the existing studies is complex, and there is also a lack of quantitative standards for the analysis of machining accuracy and machining errors. Therefore, it is necessary to propose an efficient method for beveloid gear hobbing and improve the analysis system of the calculation of machining parameters and machining errors. To realize the efficient and high-precision machining of external helical involute beveloid gears, we proposed a hobbing method based on the modification of an ordinary gear hobbing machine. This study can realize the hobbing beveloid gears with specific parameters. It is helpful to the mass production and industrial application of beveloid gears due to its high efficiency and accuracy and low cost. The concept embodied in the study can realize the manufacturing of new products via appropriate transformation, which has great industrial value and is also of great significance in terms of economy and sustainable use of resources. The most prominent advantage of the method we proposed is that the ordinary three-axis hobbing machine can be modified to realize the four-axis linkage so that the processing of special gears, such as beveloid gears, can be completed. Compared with the processing directly by four-axis machines, the cost can be greatly reduced, which is of great significance to the mass production of beveloid gears. However, each group of hanging wheels can only process the beveloid gears of a specific taper angle; if the taper angle changes, the tooth ratio of the hanging wheel needs to be recalculated, and a new hanging wheel mechanism needs to be assembled. Due to the limited experimental conditions, no further research can be carried out, and the follow-up research can realize the adjustability of the transmission ratio by changing the manipulator handwheel to an electronic handwheel so that the versatility of the method can be expanded.

The geometric parameters of the gear blank are determined by that of the trial beveloid gear. The main parameters include the diameter of the inner and outer transverse plane (the reference standard is the taper angle of the beveloid gear), the thickness of the gear blank (corresponding to the gear tooth width), and the shaft hole diameter. Compared with traditional cylindrical gears, the cylindrical surface needs to be machined into a conical surface to form a round table. In addition, in the hobbing processing, it is necessary to set aside the machining allowance according to the characteristics of the bobbing machine. Regarding the determination of the material of the gear, it is necessary to comprehensively consider the cutting performance, mechanical properties, and material cost. In the machinery industry, 45# steel has been widely used, especially in the manufacturing of transmission parts due to its high strength, good plasticity, strong wear resistance, and good corrosion resistance. Therefore, 45# steel is selected as the material of the gear blank to complete the production of the trial beveloid gear [43].

The remaining parts of the paper are presented as follows: In Section 2, the relative position and motion relationship between the cutter and the gear blank is analyzed, and the hobbing processing parameters for beveloid gears are calculated. In Section 3, the transmission principle of the hobbing machine is analyzed, and the modification method of the hobbing machine is completed. In Section 4, the actual hobbing of the beveloid gear is carried out, and the error of the hobbing is analyzed. In Section 5, the main conclusions are summarized.

2. Analysis of Hobbing Processing Parameters for Beveloid Gears

2.1. Analysis of the Relative Position and Motion Relationship between the Cutter and Gear Blank

In the hobbing process of involute cylindrical gears, the hobbing cutter is fed along the gear axial direction. When a pass cutting is completed, the main motion of the gear hobbing machine includes the rotary motion of the cutter around its axis to form the cutting motion of the cutting edge, the rotary motion of the workpiece around its axis, and the feed motion of the cutter along the gear axis, the two form a generating motion. In the hobbing processing of involute helical cylindrical gears, in addition to meeting the above motion requirements, it is also necessary to meet the differential relationship between the feed motion of the cutter along the axial direction and the rotary motion of the workpiece. Compared to helical cylindrical gears, in the hobbing process of helical beveloid gears, there is a taper angle (δ) between the imaginary rack pitch plane and the gear axis, which is shown in Figure 1:

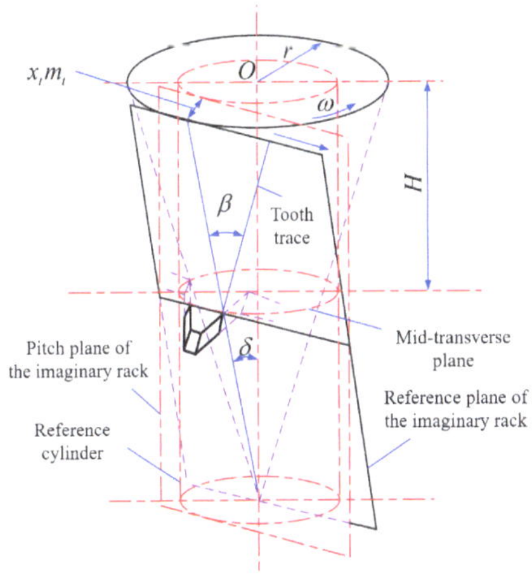

Figure 1. The schematic diagram of the helical beveloid gear's structure. H—the distance between the outer transverse plane and the mid-transverse plane; β—the spiral angle; δ—the taper angle; $x_t m_t$—addendum modification of the transverse plane; x_t—transverse modification coefficient; m_t—transverse module.

Therefore, in the hobbing process of helical beveloid gears, in addition to meeting the motion requirements of traditional gear processing, it is necessary to add the feed movement of the gear hobbing cutter along the gear radial. At the same time, the relationship between the axial feed rate of the gear hobbing cutter and the radial feed rate along the gear should also be considered to complete the setting of the motion parameters. Taking the right-handed involute beveloid gear as the main research object, in the hobbing process, the relative motion and the relative position between the cutter and the gear are shown in Figure 2.

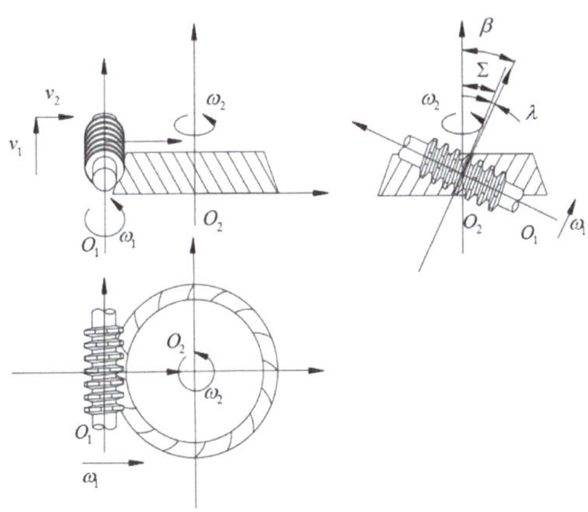

Figure 2. The relative motion and the relative position between the cutter and the gear. Σ—the installation angle of the cutter; λ—the lead angle of the cutter.

According to Figure 2, the main motion of the cutter includes the rotary motion around the axis with the angular velocity ω_1, the vertical feed motion along the gear axis direction with the velocity v_1, the parallel feed motion along the radial direction of the gear with the velocity v_2. The gear blank makes the rotary motion around its axis with the angular velocity ω_2. The feed rate v_1 and v_2 of the cutter are determined according to the taper angle of the beveloid gear (δ) that is designed. The installation angle of the cutter (Σ) is determined according to the spiral angle of the beveloid gear (β), the taper angle of the beveloid gear (δ), and the lead angle of the cutter (λ). The rotation speed ω_2 of the gear blank is determined according to the rotation speed ω_1 of the cutter, the spiral angle of the beveloid gear (β), and the transmission ratio of the cutter and the gear. The schematic diagram of the relative position between the axis of the cutter and the gear blank is shown in Figure 3.

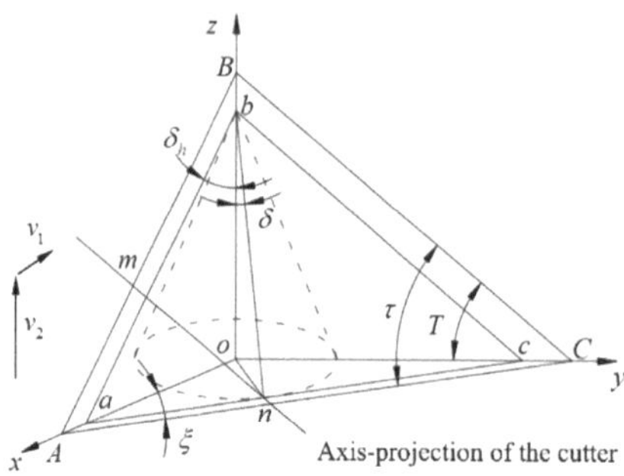

Figure 3. The relative position between the axis of the cutter and the gear blank. τ—the inclination angle of the cutter axis in the plane ABC; δ_h—the motion taper angle of the cutter.

From Figure 3, the motion of the cutter axis mn produces the envelope plane ABC. The plane abc is the nominal cylindrical envelope plane of the cutter; it is also the reference plane of the counterpart rack. The plane abc is parallel to the plane ABC, and the angle between the plane abc and the z-axis is δ, which is the taper angle of the beveloid gear. According to the definition of the spiral angle of the beveloid gear, the inclination angle of the cutter axis (τ) can be calculated by Equations (1) and (2).

$$\tau = \beta - \lambda \tag{1}$$

$$\sin\tau = \frac{\overline{Ao}/\cos\delta}{\overline{on}} = \frac{\overline{Ao}}{\overline{on}}\frac{1}{\cos\delta} \tag{2}$$

The angle between line BC and the y-axis is the installation angle of the cutter (Σ). Based on the spatial geometric relationship, Σ can be calculated by Equations (3) and (4).

$$\sin\Sigma = \frac{\overline{Ao}}{\overline{on}} \tag{3}$$

$$\Sigma = \arcsin(\sin\tau\cos\delta) = \arcsin[\sin(\beta-\lambda)\cos\delta] \tag{4}$$

The motion taper angle of the cutter (δ_h) is inconsistent with the taper angle of the beveloid gear (δ). δ_h can be calculated based on the spatial geometric relationship, so that the relationship between the axial feed motion and the radial feed motion of the cutter can be determined. The derivation process of δ_h is shown in Equations (5)–(7).

$$\tan\xi = \frac{\overline{on}}{\overline{an}} = \frac{\overline{cn}}{\overline{on}} = \frac{\overline{cn}}{\overline{bn}}\frac{\overline{bn}}{\overline{on}} = \frac{1}{\tan\tau}\frac{1}{\sin\delta} \tag{5}$$

$$\tan\delta_h = \frac{\overline{ao}}{\overline{bo}} = \frac{\overline{on}}{\overline{bo}}\frac{\overline{ao}}{\overline{on}} = \frac{\tan\delta}{\sin\xi} \tag{6}$$

$$\delta_h = \arctan\left(\tan\delta\sqrt{1+\tan^2(\beta-\lambda)\sin^2\delta}\right) \tag{7}$$

In the process of hobbing the beveloid gear, the differential motion between the gear blank and the hob forms the spiral angle. The relative motion between the cutter axis and the gear blank in the differential motion is shown in Figure 4.

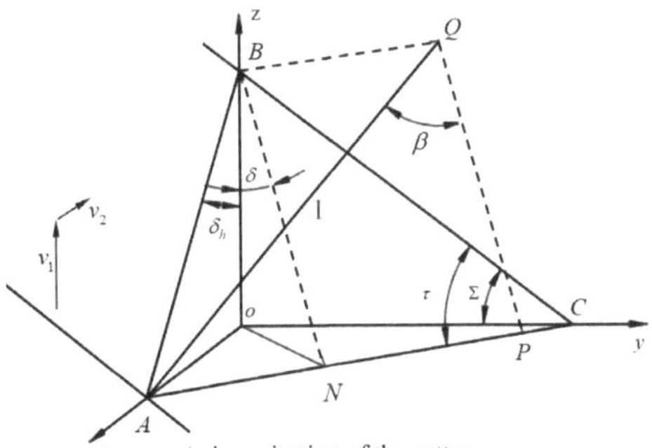

Figure 4. The relative motion between the cutter axis and the gear blank in the differential motion.

To form the spiral angle of the beveloid gear, in other words, the envelope plane of the cutter tooth that is parallel to the tooth trace (I) of the counterpart rack, the cutter needs to translate by $\overline{BQ}$ tangentially along the gear blank when point A on the axis of the cutter moves to point B with the movement of the cutter, so that point A moves to point Q, which converts the relative motion into the differential rotary motion between the cutter and the gear blank. The spiral angle of the differential motion of the beveloid gear in hobbing is defined as β', and the derivation process of β' is shown in Equations (8)–(10).

$$\overline{BQ} = \overline{AP} - \overline{AN} = \overline{BN}\tan\beta - \overline{BN}\sin^2\delta\tan\tau \tag{8}$$

$$\tan\beta' = \frac{\overline{BQ}}{\overline{BO}} = \frac{\overline{BQ}}{\cos\delta\overline{BN}} = \frac{\tan\beta - \sin^2\delta\tan\tau}{\cos\delta} \tag{9}$$

$$\beta' = \arctan\left[\frac{\tan\beta - \sin^2\delta\tan(\beta-\lambda)}{\cos\delta}\right] \tag{10}$$

2.2. Calculation of Hobbing Processing Parameters for Beveloid Gears

The dividing tooth motion and differential motion of ordinary hobbing machines are realized by the hanging wheel. The hanging wheel belongs to the external transmission chain of the hobbing machine, and different hobbing machines have different internal transmission chains in general. To realize the calculated transmission ratio, it is necessary to combine the transmission ratio of both the internal and the external transmission chain.

Hobbing the beveloid gear, the calculation method of the transmission ratio of the hobbing machine's dividing tooth motion of the hanging wheel is the same as that of the cylindrical gear. However, the transmission ratio of the differential hanging wheel is different from that of the cylindrical gear. The transmission ratio of the hobbing machine's dividing tooth motion can be calculated by Equation (11).

$$i_{\mathrm{DIV}} = \frac{KC_1}{Z} \tag{11}$$

K—number of starts of the cutter;
C_1—dividing constant of the hobbing machine;
Z—number of teeth of gear to be machined.

The differential motion transmission ratio can be calculated by Equation (12).

$$i_{\text{DIFF}} = \frac{C_2}{m_n}\sin\left\{\arctan\left[\frac{\tan\beta - \sin^2\delta\tan(\beta-\lambda)}{\cos\delta}\right]\right\} \tag{12}$$

When hobbing the gear blank, the shaft hole of the gear blank is selected as the clamping positioning datum, and the outer circle of the gear blank is selected as the cutter setting datum. It is necessary to calculate the size of the gear blank before the hobbing process. The structure of the gear blank is shown in Figure 5.

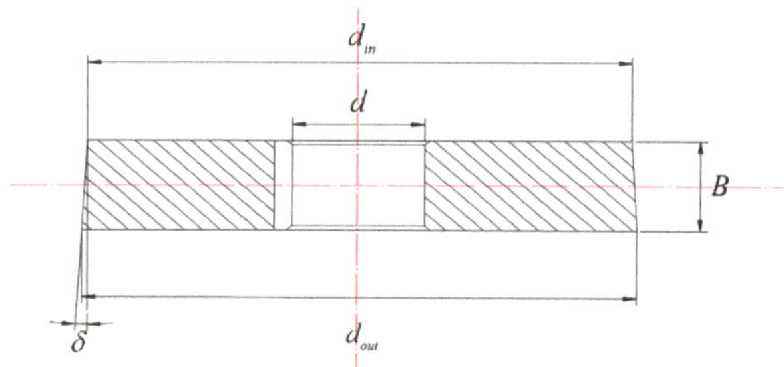

Figure 5. The structure of the gear blank.

The thickness (B) and taper angle (δ) of the gear blank are consistent with those of the gear, and the diameter of the shaft hole (d) is determined according to the shaft diameter. The diameter of the outer transverse plane of the gear blank can be calculated by Equation (13).

$$d_{out} = 2\left(\frac{m_n z}{\cos\beta} + \frac{m_n x}{\cos\beta} + \frac{m_n h_{an}^*}{\cos\delta}\right) \tag{13}$$

The diameter of the inner transverse plane of the gear blank can be calculated by Equation (14).

$$d_{in} = 2\left(\frac{m_n z}{\cos\beta} + \frac{m_n x}{\cos\beta} + \frac{m_n h_{an}^*}{\cos\delta} - B\tan\delta\right) \tag{14}$$

In the hobbing process of cylindrical gears, the cylinder of the gear blank is selected as the cutter aligning datum. After the cutter is aligned, the cutter is moved to the top of the gear blank along the axial direction. Rotating the radial feed handwheel of the gear hobbing machine, the cutter moved for the whole depth of the gear (h) along the radial direction of the gear blank.

The whole depth of the gear (h) can be calculated by Equation (15).

$$h = m_n(2h_{an}^* + c_n^*) \tag{15}$$

The cutter aligning datum of the beveloid gears' hobbing process is similar to that of cylindrical gears. Selecting the round table surface of the gear blank as the cutter setting datum, and the schematic diagram of the cutter setting is shown in Figure 6. O_1, O_2, and O_3 represent the axes of the cutter, and the motion path of the cutter is $O_1 \to O_2 \to O_3$.

E—the moving distance of the cutter after the cutter aligns. It can be calculated by Equation (16).

$$E = \frac{m_n(2h_{an}^* + c_n^*)}{\cos\delta\sin\xi} = \frac{m_n(2h_{an}^* + c_n^*)\sqrt{1+\tan^2(\beta-\lambda)\sin^2\delta}}{\cos\delta} \tag{16}$$

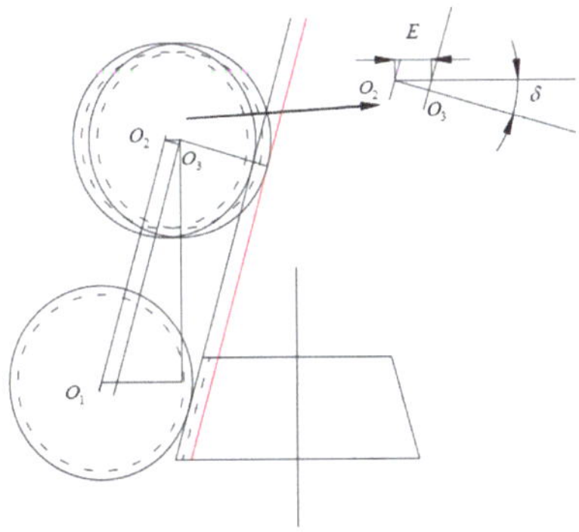

Figure 6. The schematic diagram of the cutter setting.

In the hobbing process of beveloid gears, the axial feed of the cutter is expressed by the axial feed distance of the cutter holder during one rotation of the worktable. The radial movement of the cutter holder can be realized by the rotation of the handwheel, and the radial feed distance of the cutter holder during one rotation of the handwheel is defined as l_1. The axial movement of the cutter holder can also be realized by the rotation of the handwheel, and the axial feed distance of the cutter holder during one rotation of the handwheel is defined as l_2. To form the taper angle of the beveloid gear, the transmission ratio between the axial feed handwheel and the radial feed handwheel of the hobbing machine (i_{AR}) needs to satisfy Equation (17).

$$i_{AR} = \frac{l_1}{\tan\delta_h l_2} = \frac{l_1}{l_2 \arctan\left[\tan\delta\sqrt{1+\tan^2(\beta-\lambda)\sin^2\delta}\right]} \tag{17}$$

The processing parameters and calculation formulas are summarized in Table 1. In addition, the gear rotation direction can be changed by increasing or decreasing differential hanging wheels.

Table 1. The processing parameters and calculation formulas.

Parameters	Symbol	Calculation Formula
Rotate speed of the cutter	ω_1	According to the allowable speed of hobbing
Feed	S	According to the requirements of machining accuracy and efficiency
Dividing tooth motion transmission ratio	i_{DIV}	$\frac{KC_1}{Z}$
Differential motion transmission ratio	i_{DIFF}	$\frac{C_2}{m_n}\sin\left\{\arctan\left[\frac{\tan\beta-\sin^2\delta\tan(\beta-\lambda)}{\cos\delta}\right]\right\}$
The transmission ratio between the axial feed handwheel and the radial feed handwheel	i_{AR}	$\frac{l_1}{l_2\tan\delta\sqrt{1+\tan^2(\beta-\lambda)\sin^2\delta}}$
Cutter install angle	Σ	$\arcsin[\sin(\beta-\lambda)\cos\delta]$
Moving distance of the cutter after the cutter aligning	E	$\frac{m_n(2h_{an}^*+c_n^*)\sqrt{1+\tan^2(\beta-\lambda)\sin^2\delta}}{\cos\delta}$

3. Modification of Hobbing Machine

3.1. Analysis of the Transmission Principle of Hobbing Machine

In the hobbing process of helical cylindrical gears, the diagram of the transmission principle is shown in Figure 7.

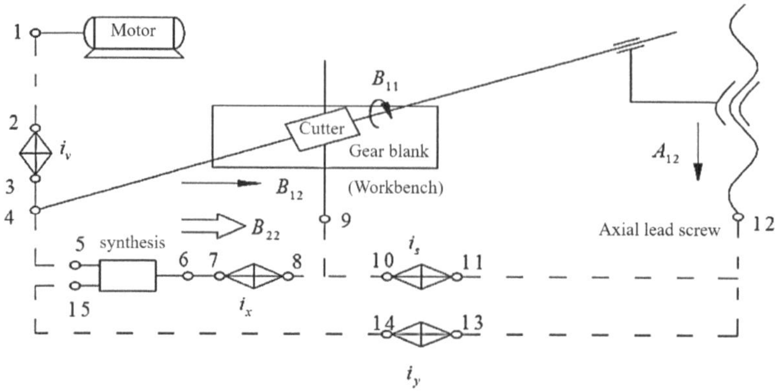

Figure 7. The transmission principle of helical cylindrical gears hobbing process.

As is shown in Figure 7. The transmission chain includes the main motion chain, the dividing tooth motion chain, the axial feed motion chain, and the differential motion chain. In the main motion chain, the power is transmitted to the cutter from the motor so that the rotary hobbing motion of the cutter is realized. The power transmission path is $\textit{the motor} \rightarrow 1 \rightarrow 2 \xrightarrow{i_v} 3 \rightarrow 4 \rightarrow \textit{the cutter}$. The speed of the cutter can be adjusted by the transmission ratio i_v. In the dividing tooth motion chain, the power is transmitted to the workbench from the cutter so that the rotary dividing tooth motion of the workbench is realized. The power transmission path is $\textit{the cutter} \rightarrow 5 \rightarrow 6 \rightarrow 7 \xrightarrow{i_x} 8 \rightarrow 9 \rightarrow \textit{the workbench}$. The number of gear teeth can be adjusted by the transmission ratio i_x. In the axial feed motion chain, the power is transmitted to the axial lead screw from the workbench. The axial feed motion in the hobbing process is realized so that the complete tooth surface of the gear is processed. The power transmission path is $\textit{the workbench} \rightarrow 10 \xrightarrow{i_s} 11 \rightarrow 12 \rightarrow \textit{the axial lead screw}$. The speed of the axial feed motion can be adjusted by the transmission ratio i_s. In the differential motion chain, the power is transmitted to the workbench from the cutter holder screw. Based on the dividing tooth motion, The workbench compensates for the differential motion relative to the cutter, and the spiral angle of the helical cylindrical gear is formed. The power transmission path is $\textit{the axial lead screw} \rightarrow 13 \xrightarrow{i_y} 14 \rightarrow 15 \rightarrow \textit{the workbench}$. The spiral angle of the gear can be adjusted by the transmission ratio i_y.

In the hobbing process of helical beveloid gears, in addition to the axial feed motion relative to the gear blank, the cutter also needs to have a radial feed motion relative to the gear blank to form a taper angle δ. The relationship between the radial feed and the axial feed of the cutter relative to the gear blank should be satisfied by Equation (18).

$$S_R/S_A = \tan\delta_h \tag{18}$$

S_R—the radial feed of the cutter during one rotation of the worktable (mm);
S_A—the axial feed of the cutter during one rotation of the worktable (mm).

To establish the connection between the radial feed and the axial feed of the cutter of the hobbing machine, it is necessary to establish the transmission chain between the axial feed screw and the radial feed screw of the hobbing machine. To achieve these above, the taper angle transmission chain is added to the transmission chain of helical cylindrical gears. And the transmission principle of the helical beveloid gears hobbing process is shown in Figure 8.

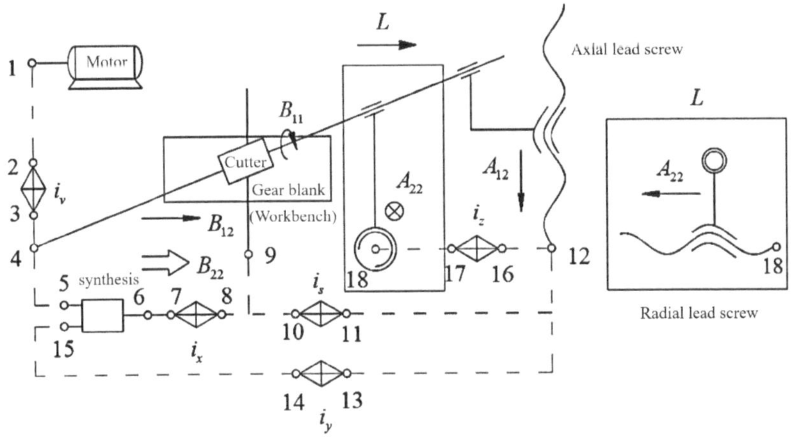

Figure 8. The transmission principle of helical beveloid gears hobbing process.

The power transmission path of the taper angle (δ) transmission chain is *the axial lead screw* $\rightarrow 16 \xrightarrow{i_z} 17 \rightarrow 18 \rightarrow$ *the radial lead screw*. The taper angle (δ) can be adjusted by the transmission ratio i_z, which can be calculated by Equation (19).

$$i_z = \frac{l_1}{\tan \delta_h l_2} \tag{19}$$

l_1—the radial feed of the cutter during one rotation of the radial feed handwheel (mm);
l_2—the axial feed of the cutter during one rotation of the axial feed handwheel (mm).

3.2. Modification Design and Assembly of the Hobbing Machine

Realizing the linkage between the axial feed and the radial feed of the cutter is the purpose of modifying the hobbing machine. In our study, the Y-38 gear hobbing machine was selected for modification, and the structure schematic diagram of the Y-38 gear hobbing machine is shown in Figure 9.

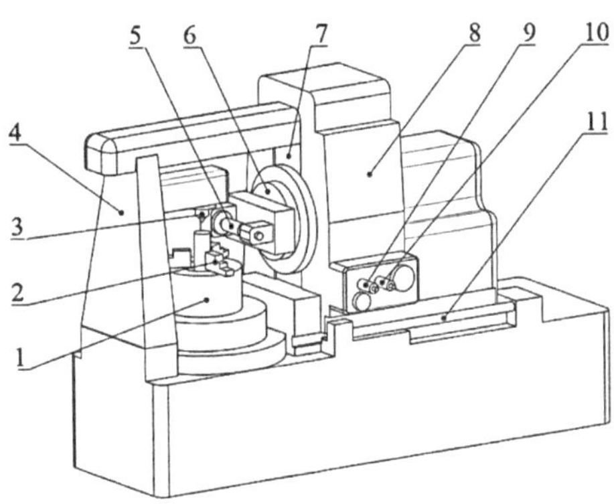

Figure 9. The structure schematic diagram of the Y-38 gear hobbing machine: 1—rotary workbench; 2—loading fixture of the gear blank; 3—live center; 4—hobbing machine body; 5—cutter; 6—cutter holder and its turntable; 7—radial rail; 8—axial workbench; 9—axial feed handwheel shaft; 10—radial feed handwheel shaft 1-axial rail.

It is necessary to measure the structural dimensions of the hobbing machine accurately and determine the feed motion parameters before installing the hanging wheel mechanism. The schematic diagram of the installation position of the hanging wheel mechanism relative to the hobbing machine is shown in Figure 10.

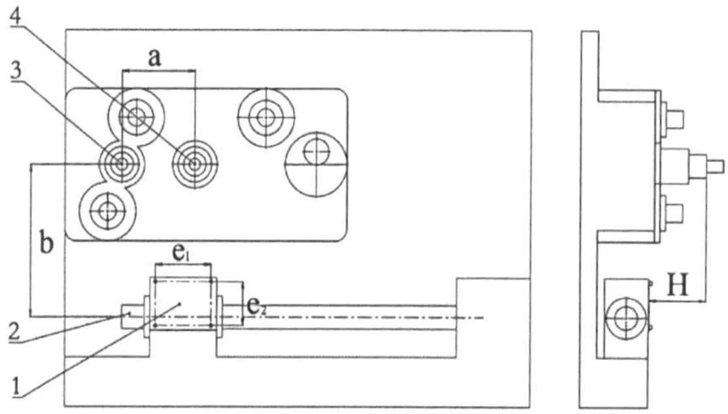

Figure 10. The schematic diagram of the installation position of the hanging wheel mechanism: 1—cover plate; 2—axial lead screw rail; 3—axial feed handwheel shaft; 4—radial feed handwheel shaft; a—the distance between the two handwheel shafts; b—the height difference between the center of the cover plate and the axial feed handwheel shaft axis; e_1 and e_2—the distance between the fixed bolt of the cover plate; H—the height difference between the cover plate transverse plane and the handwheel install plane.

After measurement, it can be obtained that a = 130 mm, b = 260 mm, e_1 = 100 mm, e_2 = 75 mm, H = 100 mm, the size of the cover plate is 90 mm × 120 mm, and four fixed bolts with the specification of M6 are used to fix the cover plate.

The radial and axial feed parameters of the hobbing machine were measured using a dial gauge, which was installed on the radial rail and the axial rail, respectively. After calculation, the radial feed handwheel rotates for one rotation, $l_1 = 2$ mm (on average). The axial feed handwheel rotates for one rotation, $l_2 = 1.7$ mm (on average).

There is a gap between the radial feed lead screw and the axial feed lead screw of the hobbing machine; although the rotating error of the axial feed motion can be eliminated by the gravity of the cutter holder, the gap of the radial feed lead screw still has an impact on the hobbing beveloid gears. After measurement, the rotating error of the radial feed motion is 0.25 mm, which needs to be eliminated during the hobbing process.

According to the axial feed motion parameters and radial feed motion parameters of the hobbing machine, the transmission ratio (i_{AR}) between the axial feed handwheel and the radial feed handwheel was preliminarily calculated. Different taper angles of beveloid gears were selected, and the transmission ratio (i_{AR}) was calculated as shown in Table 2.

Table 2. The transmission ratio of the handwheel for different taper angles of beveloid gears.

The Taper Angle of the Beveloid Gear δ/°	The Transmission Ratio of the Handwheel i_{AR}
1	67.3998
2	33.6894
3	22.4479
4	16.8236
5	13.4462
6	11.1923
7	9.5803
8	8.371
9	7.427
10	6.6721

To facilitate the experimental design and description, the taper angle of the beveloid gear for trial hobbing was selected as $\delta = 3.06°$.

Due to the RPM ratio between the radial handwheel and the axial handwheel of the hobbing machine being relatively large and the center distance between the two handwheel shafts being close, it is difficult for the primary transmission to achieve a large transmission ratio. At the same time, the double-geared transmission needs to change the body of the hobbing machine greatly. For these above, the two-stage synchronous belt transmission was selected. Due to the limited space, it is inconvenient to arrange the tensioning device for the synchronous belts. In addition, due to the small torque transmitted by the hanging wheel mechanism and the slow transmission speed, the design can meet the processing requirements. The diagram of hanging wheel transmission in synchronous belt type is shown in Figure 11.

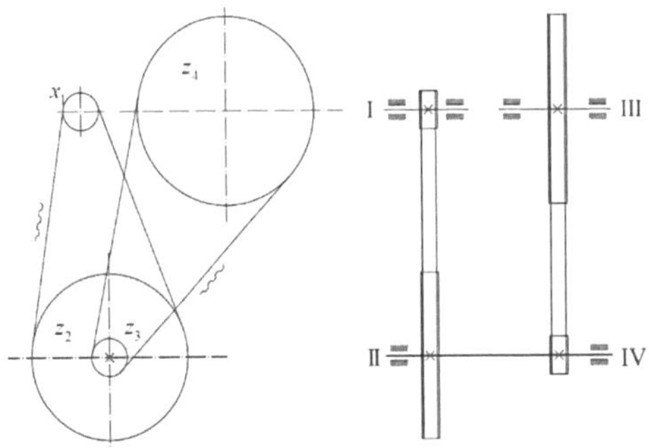

Figure 11. The diagram of hanging wheel transmission with synchronous belt type.

According to Equation (17), i_{AR} can be calculated as follows:

$$i_{AR} = \frac{l_1}{l_2 \arctan\left[\tan\delta\sqrt{1+\tan^2(\beta-\lambda)\sin^2\delta}\right]} = 22.0069 \approx 22$$

To facilitate the selection of the synchronous belt, the structure of the hanging wheel should be compacted, and the center distance of the synchronous belt should be closed. The transmission ratio was assigned as $i_{12} = 4.4$, $i_{34} = 5$. According to the install position and the diameter of the install shaft, the number of axial synchronous belt wheel teeth was determined as $z_1 = 20$, the number of radial synchronous belt wheel teeth was determined as $z_4 = 100$, the number of the large center synchronous belt wheel teeth was determined as $z_2 = 88$, the number of the small center synchronous belt wheel teeth was determined as $z_3 = 20$.

The actual transmission ratio of the synchronous belt hanging wheel was calculated by Equation (20).

$$i_{14} = \frac{z_4}{z_3}\frac{z_2}{z_1} = 22 \quad (20)$$

The actual taper angle was calculated by Equation (21).

$$\delta\prime = \arctan\frac{l_1}{i_{14}l_2} = \arctan\frac{2}{22 \cdot 1.7} = 3.061° \quad (21)$$

The deviation between the actual taper angle and the design taper angle was calculated by Equation (22).

$$\Delta\delta = \delta\prime - \delta = 3.061 - 3.06 = 0.001° \quad (22)$$

From Equation (22), the deviation was within the allowable range and met the processing standard. If the design taper angle is another angle, it can be achieved by changing the number of the large center synchronous belt wheel teeth (z_2).

According to the model and the teeth number of the synchronous belt wheel selected, the three-dimensional assembly drawing of the hanging wheel mechanism is shown in Figure 12.

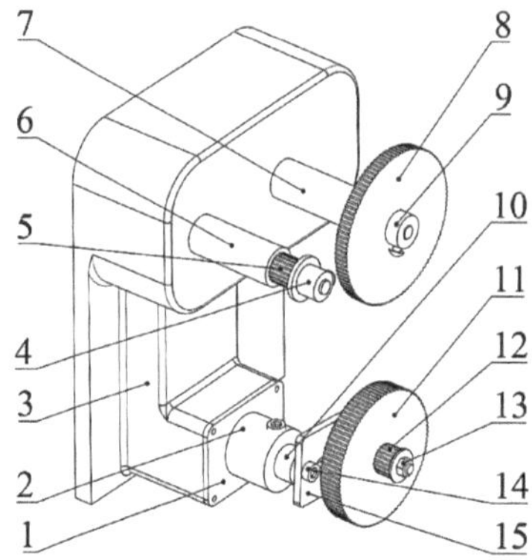

Figure 12. The three-dimensional assembly drawing of the hanging wheel mechanism: 1—cover plate; 2—shaft seat; 3—hobbing machine body; 4—fix bolt of axial synchronous belt wheel; 5—axial synchronous belt wheel; 6—axial rotary shaft 7—radial rotary shaft; 8—radial synchronous belt wheel; 9—fix bolt of radial synchronous belt wheel; 10—center rotary shaft; 11—large center synchronous belt wheel; 12—small center synchronous belt wheel; 13—fix bolt of the hanging wheel; 14—connecting plate screws; 15—connecting plate.

The center distance of the synchronous belt wheels was measured, and the synchronous belt model was selected as M5-830. In addition, the tensioning of the synchronous belt can be realized by adjusting the connecting plate. As shown in Figure 12, there is a guide groove on the connecting plate, and the spatial relative position of the center synchronous belt wheel can be adjusted by loosening the connecting plate screws so that the preload of the synchronous belt can be adjusted.

4. Hobbing Beveloid Gears

4.1. Calculation of Hobbing Parameters and Hobbing Process

The process of hobbing beveloid gears mainly includes determining the basic parameters of beveloid gears, calculating the hobbing parameters based on the parameters of beveloid gears and cutters, determining the number of hanging wheel teeth by checking the table, assembling the hanging wheels on the hobbing machine, clamping the gear blank, setting the cutter, retracting the cutter, feeding the cutter and hobbing gears.

According to the study in Sections 2 and 3, taking the involute helical beveloid gear with specific parameters as an example, the process of hobbing the beveloid gear is introduced as follows.

The basic parameters of the beveloid gear are set as shown in Table 3.

The basic parameters of the cutter are set, as shown in Table 4.

The calculation of the basic parameters of the gear blank are shown in Table 5.

The calculation results of the hobbing parameters are shown in Table 6.

Hobbing the outer circle, transverse plane, and shaft hole of the gear blank. The gear blank hobbing was completed, as shown in Figure 13.

Table 3. The basic parameters of the beveloid gear.

Parameter	Symbol of the Parameter	Unit	Symbol of the Unit	Size
Normal modulus	m_n	millimetre	mm	2
Normal pressure angle	α_n	degree	°	20
Spiral angle	β	degree	°	8.5
Taper angle	δ	degree	°	3.06
Tooth width	B	millimetre	°	25
Tooth number	z	—	—	38
Modification coefficient of outer transverse plane	x	—	—	+0.33
Normal addendum coefficient	h^*_{an}	—	—	1
Normal tip clearance coefficient	C^*_n	—	—	0.25
Rotation direction	L/R	—	—	Right-hand

Table 4. The basic parameters of the cutter.

Parameter	Symbol of the Parameter	Unit	Symbol of the Unit	Size
Normal modulus	m_n	millimetre	mm	2
Normal pressure angle	α_n	degree	°	20
Lead angle	λ	degree	°	2.21
Normal addendum coefficient	h^*_{an}	—	—	1
Normal tip clearance coefficient	C^*_n	—	—	0.25
Number of starts	K	—	—	1
Number of slots	N	—	—	12
Rotation direction	L/R	—	—	Right-hand

Table 5. The basic parameters of the gear blank.

Parameter	Symbol of the Parameter	Unit	Symbol of the Unit	Size
Diameter of outer transverse plane	d_d	millimetre	mm	82.8
Width	B	millimetre	mm	25
Taper angle	δ	degree	°	3.06
Diameter of shaft hole	d	millimetre	mm	30
Size of keyway	$b \times h$	millimetre	mm	8×7

Figure 13. The gear blank before gear hobbing.

Table 6. The hobbing parameters.

Parameter	Symbol of the Parameter	Unit	Symbol of the Unit	Size
Dividing tooth motion transmission ratio	i_{DIV}	—	—	1.5833
Differential motion transmission ratio	i_{DIFF}	—	—	0.5873
Cutter installation angle	Σ	degree	°	6.29
Cutter moving distance	E	millimetre	mm	4.506
Radial feed	S	millimetre/r	mm/r	0.5
The transmission ratio between the axial and radial feed handwheel	i_{AR}	—	—	22.0074

According to the calculated hanging wheel transmission ratio, assembling the dividing tooth hanging wheel and the differential hanging wheel. The machining efficiency and tooth surface processing quality should be considered before selecting and assembling the appropriate axial feed hanging wheel. The diagram of the hanging wheel mechanism assembling is shown in Figure 14.

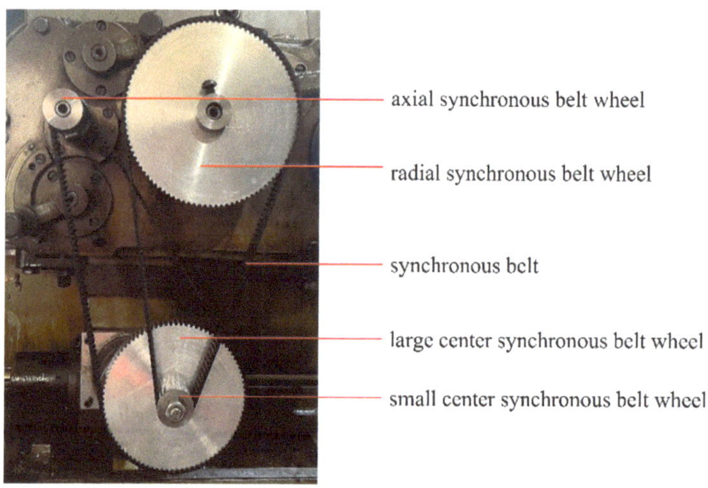

Figure 14. The diagram of the hanging wheel mechanism assembling.

According to the lead angle of the cutter and the spiral angle of the beveloid gear to be processed, the cutter installation angle was determined, and the cutter was assembled. We clamped the gear blank on the loading fixture. Then, we measured the radial runout of the gear blank and fine-tuned the gear blank until the hand of the dial gauge was stable. The gear blank after clamping is shown in Figure 15.

Take the side of the gear blank as the reference, rotating the radial feed handwheel slowly until the tip of the cutter touches the side of the gear blank, as shown in Figure 16.

It is necessary to rotate the radial feed handwheel to keep the hob away from the gear blank if the knife marks are too deep. Then, rotating the radial feed handwheel in reverse to make the cutter close to the gear blank so that the handwheel rotation error can be eliminated. After the cutter aligned, we installed the synchronous belt of the hanging wheel, raised the cutter to the top of the gear blank, and then removed the synchronous belt. Next, we rotated the radial feed handwheel to make the cutter move to the axis of the workbench. During the cutter alignment process, it is necessary to consider the radial feed lead screw gap so that the rotation error can be eliminated. Aligning the cutter and installing the synchronous belt of the hanging wheel again, starting the machine, and the hobbing process of the beveloid gear was completed.

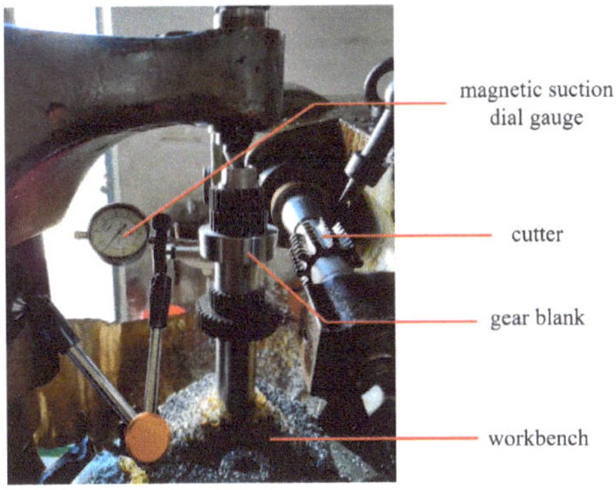

Figure 15. The diagram of the gear blank after clamping.

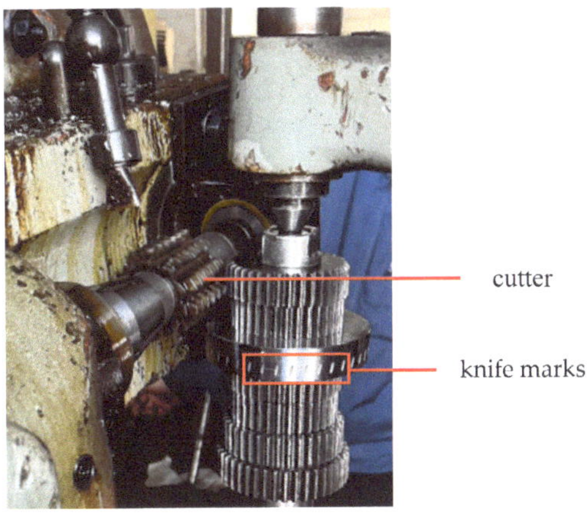

Figure 16. The diagram of the gear blank after the cutter aligning.

4.2. Error Measurement and Analysis

The accuracy grade of the gear has a significant impact on the transmission efficiency. Improving the accuracy grade of the gear, the load distribution of the gear is more uniform, the accuracy and stability of the transmission motion are better, the energy loss caused by the vibration and impact of the gear is reduced, and the transmission efficiency is improved. In the industry standard, the 6-level accuracy has requirements for the size, tooth pitch error, tooth profile deviation, etc. According to the actual situation of our experiment, only some of the parameters are calculated and compared, and the corresponding industry standards are also listed in the paper. Error detection was performed on the machined beveloid gear. The test items include the distance from the root circle of the outer transverse plane to the top diameter of the shaft hole (d_1), the distance from the root circle of the inner transverse plane to the top diameter of the shaft hole (d_1'), the chordal thickness of the reference circle (l), the total deviation of the left tooth surface reference circle helix (Δ_L) and the total deviation of the right tooth surface reference circle helix (Δ_R). The schematic diagram of the measurement is shown in Figure 17.

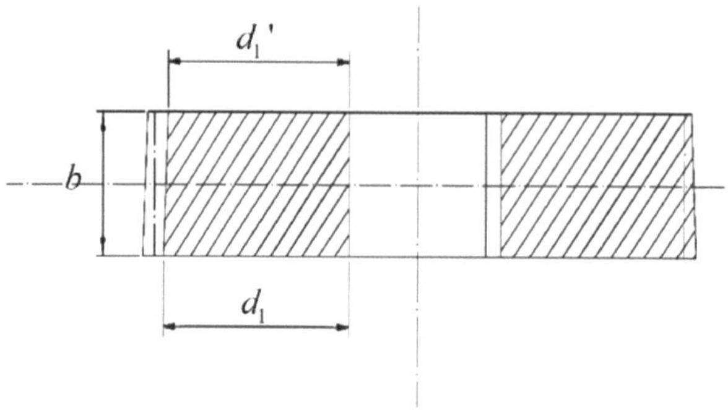

Figure 17. The schematic diagram of the measurement.

The machined taper angle (δ') can be calculated indirectly by the result of measuring the distance from the root circle of the outer transverse plane to the top diameter of the shaft hole (d_1), the distance from the root circle of the inner transverse plane to the top diameter of the shaft hole (d_1') and teeth width (b). The machined taper angle (δ') can be calculated by Equation (23).

$$\delta\prime = \arctan\frac{d_1 - d_1\prime}{b} \tag{23}$$

The taper angle deviation ($\Delta\delta$) can be calculated by Equation (24).

$$\Delta\delta = \delta\prime - \delta \tag{24}$$

Due to the different radii of the base circle of the beveloid gear's left and right tooth surfaces, the tooth thickness error cannot be detected by measuring the common normal. Instead, we measured the chordal thickness of the reference circle (l) and that in theoretical (l') to analyze the error. The relationship between the chordal thickness of the reference circle (l) and the tooth thickness of the reference circle is shown in Figure 18.

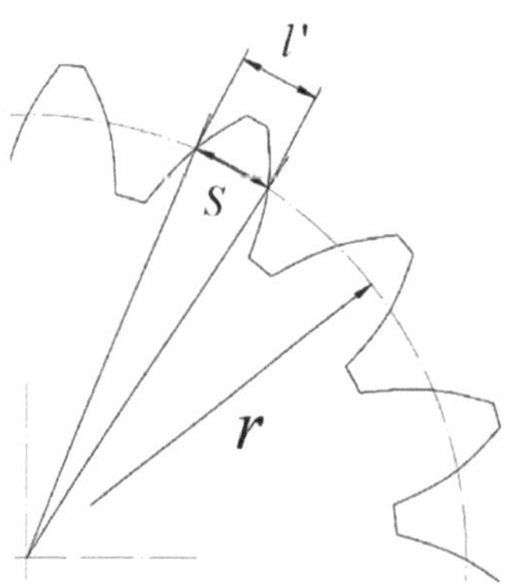

Figure 18. The diagram of the relationship between the chordal thickness and the tooth thickness of the reference circle.

On the outer transverse plane of the beveloid gear, the tooth thickness of the reference circle (S) can be calculated by Equation (25). and the circular thickness on the diametral pitch (DP) of the reference circle in theoretical (l') can be calculated by Equation (26).

$$S = \frac{m_n}{\cos\beta}\left(\frac{\pi}{2} + \frac{2x\tan\alpha_n\cos\delta}{\cos\beta}\right) \tag{25}$$

$$l' = 2r\sin\frac{S}{2r} \tag{26}$$

The reference circle helix of the beveloid gear was selected as the measurement object. To analyze the total deviation of the helix of the beveloid gear, it is necessary to calculate the reference circular spiral angle of the beveloid gear separately. The spiral angle of the left tooth surface of the beveloid gear (β_L) can be calculated by Equation (27), and the spiral angle of the right tooth surface of the beveloid gear (β_R) can be calculated by Equation (28).

$$\beta_L = \arctan\left(\frac{\tan\alpha_n\cos\delta}{\cos\beta} + \sin\delta\tan\beta\right) \tag{27}$$

$$\beta_R = \arctan\left(\frac{\tan\alpha_n\cos\delta}{\cos\beta} - \sin\delta\tan\beta\right) \tag{28}$$

According to the size of the spiral angle, the ratio of the differential hanging wheel of the hobbing machine was calculated. After selecting the appropriate hanging wheel and assembling the mechanism, we adjusted the clutch of the hobbing machine so that the axial motion of the cutter holder and the rotary motion of the workbench could be realized. In addition, to measure the deviation of the helix of the beveloid gear, the magnetic suction gauge was placed on the cutter holder of the hobbing machine and moved with the tool holder. At the same time, the contact head of the magnetic suction gauge was in contact with the reference circle of the tooth surface to be measured. The measurement results of the beveloid gear size of the hobbing processing are shown in Table 7.

Table 7. The measurement results of the beveloid gear size.

Serial Number	d_1/mm	d_1'/mm	b/mm	l/mm
1	21.59	20.27	24.99	3.66
2	21.58	20.26	24.98	3.67
3	21.58	20.25	24.98	3.66
4	21.60	20.26	25.00	3.65
5	21.59	20.25	25.01	3.67
6	21.57	20.24	24.98	3.68
7	21.57	20.25	24.99	3.66
8	21.58	20.25	24.99	3.67
9	21.60	20.27	24.98	3.65
10	21.59	20.26	24.99	3.67
Average value	21.585	20.256	24.989	3.664

Due to the structural limitation of the hobbing machine, the synchronous belt tensioning device was not installed, which led to the radial feed handwheel rotating in a crawling state and reducing the machining quality of the gear tooth surface. After reducing the axial feed, although the crawling state of the radial feed handwheel cannot be eliminated, the machining quality of the tooth surface was significantly improved. For the above, we not only reduced the axial feed but also hobbed the gear twice so that the crawling state of the radial feed handwheel was eliminated and the quality of tooth surface machining was improved.

When the axial feed was selected as 1 mm/r, the state of the tooth surface for single hobbing is shown in Figure 19a.

When the axial feed was selected as 0.5 mm/r, the state of the tooth surface for single hobbing is shown in Figure 19b.

When the axial feed was selected as 0.5 mm/r, the state of the tooth surface for double hobbing is shown in Figure 19c.

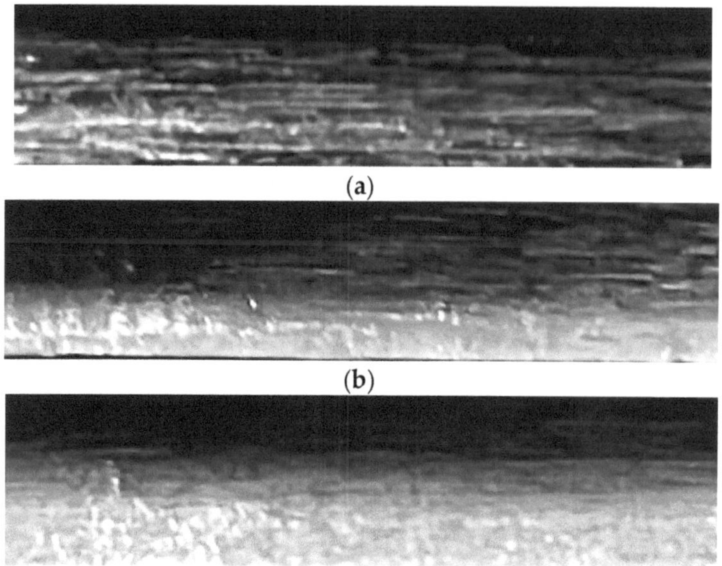

Figure 19. (a) The state of the tooth surface for single hobbing with the axial feed of 1 mm/r. (b) The state of the tooth surface for single hobbing with the axial feed of 0.5 mm/r. (c) The state of the tooth surface for double hobbing with the axial feed of 0.5 mm/r.

Based on the 6-level accuracy standard of GB/T11334-2005 "Geometrical Product Specifications (GPS) Taper Tolerance" [44], the taper tolerance of the beveloid gear was selected as 0.02° according to the table. The deviation between the actual value of the beveloid gear taper angle and the theoretical value was calculated as Equation (29).

$$\Delta\delta = \delta\prime - \delta = 3.044 - 3.06 = -0.016^{\circ} \tag{29}$$

From the result of Equation (29), it can be found that the taper deviation of the beveloid gear was within the tolerance range and met the 6-level accuracy standard. The deviation of the taper angle of the beveloid gear is mainly caused by the gap of the feed screw of the hobbing machine, the teeth number rounding of the synchronous belt wheel in the synchronous belt hanging wheel, and the transmission error of the internal transmission chain of the hobbing machine. To improve the accuracy of the taper angle of the beveloid gear, an improvement scheme can be proposed from the above aspects.

According to GB/T10095.1-2022 [45] "Cylindrical gears—ISO system of flank tolerance classification—Part 1: Definitions and allowable values of deviations relevant to flanks of gear teeth" [45] and referring to the 6-level accuracy grade standard for cylindrical gears, the tolerance of the circular thickness on the DP of the reference circle was obtained as 0.005 mm by checking the table. And the total deviation of the helix of the beveloid gear was obtained as 0.011 mm. It is calculated that the theoretical chordal thickness of the reference circle of the outer transverse plane of the beveloid gear (l') is 3.667 mm. The deviation of the chordal thickness was calculated by Equation (30).

$$\Delta l = l - l\prime = 3.664 - 3.667 = -0.003 \text{ mm} \tag{30}$$

From Equation (30), the result meets the 6-level accuracy standard within the allowable tolerance range. The chordal thickness error of the beveloid gear reference circle is mainly caused by the size error of the cutter and the transmission error of the hanging wheel of the hobbing machine. The accuracy of the chordal thickness of the beveloid gear can be improved from these two aspects.

Finally, we measured the total deviation of the left and right tooth surface reference circle helix (Δ_L and Δ_R) of the beveloid gear. The results showed that the hand of the dial gauge swung around the 0 lines, and the maximum runout did not exceed 0.01 mm. According to these above, the beveloid gears hobbed met the 6-level accuracy standard.

5. Conclusions

To improve the machining efficiency and accuracy of processing beveloid gears, reduce processing costs, and promote the mass production and industrial application of beveloid gears, we proposed a beveloid gear hobbing processing method based on the modification of ordinary gear hobbing machine and verified the feasibility and accuracy of the method via actual processing.

Firstly, according to the structural characteristics of beveloid gears, we analyzed the relative position and relative motion relationship between the cutter and the gear blank during the hobbing process based on the principle and method of gear hobbing. And the calculation method of the relevant processing parameters such as feed, cutter installation angle, and feed distance of the beveloid gears was derived.

Next, we designed the transmission chain between the axial feed and the radial feed of the hobbing machine and analyzed the transmission relationship between them by combining the transmission principle of helical cylindrical gears in the gear hobbing machine and the hobbing requirements of beveloid gears. After that, we analyzed the structural dimensions and feed motion parameters of the hobbing machine and then designed the modification scheme of the gear hobbing machine. We realized the linkage between the axial feed and the radial feed of the hobbing machine via the designed hanging wheel mechanism in synchronous belt type so that the hobbing processing requirements of beveloid gears were satisfied.

Finally, we completed the actual hobbing processing of the beveloid gear and verified the feasibility of the method proposed before. Then, the error measurement and analysis of the trial-produced beveloid gear were carried out. The results showed that the accuracy of the trial-produced beveloid gear met the 6-level standard, which meets the accuracy requirements of machining, and the accuracy of the processing method proposed in this study was verified.

Author Contributions: J.W. conceived and designed the study and applied for the funding which financially supported the study. H.Y. conducted data extraction, performed the analyses, and wrote the article. B.Y. and H.L. provided their comments and suggestions during discussions and by writing, especially on the data extraction method and paper formatting. All authors have read and agreed to the published version of the manuscript.

Funding: This research was funded by the National Natural Science Foundation of China under Grant Nos. 52075116 and 52175082 and the Natural Science Foundation of Shandong Province under Grant No. ZR2021ME025.

Data Availability Statement: Data are not available due to restrictions privacy or ethical, confidentiality.

Conflicts of Interest: The authors declare no conflicts of interest.

References

1. Yu, D. Research on the evaluation index system of comparative advantage of waterway transportation. *Transp. Res.* **2018**, 4, 1–6. [CrossRef]
2. Dong, S.; Wo, M. Advantages of large-scale equipment for waterway transportation. *Water Transp. Sci. Technol. Inf.* **1998**, 4, 8–9.
3. Fu, X. Opportunities and Challenges Faced by Waterway Transport in the New Era. *Pearl River Water Transp.* **2023**, 11, 21–23. [CrossRef]
4. Han, J. Exploring the economic development advantages of low-carbon economy sewer transportation. *China Storage Transp.* **2023**, 10, 104–105. [CrossRef]

5. Liu, J.; Chen, R.; Li, S.; Hu, X. Development and Application of Intelligent Waterway Transportation System: Case Study and Prospect. *Traffic Inf. Secur.* **2023**, *41*, 175–181.
6. Li, M. Water transport under the new economic situation. *China Shipp. Wkly.* **2021**, *12*, 46–47.
7. Wang, Y. Analysis of ship development trend. *Heilongjiang Sci.* **2016**, *7*, 108–109.
8. Li, Y.; Zheng, Z.; Wu, T.; Dou, Z. Development Trend and Challenges of Intelligent Ships in the Era of Industry 4.0. *Ship Eng.* **2023**, *45* (Suppl. S1), 224–229. [CrossRef]
9. Xu, W.; Qu, R.; Xue, G.; Xu, M.; Zhu, X.; Liu, Z.; Zhang, Q. Research status and development trend of intelligent ship system. *Ship* **2023**, *34*, 46–55. [CrossRef]
10. Wu, Q. Research status and future development trend of intelligent ships. *China Shipp. Wkly.* **2023**, *22*, 51–53.
11. Chang, S. Technology development status and the prospect of ship high-power gear transmission device. *Ship Sci. Technol.* **2010**, *32*, 17–22+45.
12. Zhang, L. Development of ship gear device. *Mech. Electr. Equip.* **1992**, *6*, 26–28.
13. Wang, H. Study on Contact Analysis and Dynamic Characteristics of Small Inclination Marine Gearbox. Ph.D. Thesis, Chongqing University, Chongqing, China, 2010.
14. Yang, T. Study on Dynamic Characteristics of New Marine RV Transmission Gearbox. Master's Thesis, Harbin Institute of Technology, Harbin, China, 2016.
15. Zhu, C. *Research and Application of Key Technologies of Spatial Variable Tooth Thickness Transmission*; Chongqing University: Chongqing, China, 1 December 2013.
16. Wu, J.; Li, G.; Wang, Q.; Sun, Z.; Li, H. Analysis and calculation of the differential difference of RV reducer for robot beveloid gear. *Mech. Des.* **2000**, *3*, 24–26+28–29+47. [CrossRef]
17. Beam, A.S. Beveloid gearing. *Mach. Des.* **1954**, *26*, 220–238.
18. Li, G.; Wu, J.; Li, H.; Qi, Y.; Lin, S.; Chen, X. Design and calculation of meshing involute thickening gears in parallel shafts. *China Mech. Eng.* **2000**, *8*, 52–55+53.
19. Du, X.; Zhu, C.; Song, D.; Xu, X. Current situation and development trend of interleaved shaft helical gear transmission variable tooth thickness technology. *Mech. Des.* **2012**, *29*, 1–7. [CrossRef]
20. Liu, H.; Lou, Y.; Liu, Q.; Chan, C.Y. Forming method to process drum gears. *China Heavy Equip.* **2008**, *2*, 20–24. [CrossRef]
21. Wang, Y. Discussion on the technology of forming gear processing. *Heavy Mach. Sci. Technol.* **2002**, *3*, 11–14. [CrossRef]
22. Wang, H. Forming Tooth Grinding. *Automot. Technol. Mater.* **2008**, *1*, 56–58.
23. Zhang, H.; Lou, Z.; Zhang, X.; Gao, B.; Nan, B. Research on the application of straight bevel gear milling technology//Chinese Society of Aeronautics and Astronautics. In Proceedings of the 4th China Aviation Science and Technology Conference in 2019, Shenyang, China, 15 August 2019; China Aviation Publishing & Media Co.: Beijing, China, 2019; p. 8.
24. Zhu, M. Hobbing of spur bevel gear. *Shanghai Mach.* **1965**, *7*, 15–18.
25. Zhang, S. *The Gear Shaping Processing Method Using Gear Shaping Machine*; Xi'an Aviation Power Co., Ltd.: Xi'an, China, 1 August 2010.
26. Yang, H. Research on shaving processing technology and grinding of shaving cutter. *Sci. Technol. Innov. Appl.* **2017**, *15*, 112.
27. Hong, D. Honing the large gear on the vertical lathe. *Mach. Manuf.* **1988**, *5*, 8–10.
28. Li, R.; Huang, L.; Chen, C. Research on design and modeling method of spiral bevel gear based on general CNC machining. *Sci. Technol. Innov.* **2023**, *5*, 23–27.
29. Yao, L. Design Theory and CNC Machining Method of Concave-Convex Arc Gear. Ph.D. Thesis, South China University of Technology, Guangzhou, China, 2018.
30. Wang, M. Research on CNC Machining Method of Linear Contact Arc Bevel Gear. Master's Thesis, Tianjin University, Tianjin, China, 2022. [CrossRef]
31. Shen, Y.; Sun, R. Application and processing of variable tooth beveloid involute gear. *Mech. Des. Res.* **2003**, *3*, 83–84+55-9. [CrossRef]
32. Zhang, Q. CNC transformation of gear hobbing machine. *J. Anhui Metall. Sci. Technol. Vocat. Coll.* **2004**, *2*, 35–37.
33. He, J.; Wu, X. Principle of involute beveloid gear hobbing. *Mech. Sci. Technol.* **2003**, *5*, 751–753.
34. Huang, J. Research on the Processing Method of Spatially Beveloid Gear Pair. Master's Thesis, Harbin Institute of Technology, Harbin, China, 2015.
35. Wu, J.; Li, G.; Li, H. Research on the diagonal inserting process of internal meshing beveloid gear pair. *J. Xi'an Pet. Inst. (Nat. Sci. Ed.)* **2000**, *3*, 45–48+0.
36. Hu, R. Design and Research of Precision Inserting Involute Variable Tooth Thickness Internal Gear Shaping Cutter. Master's Thesis, Chongqing University, Chongqing, China, 2022. [CrossRef]
37. Brecher, C.; Brumm, M.; Hübner, F.; Henser, J. Influence of the manufacturing method on the running behavior of beveloid gears. *Prod. Eng.* **2013**, *7*, 265–274. [CrossRef]
38. Wen, J.; Li, G.; Li, X.; Zhang, X.; Liu, Y. Research on modification method of non-involute beveloid gear. *J. Harbin Eng. Univ.* **2003**, *6*, 660–663.
39. Jiang, P. Research on Grinding Method of Conical Worm Grinding Wheel for Beveloid Gear. Master's Thesis, Chongqing University, Chongqing, China, 2019.

40. Zhang, Y. Research on the Grinding Processing Method of Involute Beveloid Gear. Master's Thesis, Lanzhou University of Technology, Lanzhou, China, 2022.
41. Liao, Y.; Zhu, P.; Zhao, W.; Lu, J.; Han, F. Development and application of CNC machining and system of variable tooth thick gear. *Mod. Mach.* **1997**, *3*, 20–22.
42. Wang, Z.; Shen, Y.; Xu, F.; Zhou, L. Honing of beveloid gear. *Mech. Des. Res.* **2007**, *3*, 106–108. [CrossRef]
43. Sun, H. Analysis of heat treatment process of 45 steel for gear. *Sci. Technol. Innov.* **2014**, *3*, 31. [CrossRef]
44. *GB/T 11334-2005*; Geometrical Product Specifications (GPS) Taper Tolerance. Standards Press of China: Beijing, China, 2005.
45. *GB/T10095.1-2022*; Cylindrical Gears-ISO System of Flank Tolerance Classification-Part 1: Definitions and Allowable Values of Deviations Relevant to Flanks of Gear Teeth. Standards Press of China: Beijing, China, 2022.

machines

MDPI

Article

A Semi-Analytical Loaded Contact Model and Load Tooth Contact Analysis Approach of Ease-Off Spiral Bevel Gears

Yuhui Liu, Liping Chen *, Xian Mao and Duansen Shangguan

School of Mechanical Science and Engineering, Huazhong University of Science and Technology, Wuhan 430074, China; yuhuiliu@hust.edu.cn (Y.L.); xianmao@hust.edu.cn (X.M.); ahcq1990@hust.edu.cn (D.S.)
* Correspondence: chenlp@hust.edu.cn; Tel.: +86-133-0716-2544

Abstract: This paper presents an innovative and comprehensive methodology for loaded tooth contact analysis (LTCA) of spiral bevel gears, integrating ease-off surface computation with high-precision virtual generating tooth surfaces. The methodology integrates an error-sensitivity analysis model with a semi-analytical LTCA model for spiral bevel gears based on ease-off surfaces, developed using a Universal Generation Model. By leveraging sophisticated corrections in the machining process, the desired ease-off surfaces are obtained, ensuring the accuracy of the generated tooth surfaces. This simulation ensures minimal errors between theoretical and virtual generating tooth surfaces, providing a reliable basis for LTCA. The LTCA model is formulated using CNC-generated tooth surfaces, focusing on misalignments such as pinion offset, adjustment errors, and angular position errors along the pinion and gear axis. The feasibility and effectiveness of the proposed method are verified through comparisons with LTCA software analysis results, demonstrating its high accuracy in predicting the impact of misalignments on contact patterns and load distribution. This integrated approach offers significant advancements in the design and analysis of spiral bevel gears, providing a robust tool for predicting and analyzing gear performance under various misalignment conditions. The combined methodology enhances the reliability and accuracy of LTCA, ensuring optimal performance and durability of spiral bevel gears in practical applications.

Keywords: spiral bevel gears; ease-off topography; semi-analytical loaded contact model; misalignments; error-sensitivity

1. Introduction

Spiral bevel gears are essential components in many mechanical systems, including automotive, aerospace, and industrial machinery, due to their ability to transmit power efficiently and smoothly between intersecting axes. The design and analysis of these gears are critical to ensure optimal performance, durability, and reliability. Traditional methods for analyzing spiral bevel gears have often relied on empirical data and trial-and-error approaches, which can be time-consuming and less accurate [1,2]. The concept of ease-off surfaces is fundamental in the design of spiral bevel gears. These surfaces represent the deviation of the actual tooth surface from the ideal conjugate surface, allowing for localized bearing contact and reducing edge-loading conditions to prevent early failure.

The research background for spiral bevel gears revolves around improving their load capacity and reducing the issues caused by misalignment and assembly errors. Over the years, substantial research has been conducted to enhance the performance and durability of these gears. Li [3] conducted an in-depth analysis of bending stress in orthogonal curve-face gears, proposing a method for calculating and testing the bending stress at the gear tooth root, which is crucial for understanding the gear's load-bearing capacity. Similarly, He [4] introduced a multi-step analytical approach for identifying the initial contact point in spiral bevel and hypoid gears, accounting for misalignments and using geometric kinematic transformations and a conic self-adaptive trust region algorithm, to improve tooth contact analysis accuracy and performance. Chen [5] introduced a semi-analytical Q-SLTCA method that efficiently calculates single tooth load contact characteristics and meshing stiffness, validated against finite element methods. Temirkhan [6] presented a computationally efficient quasi-static model for the three-dimensional non-conjugate contact problem between two surfaces, simplifying the conventional five-equation system to two nonlinear equations, and demonstrates improved accuracy and stability, particularly when applied to spur gears with crowned tooth surfaces. Marciniec [7] compared numerical methods for Gleason-type bevel gear contact patterns using mathematical models and finite element analysis, the results suggest mathematical analysis alone can suffice, avoiding the need for experimental verification. A real gear tooth surface modeling method was proposed in [8] to demonstrate actual tooth contact performances with manufacturing errors.

Complementing this, Simon [9] introduced a multi-objective optimization technique specifically tailored for hypoid gears, emphasizing the minimization of tooth contact pressure and transmission error, while simultaneously maximizing their mechanical efficiency. The optimization process heavily depends on LTCA for accurately predicting the distribution of tooth contact pressure and transmission errors, underscoring the critical role of precise computational methodologies in the analysis of gears. Moslem [10] used Transmission3D-Calyx software to perform loaded and unloaded tooth contact analyses of spiral bevel gears, evaluating the impact of axial and radial misalignments on mesh stiffness and gear lifespan. This real gear tooth surface modeling method was used to predict the wear tendency of gear tooth surfaces [11]. Building on these methods, Jiang [12] reviewed the development of logarithmic spiral bevel gears, emphasizing advancements in mathematical modeling, simulation analysis, and CNC machining centers. Adrian [13] outlined a method for generating gear tooth flanks using diagonal milling, introducing a virtual machine tool for precise and rapid calculation of change gears. Moreover, Li [2] investigated the design and power loss assessment of noncircular gear pairs for infinitely variable transmissions. This study presented an innovative approach using modified high-order elliptical pitch curves and introduced a method for evaluating power loss, crucial for enhancing the smoothness and efficiency of gear transmission systems. Similarly, Simon [14] explored advancements in mixed elastohydrodynamic lubrication and the performance enhancements of hypoid gears. Using a unified numerical approach, this research offers a comprehensive analysis of lubrication conditions. The findings demonstrate significant improvements in gear performance by optimizing lubrication conditions and reducing frictional losses. Liu [15] introduced a semi-analytical LTCA method for spiral bevel gears, combining analytical formulas, FEA corrections, and optimization for precise tooth deformation and contact analysis. Li [16] presented a multi-objective optimization approach for spiral bevel gears, focusing on enhancing both contact performance and meshing efficiency. This study utilized sensitivity analysis to select machine tool parameters significantly impacting the tooth surface, optimizing the gear design for higher performance and efficiency. Ding [17] presented a new method that uses optimization algorithms and operational strategies to accurately identify the initial contact point for ETCA, with its accuracy and efficiency validated through numerical examples. Fu [18] developed a mathematical model and finite element analysis for spiral bevel gears, comparing results with empirical formulas and experimental tests, showing good agreement.

Expanding on these methodologies, Ding [19] presented a novel prediction and control methodology specifically tailored for the collaborative grinding process of non-orthogonal aerospace spiral bevel gears, emphasizing improvements in geometric accuracy and load distribution through sophisticated simulation and optimization techniques. Kolivand [20] introduces a computationally efficient load distribution model for hypoid gears that uses ease-off topography and Rayleigh-Ritz shell models for accurate contact analysis, requiring less computational effort than finite element methods. Building on this, Simon [21] devised a multi-objective optimization technique aimed at optimizing the production of face-milled hypoid gears, leveraging numerically controlled machine tools for precise and efficient manufacturing. Stanasel [22] developed a mathematical model for manufacturing high-performance cylindrical gears with curved cycloidal teeth, validated through MATLAB simulation and Solid Edge modeling for future stress analysis. Adrian Ghionea [23] proposed a new penetration-based contact model that enables more efficient and accurate computation of three-dimensional contact loads, thereby supporting system-level and dynamic analysis, addressing the challenges of contact simulation posed by the complex geometry of spiral bevel gears. Mathur [24] presents a numerical model for loaded tooth contact analysis of straight bevel gears in pericyclic transmissions, using finite strip methods to calculate tooth deflection and a variational framework for load distribution, highlighting the high-power density and noise reduction potential. Moreover, Li [25] focused on the impact of lubrication on gear performance, underscoring the importance of proper lubrication in reducing friction and wear in spiral bevel gears. The integration of these diverse research efforts provides a comprehensive understanding of the current state of the field, showcasing the multifaceted approach required to address the challenges and enhance the performance of spiral bevel gears. Despite these advancements, several bottlenecks persist in the field. One significant challenge is achieving precise tooth surface corrections during the machining process. This difficulty is compounded by the need for accurate simulation tools that can predict gear behavior under various load and misalignment scenarios.

Building on these insights into material performance and failure mechanisms, this paper tackles the challenge of accurately predicting loaded contact patterns in spiral bevel gears by introducing a novel semi-analytical model. This model integrates ease-off surfaces derived from universal gear machining settings, and Hertzian contact theory, ensuring high accuracy in generated tooth surfaces and improving gear performance and durability. The approach ensures high accuracy in generated tooth surfaces, leading to improved performance and durability of spiral bevel gears. The main bottleneck addressed by this paper is the difficulty in achieving precise tooth surface corrections during the machining process.

Consequently, this paper also focuses on the simulation and analysis of misalignments in spiral bevel gears. Using high-precision virtual generating tooth surfaces obtained using a Universal Generation Model, the study minimizes the errors between theoretical and virtual surfaces. By formulating an LTCA model using CNC-generated tooth surfaces, the finite element analysis (FEA) based LTCA model developed in this paper considers various misalignment errors, such as pinion offset and adjustment errors along the pinion and gear axis. In such a way, the research accounts for misalignments and load distribution, providing a comprehensive approach to analyzing gear performance under realistic operational conditions.

2. Generation of Gear Tooth Profiles

2.1. The Dimensional Configuration of Spiral Bevel Gears

The intricate geometry of spiral bevel gears is solely determinable through the resolution of implicit equations, which meticulously capture the intricacies of the manufacturing process, including the geometry of the cutting tool and the specific settings of the machine tool.

In addition to meeting the basic geometric and gear blank dimensions, it is essential to fulfill performance or functionality requirements concerning contact performance. Key aspects frequently evaluated in the design of spiral bevel gears include the location, size, and shape of the contact pattern on the gear tooth surfaces, as well as the motion transmission error amplitude of the gear pair. Figure 1 illustrates the pinion generation setup, which includes a head-cutter with blades, a virtual generating gear, and the pinion.

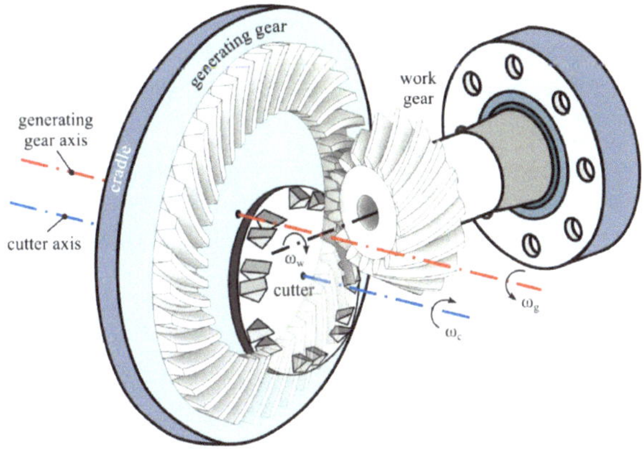

Figure 1. Generalized bevel gear generation configuration.

2.2. The Geometry of Cutting Tools and Their Relative Movement

A universal mathematical framework is employed for the generation of tooth surfaces for both pinion and gear, as shown in Figure 2. The coordinate system S_2 is firmly attached and remains stationary relative to the workpiece. The coordinate system S_{c2}, firmly attached to the generating machine cradle, describes its angular position. The coordinate system S_{m2} is attached to the cutting machine framework. S_{a2} and S_{b2} assist in work piece installment. The coordinate system S_g is firmly attached to the gear head-cutter. The angles ψ_2 and ψ_{c2} are associated with instantaneous rotational positions of the gear and the cradle, respectively. There exist six potential auxiliary kinematic variables that can be utilized to alter the tooth surface of a manufactured spiral bevel gear: $\Delta\mathbf{X}_{D2}$ represents the machine center to back, $\Delta\mathbf{E}_{m2}$ denotes the blank offset, γ_{m2} is the machine root angle, S_{r2} signifies the radial distance, $\mathbf{X}_{b2}$ indicates the sliding base, and q_2 is the cradle rotation angle. The gear roll ratio m_{2c_2} is determined by

$$m_{2c_2} = \frac{\omega_w}{\omega_c} = \frac{\mathrm{d}\psi_2}{\mathrm{d}t} / \frac{\mathrm{d}\psi_{c_2}}{\mathrm{d}t} \tag{1}$$

where ω and ψ represent angular speeds of the gear and the cradle, respectively.

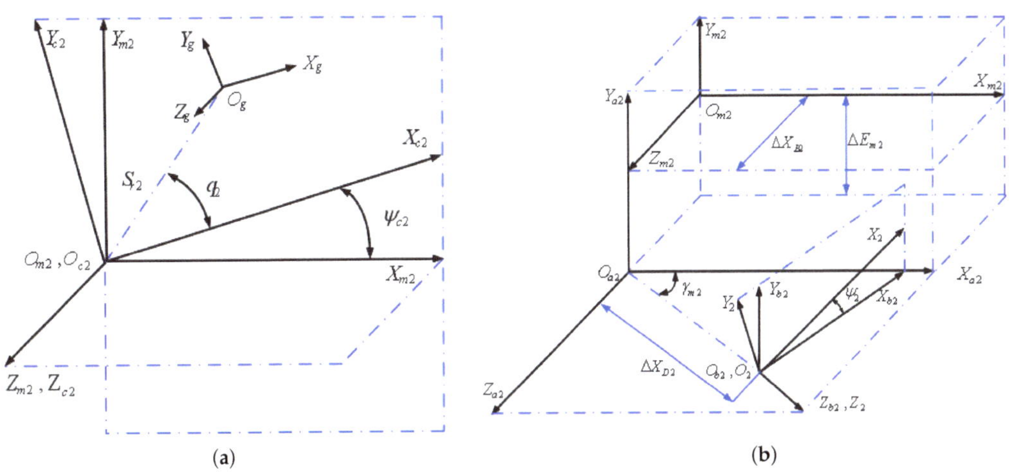

Figure 2. Coordinate systems for the generation of gear tooth profiles. (a) Depiction of the installation of the machine tool for the process of generation; and (b) Illustration of machine tool settings.

As shown in Figure 3, The cutting edge of the head-cutter blade is divided into four distinct sections: the edge, toprem, profile, and flankrem portions. The cutting surfaces are created by rotating the blade around the z_g-axis with an angle of θ_g. The primary shaping of the gear tooth surface occurs through the profile section, which is characterized by a straight line with a profile angle of α_g. The fillet on the gear tooth surface is generated by the edge section, which incorporates a corner radius of ρ_ω. Referring to Figure 3, an arbitrary point M_b located on the cutting surface of the blade is determined by parameters s_g and θ_g, as shown in Figure 4. The generating surface $\sum_g^{(a)}$ of the profile section of the head-cutter blade is expressed through the vector function $\mathbf{r}_g^{(a)}(s_g, \theta_g)$ as

$$\mathbf{r}_g{}^{(a)}(s_g, \theta_g) = \begin{bmatrix} (r_G \pm s_g \sin\alpha_g)\cos\theta_g \\ (r_G \pm s_g \sin\alpha_g)\sin\theta_g \\ -s_g \cos\alpha_g \end{bmatrix} \tag{2}$$

where α_g is the blade angle, s_g denotes tooth surface parameters, and r_G represents the head-cutter point radius. The upper signs in Equation (2) signify generation of the concave side of the gear tooth surface, whereas the lower signs correspond to the convex side.

The unit normal vector to the generating surface $\sum_g^{(a)}$ of the profile is denoted by $\mathbf{n}_g^{(a)}(\theta_g)$ and is represented as

$$\mathbf{n}_g{}^{(a)}(\theta_g) = \begin{bmatrix} \cos\alpha_g \cos\theta_g \\ \cos\alpha_g \sin\theta_g \\ \pm\sin\alpha_g \end{bmatrix} \tag{3}$$

The generating surface $\sum_g^{(b)}$ relates to the edge section of the head-cutter blade, while $\sum_g^{(a)}$ is described by the vector function $\mathbf{r}_g^{(b)}(s_g, \theta_g)$ as

$$\mathbf{r}_g{}^{(b)}(\lambda_w, \theta_g) = \begin{bmatrix} (r_w \pm \rho_w \sin\lambda_w)\cos\theta_g \\ (r_w \pm \rho_w \sin\lambda_w)\sin\theta_g \\ -\rho_w(1-\cos\lambda_w) \end{bmatrix}, 0 \le \lambda_w \le \frac{\pi}{2} - \alpha_g \tag{4}$$

where $r_w = r_G \mp \rho_w(1-\sin\alpha_w)/\cos\alpha_g$. ρ_ω is the corner radius in the edge section of head-cutter blade for the fillet of the gear.

The unit normal to the generating surface $\sum_g^{(b)}$ of the edge section is denoted by $\mathbf{n}_g^{(b)}(\theta_g)$ and is represented as

$$\mathbf{n}_g{}^{(b)}(\theta_g) = \begin{bmatrix} \sin\lambda_w \cos\theta_g \\ \sin\lambda_w \sin\theta_g \\ \pm\cos\lambda_w \end{bmatrix} \tag{5}$$

The surface of the generated gear tooth, denoted as $\sum_2$, is represented as

$$\begin{cases} r_2{}^{(a)}(s_g, \theta_g, \psi_2) = M_{2g}(\psi_2) r_g{}^{(a)}(s_g, \theta_g) \\ r_2{}^{(b)}(\lambda_w, \theta_g, \psi_2) = M_{2g}(\psi_2) r_g{}^{(b)}(\lambda_w, \theta_g) \end{cases} \tag{6}$$

where ψ_2 is the rotation angle of the gear, and $\mathbf{M}_{2g}$ represents the transformation matrix from coordinate S_g to S_2.

$$\mathbf{M}_{2g}(\psi_2) = \mathbf{M}_{2b_2}\mathbf{M}_{b_2a_2}\mathbf{M}_{a_2m_2}\mathbf{M}_{m_2c_2}\mathbf{M}_{c_2g} \tag{7}$$

here,

$$\mathbf{M}_{c_2g} = \begin{bmatrix} 1 & 0 & 0 & S_{r_2}\cos q_2 \\ 0 & 1 & 0 & S_{r_2}\sin q_2 \\ 0 & 0 & 1 & 0 \\ 0 & 0 & 0 & 1 \end{bmatrix},$$

$$\mathbf{M}_{m_2c_2} = \begin{bmatrix} \cos\psi_{c_2} & -\sin\psi_{c_2} & 0 & 0 \\ \sin\psi_{c_2} & \cos\psi_{c_2} & 0 & 0 \\ 0 & 0 & 1 & 0 \\ 0 & 0 & 0 & 1 \end{bmatrix},$$

$$\mathbf{M}_{a_2m_2} = \begin{bmatrix} 1 & 0 & 0 & 0 \\ 0 & 1 & 0 & \Delta E_{m_2} \\ 0 & 0 & 1 & -\Delta X_{B_2} \\ 0 & 0 & 0 & 1 \end{bmatrix},$$

$$\mathbf{M}_{m_2c_2} = \begin{bmatrix} \sin\gamma_{m_2} & 0 & -\cos\gamma_{m_2} & 0 \\ 0 & 1 & 0 & 0 \\ \cos\gamma_{m_2} & 0 & \sin\gamma_{m_2} & -\Delta X_{D_2} \\ 0 & 0 & 0 & 1 \end{bmatrix},$$

$$\mathbf{M}_{2b_2} = \begin{bmatrix} \cos\psi_2 & \sin\psi_2 & 0 & 0 \\ -\sin\psi_2 & \cos\psi_2 & 0 & 0 \\ 0 & 0 & 1 & 0 \\ 0 & 0 & 0 & 1 \end{bmatrix}.$$

$$\begin{cases} \mathbf{n}_g{}^{(a)} \cdot \mathbf{v}_{cg} = f_g{}^{(a)}(s_g, \theta_g, \psi_2) = 0 \\ \mathbf{n}_g{}^{(b)} \cdot \mathbf{v}_{cg} = f_g{}^{(b)}(\lambda_w, \theta_g, \psi_2) = 0 \end{cases} \quad (8)$$

where $\mathbf{v}_{cg}$ is the relative speed vector of the head-cutter to the gear, and $\mathbf{n}_g^{(a)}$, $\mathbf{n}_g^{(b)}$ are unit normal vectors of the head-cutter.

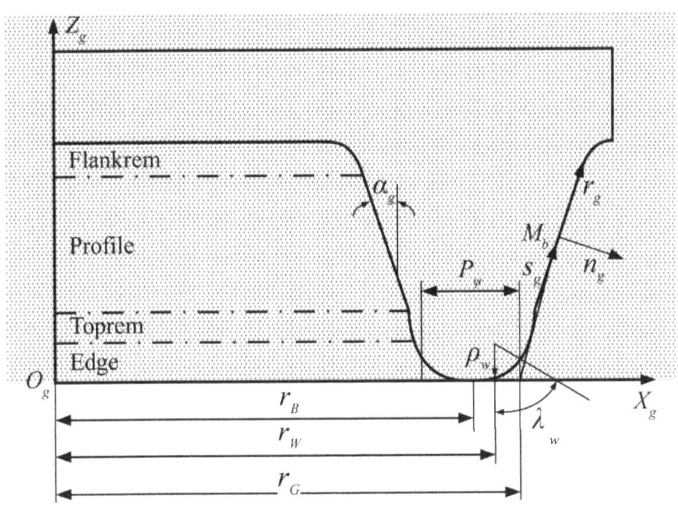

Figure 3. The four distinct segments of blades.

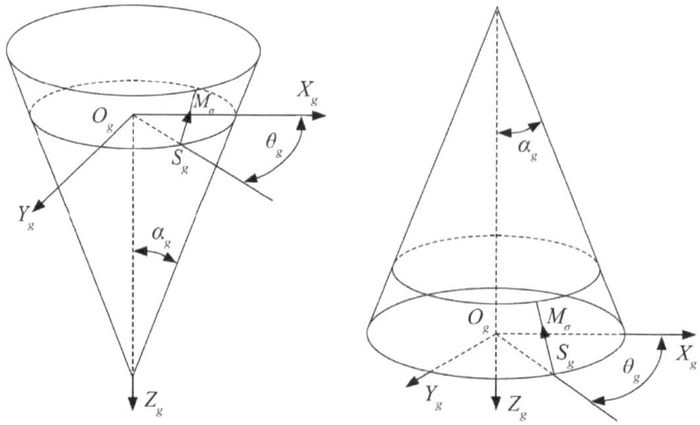

Figure 4. Blades and generating cones utilized in the gear generating tool: Specific generating tool cones designed for the concave and convex sides.

2.3. *Kinematics of Universal Motion Concept (UMC)*

The advent of modern CNC spiral bevel gear generating machines has significantly advanced the generation concept [26,27]. The UMC, which has been developed and seamlessly integrated with machine control software [28], greatly enhances the capabilities of free-form CNC spiral bevel generators. Figure 5 depicts the kinematic representation of the machine tool configurations for a UMC spiral bevel generator model. Each configuration is visualized as a moving element, and its motion is mathematically described through a polynomial function that is dependent on the cradle's rotation angle. Despite the kinematic model's foundation in mechanical machinery, we can incorporate the advanced capabilities of modern CNC machines by expressing the machine tool configurations as functions that vary with the cradle's rotational increments, thereby eschewing the assumption of static, fundamental settings. namely,

$$R_a = R_{a0} + R_{a1}q + R_{a2}q^2 + \cdots + R_{a5}q^5 + R_{a6}q^6 \tag{9}$$

$$X_b = X_{b0} + X_{b1}q + X_{b2}q^2 + \cdots + X_{b5}q^5 + X_{b6}q^6 \tag{10}$$

$$s = s_0 + s_1 q + s_2 q^2 + \cdots + s_5 q^5 + s_6 q^6 \tag{11}$$

$$E_m = E_{m0} + E_{m1}q + E_{m2}q^2 + \cdots + E_{m5}q^5 + E_{m6}q^6 \tag{12}$$

$$X_p = X_{p0} + X_{p1}q + X_{p2}q^2 + \cdots + X_{p5}q^5 + X_{p6}q^6 \tag{13}$$

$$\gamma_m = \gamma_{m0} + \gamma_{m1}q + \gamma_{m2}q^2 + \cdots + \gamma_{m5}q^5 + \gamma_{m6}q^6 \tag{14}$$

$$j = j_0 + j_1 q + j_2 q^2 + \cdots j_5 q^5 + j_6 q^6 \tag{15}$$

$$i = i_0 + i_1 q + i_2 q^2 + \cdots + i_5 q^5 + i_6 q^6 \tag{16}$$

Equations (9)–(16) encapsulate both the static, fundamental machine tool configurations and the dynamic adjustments in those configurations, which may be contingent upon the cradle's rotation parameter q. These motions can be precisely executed by CNC machines via computer programming.

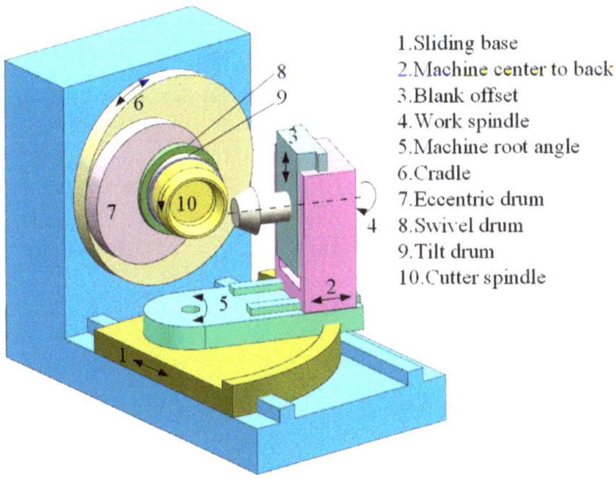

Figure 5. A machine tool model of spiral bevel gears.

2.4. The Simulation of Generation Process

As described, both face hobbing and face milling procedures can be virtually simulated utilizing the mechanical UMC spiral bevel generator shown in Figure 5. In contrast to cradle-style machines, which are predicated on the spiral bevel gear generation principle involving a generating gear, CNC spiral bevel generators boast a more adaptable design. Nonetheless, the design of spiral bevel gears can be correlated to UMC spiral bevel generators by determining the appropriate machine tool configurations. Furthermore, the generating motions specified for UMC generators can be translated into the motions executed by CNC hypoid generators via computer programming. The six axes of CNC hypoid generators are actuated directly by servo motors, empowering them to execute precise motion functions. As a result, the machine tool configurations can be dynamically characterized as functions that vary with the cradle's incremental angle q. To accurately portray the generation process, we dissect the relative motion components within the kinematic model depicted in Figure 5 and assign a unique coordinate system to each element. Given the generating roll increment angle q and using Equation (16), we may determine the generation process, as shown in Figure 6. The envelope of $\mathbf{r}_g$ with three independent variables s_g, θ_g and q is expressed as

$$\left(\frac{\partial \mathbf{r}_g}{\partial s_g} \times \frac{\partial \mathbf{r}_g}{\partial \theta_g}\right) \cdot \frac{\partial \mathbf{r}_g}{\partial q} = 0 \tag{17}$$

The generating roll is used in UMC processes and is associated with the rotations of the machine cradle and the work. It can be represented as

$$\omega_w = R_a \omega_c \tag{18}$$

where ω_w and ω_c denote the angular speeds of the workpiece (the gear) and the cradle, respectively. R_a is the ratio of roll.

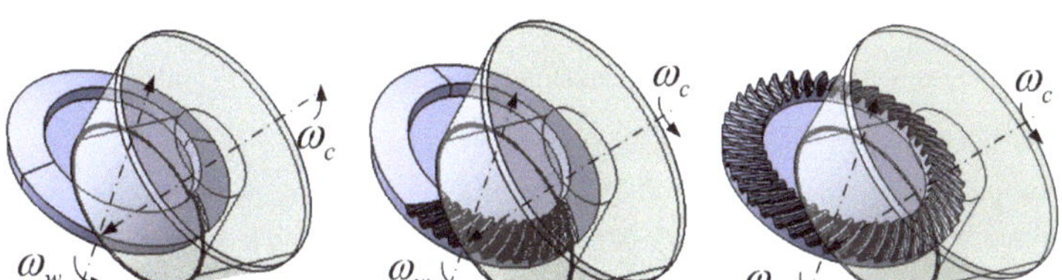

Figure 6. The generation process of the spiral bevel gear.

3. Semi-Analytical Loaded Contact Model Based on Ease-Off Topography

3.1. Theoretical Ease-Off Topography

The ease-off value represents the perpendicular difference between the ease-off surface and the conjugate tooth surface. Figure 7 illustrates examples of gear ease-off topographies for the gear's convex side, where the ease-off grids are based on areas of the gear tooth surfaces produced by higher-order motions described in Equations (19)–(22). If the gear tooth surface is produced by the pinion tooth surface under a constant rolling condition, as determined by the tooth numbers of both the pinion and the gear, the pinion ease-off can be defined as

$$\Delta \mathbf{r}_p = \left[\mathbf{r}_p(s_p, \theta_p, \psi_1) - \mathbf{r}_{pc}(s_g, \theta_g, \psi_2, \phi_1)\right] \cdot \mathbf{n}_p(s_p, \theta_p, \psi_1) \tag{19}$$

where ϕ_1 is the generating motion parameter of the pinion; $\mathbf{r}_{pc}$ represents the conjugate pinion surface generated by the gear tooth, which is determined by

$$\left\{\begin{array}{l} \mathbf{r}_{pc} = \mathbf{M}_{pg}(\phi_1)\mathbf{r}_g(u_g, \theta_g, \psi_2) \\ \mathbf{n}_{pc} = \mathbf{L}_{pg}(\phi_1)\mathbf{n}_g(u_g, \theta_g, \psi_2) \\ f_{pc}(u_g, \theta_g, \psi_2, \phi_1) = \frac{\mathrm{d}\mathbf{r}_{pc}}{\mathrm{d}\phi_1} \cdot \mathbf{n}_{pc} = 0 \end{array}\right] \tag{20}$$

where $f_{pc}(u_g, \theta_g, \psi_2, \phi_1)$ is the equation of meshing, and $\mathbf{M}_{pg}$ and $\mathbf{L}_{pg}$ are the coordinate transformation matrices from $S_p(x_p, y_p, z_p)$ to $S_g(x_g, y_g, z_g)$. Similarly, the gear ease-off tooth surface can be defined by

$$\Delta \mathbf{r}_g = \left[\mathbf{r}_g(s_g, \theta_g, \psi_2) - \mathbf{r}_{gc}(s_p, \theta_p, \psi_1, \phi_1)\right] \cdot \mathbf{n}_g(s_g, \theta_g, \psi_2) \tag{21}$$

where $\mathbf{r}_{gc}$ is the conjugate gear surface that is generated by the pinion tooth and is determined by

$$\left\{\begin{array}{c} \mathbf{r}_{gc} = \mathbf{M}_{gp}(\phi_1)\mathbf{r}_p(u_p, \theta_p, \psi_1) \\ \mathbf{n}_{gc} = \mathbf{L}_{gp}(\phi_1)\mathbf{n}_p(u_p, \theta_p, \psi_1) \\ f_{gc}(u_g, \theta_g, \psi_2, \phi_1) = \frac{\mathrm{d}\mathbf{r}_{gc}}{\mathrm{d}\phi_1} \cdot \mathbf{n}_{gc} = 0 \end{array}\right] \tag{22}$$

where $\mathbf{M}_{gp}$ and $\mathbf{L}_{gp}$ are coordinate transformation matrices from $S_g(x_g, y_g, z_g)$ to $S_p(x_p, y_p, z_p)$.

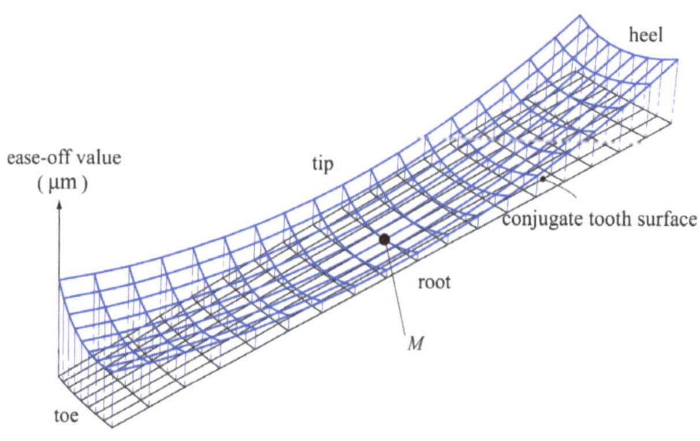

Figure 7. The ease-off topographies of the gear convex side.

3.2. Calculation of the Contact Line Direction

The principal axis of the contact ellipse, which arises during tooth contact under load, aligns itself with the direction of the contact line precisely at the point of contact. To ascertain the orientation of the principal axis of the contact ellipse, a surface Σ^{pg} is defined. The tangent plane Σ^{pg} contacts both gear tooth and pinion surface at M_m. The normal curvature, which quantifies the rate of change of the surface's direction to a given vector traversing the surface **t** on Σ^{pg} is defined by

$$\kappa_t^{pg} = \kappa_t^p - \kappa_t^g \tag{23}$$

where κ_t^p and κ_t^g are normal curvatures along vector **t**. The orientations of the major and minor axes of the contact ellipse can be delineated on the surface Σ^{pg} by identifying the principal directions,$\mathbf{t}_1$ and $\mathbf{t}_2$. The principal directions, specifically $\mathbf{t}_1$ and $\mathbf{t}_2$, can be identified on the surface Σ^{pg} to determine the major and minor axis orientations of the contact ellipse. As depicted in Figure 8, the principal orientations on the surface Σ^{pg} signify the extreme curvature values that occur as the direction of an arbitrary vector **t** is varied. Using Euler's equation in Equation (17), the normal curvature along any vector **t** on a given surface can be expressed in terms of its principal curvatures, κ_1 and κ_2.

$$\kappa_t^p = \kappa_1^p \cos^2\lambda + \kappa_2^p \sin^2\lambda \tag{24}$$

$$\kappa_t^g = \kappa_1^g \cos^2(\lambda - \varphi) + \kappa_2^g \sin^2(\lambda - \varphi) \tag{25}$$

where φ is the angle between $\mathbf{e}_1^g$ and $\mathbf{e}_1^p$, and λ is the angle between $\mathbf{e}_1^p$ and arbitrary vector **t**, with both angles measured positively in the counter-clockwise direction. Consequently, the relative curvature along the vector **t** can be expressed as

$$\kappa_t^{pg} = \kappa_1^g \cos^2(\lambda - \varphi) + \kappa_2^g \sin^2(\lambda - \varphi) - \kappa_1^p \cos^2\lambda - \kappa_2^p \sin^2\lambda \tag{26}$$

In Figure 8, to identify one of the principal directions that signify the direction of the contact line at the contact point M_m on the surface Σ^{pg}, we search for the angles that correspond to the limiting curvature values κ_t^{pg} can be obtained by setting $\mathrm{d}\kappa_t^{pg} \big/ \mathrm{d}\lambda = 0$. The angle λ corresponding to $\mathbf{t}_1$ is denoted by φ,

$$\lambda = \frac{1}{2}\tan^{-1}\left[\frac{\sin(2\varphi)}{\kappa_{12}^{pg} - \cos(2\varphi)}\right] \tag{27}$$

where $\kappa_{12}^{pg} = \frac{\kappa_1^p - \kappa_2^p}{\kappa_1^g - \kappa_2^g}$.

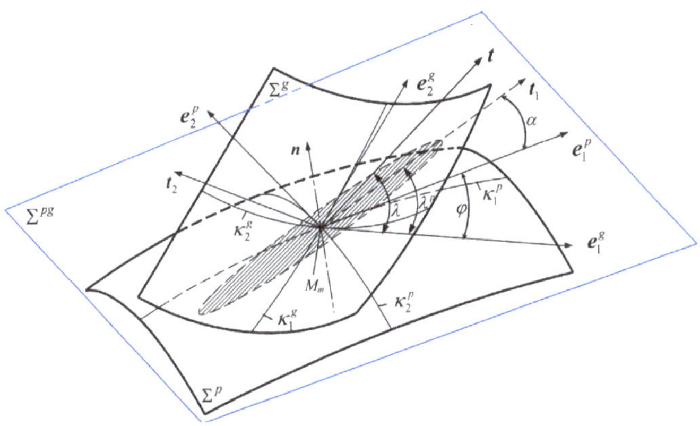

Figure 8. A depiction of the local coordinate systems on the surface at a given instant of contact.

3.3. Deformation Coordination Equation of a Contact Ellipse

During the meshing cycle of spiral bevel gears, the loaded tooth contact analysis is reiterated for each meshing position, enabling the determination of the distribution of instantaneous contact load across every contact ellipse. The elastic deformation is divided into two categories contact deformation and bending deformation. To ensure the meshing continuity between the pinion and the gear, the total deformation is constant at various contact locations. The deformation coordination equation of an arbitrary contact ellipse Ω_ℓ is defined by

$$\int_{\Omega_\ell} (f_B)_{\eta,\xi} \cdot (P_H)_\xi \mathrm{d}\xi + (f_C)_\eta \cdot (P_H)_\eta = \delta(\eta,\Delta) \cdot eR(\eta) \tag{28}$$

where $\delta(\eta,\Delta) = \begin{cases} \Delta - [\Delta r_p(\eta) + \Delta r_g(\eta)] & \text{for} \quad \delta(\eta,\Delta) > 0 \\ 0 \quad \text{for} \quad \delta(\eta,\Delta) \le 0 \end{cases}$, ℓ is the index of meshing members, $\ell = 1, 2, \cdots, n_\ell$, n_ℓ is the quantity of meshing members at the present contact location, η is the deformation measurement point coordinate, ξ is the contact point coordinate, $(P_H)_\xi$ is the contact pressure at the contact point ξ, $(f_B)_{\eta,\xi}$ is the bending flexibility of point η correspond to contact point ξ, $(f_C)_\eta$ is the contact flexibility of point $\eta = \xi$ on the meshing surface, $\delta(\eta,\Delta)$ is the deformation at point η, Δ is the passive gear rotation angle relative to the drive gear, and $eR(\eta)$ is the coefficient that transformation angle deformation to tooth coordinate, which can be represented as

$$eR(\eta) = \left[x_g(\eta) \cdot \mathbf{e}_2^g(\eta) - y_g(\eta) \cdot \mathbf{e}_1^g(\eta) \right] \tag{29}$$

Due to the total deformation of meshing teeth cannot be determined; the deformation coordination Equation (28) is statically indeterminate. It cannot be strictly solved. The total normal load of all contact points is the tangential load of gear transmission.

$$T = \sum_{\ell=1}^{n_\ell} \int_{\Omega_\ell} (P_H^*)_\xi \cdot r(\xi) \mathrm{d}\xi \tag{30}$$

where $(P_H^*)_\xi$ is tangential pressure of the contact point ξ, and $r(\xi)$ is the radius of contact point ξ around the drive gear axis. Substituting Equations (28) and (29) into Equation (30), one has

$$\begin{bmatrix} (f_{BC})_{1,1} & (f_{BC})_{1,2} & \cdots & (f_{BC})_{1,n_\ell} \\ (f_{BC})_{2,1} & (f_{BC})_{2,2} & \cdots & (f_{BC})_{2,n_\ell} \\ \cdots & \cdots & \cdots & \cdots \\ (f_{BC})_{n_\ell,1} & (f_{BC})_{n_\ell,2} & \cdots & (f_{BC})_{n_\ell,n_\ell} \end{bmatrix} \begin{bmatrix} (P_H)_1 \\ (P_H)_2 \\ \cdots \\ (P_H)_{n_\ell} \end{bmatrix} = \begin{bmatrix} \{\Delta - [\Delta r_p(1) + \Delta r_g(1)]\} \cdot eR(1) \\ \{\Delta - [\Delta r_p(2) + \Delta r_g(2)]\} \cdot eR(2) \\ \cdots \\ \{\Delta - [\Delta r_p(n_\ell) + \Delta r_g(n_\ell)]\} \cdot eR(n_\ell) \end{bmatrix} \tag{31}$$

where

$$(f_{BC})_{\eta,\xi} = \begin{cases} (f_B)_{\eta,\xi} \text{ for } \eta \ne \xi \\ (f_B)_{\eta,\xi} + (f_C)_\eta \text{ for } \eta = \xi \end{cases} \tag{32}$$

Based on the gear engagement at various contact locations, a composite flexibility matrix $\mathbf{f}_{BC}$ is constituted at the contact points on the tooth surface [29].

3.4. Semi-Analytical Contact Model for Loaded Tooth Contact Analysis

To accurately simulate the position and attributes of the contact pattern on the ease-off surface utilizing the semi-analytical loaded contact model, it is imperative to initially distribute the applied load along the major axis of the contact ellipse, spanning the entire actual contact width. On any contact ellipse Ω_ℓ, the half contact width, denoted as b_{M_m}, of this contact ellipse is determined using Hertzian theory.

$$b_{M_m} = \sqrt{\frac{4\rho_{M_m}(P_H)_{M_m}}{\pi b} \cdot \frac{1}{\left(\frac{1-v_p^2}{E_p} + \frac{1-v_g^2}{E_g}\right)}} \tag{33}$$

where v_g and v_p are the gear and Poisson's ratios of the pinion, respectively. $(P_H)_{M_m}$ is the pressure at the contact point M_m. b is tooth width. E_p and E_g are the elasticity moduli for materials of the pinion and gear, respectively. ρ_{M_m} represents the equivalent radius of curvature that corresponds to the direction aligned with the major axis of the contact ellipse, which is defined as

$$\rho_{M_m} = \left(\frac{1}{(\rho_p)_{M_m}} + \frac{1}{(\rho_g)_{M_m}} \right)^{-1} \tag{34}$$

where $(\rho_p)_{M_m}$ and $(\rho_g)_{M_m}$ represent the radii of curvature of the pinion and the gear at the location M_m along the direction of the contact line, respectively, and $(P_H)_{M_m}$ is the pressure at the contact point M_m that is determined by Hertzian theory.

$$(p_H)_{M_m} = \sqrt{\frac{(F_n)_{\Omega_\ell}}{a_{M_m}} \cdot \frac{1}{\rho_{M_m}} \cdot \frac{1}{\pi\left(\frac{1-v_p^2}{E_p} + \frac{1-v_g^2}{E_g}\right)}} \tag{35}$$

where a_{M_m} is the length of this contact ellipse, $a_{M_m} = \sqrt{2\tau/\kappa_t^{pg}}$, in which τ is the unload separation distance, and $(F_n)_{\Omega_\ell}$ is the normal load carried by this contact ellipse, $(F_n)_{\Omega_\ell} = \int_{\Omega_\ell} (P_H^*)_\xi \mathrm{d}\xi$.

4. Validation of the LTCA Model by Using CNC-Generated Tooth Surfaces

Typically, the installation state of gears does not perfectly align with the design specifications, leading to misalignments. These misalignments result in a "mismatched" spiral bevel gear pair, introducing uncertainty regarding the path of contact between the meshing tooth surfaces and the actual load distribution. This methodology is utilized to investigate the consequences of tooth inaccuracies and misalignments on various aspects, including the location of the contact path, the potential extent of the contact area, any separations within this potential contact area, and the angular deviations exhibited by the driven gear member from its theoretically precise position, as dictated by the gear tooth ratio.

4.1. Misalignments and Load Distribution of Spiral Bevel Gears

As is shown in Figure 9, the meshing of tooth surface is considered in the fixed coordinate system S_h. In Figure 9a,b, movable coordinate system S_1 and S_2 are rigidly connected to pinion and gear, respectively. And both of the pinion and gear are around the Z-axis with a rotation angle. Auxiliary coordinate system S_{b1} and S_{b2} are on behalf of the rotation of pinion and gear, respectively. The misalignments of this system are calculated through coordinate system S_{b1} and S_{b2} relative to a fixed coordinate system S_h, which is shown in Figure 9c. In Figure 9, ΔA_1 and ΔA_2 are axial displacement of pinion and gear, respectively, $\Delta\Sigma$ is a change of shaft angle, ΔE is the shortest distance between the two axis of pinion and gear when they are crossed but not intersected.

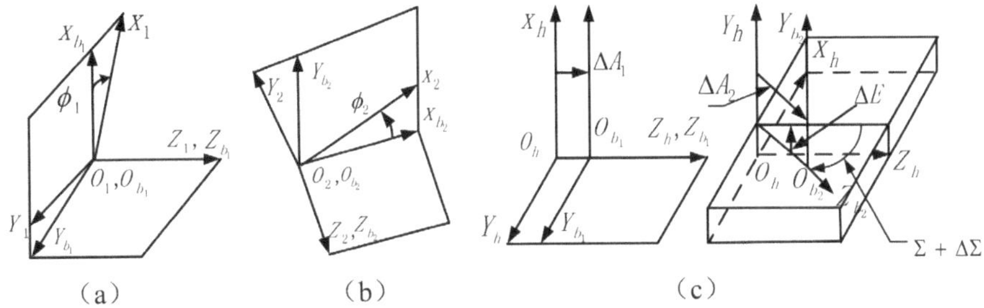

Figure 9. Mesh coordinate system with misalignments.(**a**) Corner of pinion; and (**b**) Coner of gear; and (**c**) ΔA_1, ΔA_2, ΔE and $\Delta\Sigma$ with misalignments.

4.2. Validation of the LTCA Model by Simulating Various Misalignment Scenarios and Comparing Results with LTCA Software Analysis

Based on the core parameters in Table 1, a 3D solid model is constructed in CAD and then imported into the finite element pre-processing software to obtain the finite element mesh model. The finite element meshes of the gear system are shown in Figure 10.

Figure 10. Finite element meshes of a spiral bevel gear system.

The gear finite element analysis model preprocessing and boundary conditions are set as follows, the same material properties are assigned to both the pinion and the gear, with an elastic modulus of 209,000 MPa and a Poisson's ratio of 0.3. The elastic modulus of the gear's support structure is 197,000 MPa, with a Poisson's ratio of 0.3. The element property uses hexahedral first-order reduced integration elements (C3D8R). An implicit, static analysis algorithm is adopted with a step size of 0.02 during the analysis process, requiring 50 steps to complete the analysis. This setup helps to obtain the analysis structure at different contact positions during meshing. The tooth surfaces of the gear and pinion are defined as a contact pair, with the friction coefficient in the tangential contact properties set to 0.1. Reference points are established at the center of the bearing installation positions on both the pinion shaft and the gear support structure. Coupling constraints are created between the reference points and the bearing contact surfaces. Two perpendicular spring elements are added between the reference points and fixed spatial points, with a spring stiffness of 120,000 N/mm. Contact force, contact pressure, stress, and displacement are set as the output variables. The boundary conditions are defined by constraining all degrees of freedom of both the gear and pinion except for rotation along the axis, and setting the rotation angle of the pinion to 240°.

To ensure convergence during the gear contact analysis process, the analysis is divided into three steps: In the first analysis step, a small rotation is applied to the degree of freedom along the gear's axis while fixing the pinion. This eliminates the backlash and brings the tooth surfaces into contact, ensuring that no rigid body displacement occurs, and that the initial contact iteration converges. In the second analysis step, the small rotation of the gear is removed, and a torque is applied to the degree of freedom along its axis to analyze the contact state under the initial load. In the third analysis step, the fixed constraint on the pinion's axis is removed, and a rotational boundary condition is applied to simulate the load analysis during the pinion's rotation.

According to the regulation of national standard GB11365-89 [30], three misalignments should be controlled when installing spiral bevel gears, such as gear ring axial displacement Δf_{AM}, gear shaft spacing deviation Δf_{α}, and gear shaft angle deviation ΔE_{Σ}. According to the gear basic parameters in Table 1 (precision grade level 7), the limit deviation of Δf_{AM} is ± 0.56 mm; the limit deviation of Δf_{α} is ± 0.02 mm. The limit deviation of ΔE_{Σ} is ± 0.06 mm, and the equivalent angular deviation is $\Delta E_{\Sigma} = \pm 0.03572°$.

Table 1. The geometric parameters of spiral bevel gear system.

Parameter	Pinion	Gear
number of teeth	6	37
face angle	16°3′	78°26′
root angle	11°15′	73°31′
direction	left	right
shaft angle/	90	90
module	8.243	8.243
face width/mm	45.12	40
face apex to crossing point/mm	−3.14	1.4
pitch angle	11°43′	77°58′
pitch apex to crossing point/mm	0.71	1.4
root apex to crossing point/mm	−7.59	1.39
outer cone distance/mm	155.35	155.92
outside diameter/mm	84.07	305.52
pitch diameter/mm	304.99	304.99

4.3. *Influence of Gear Ring Axial Displacement* Δf_{AM}

Set $\Delta f_{AM} = \pm 0.056$ mm, and take two limit values in the finite element analysis. Adjust the model considering the misalignments and make it no interference between gears. The obtained FEA results are presented in Table 2. It is observable that the inclination angle of the contact path remains relatively constant, while the contact pattern shifts from the toe towards the heel of the mating teeth as the value of Δf_{AM} increases, as shown in Table 2.

Table 2. FEA results with gear ring axial displacement.

Error Value	FEA Results (Torque = 100 N·m)
$\Delta f_{AM} = -0.056$ mm	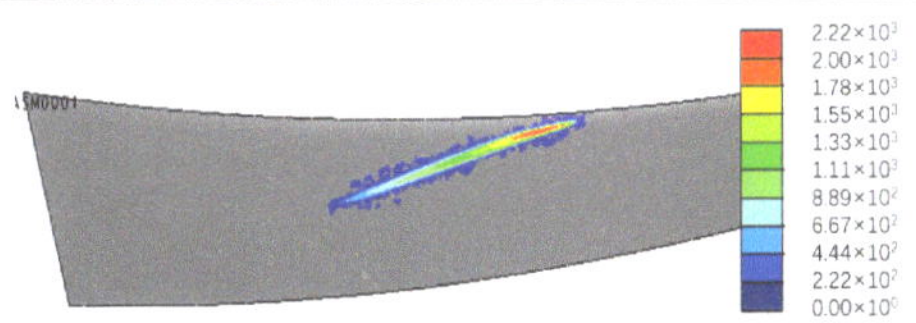
$\Delta f_{AM} = 0$ mm	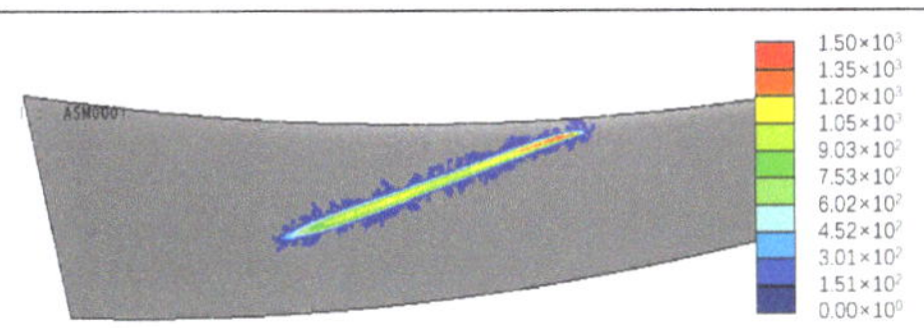
$\Delta f_{AM} = +0.056$ mm	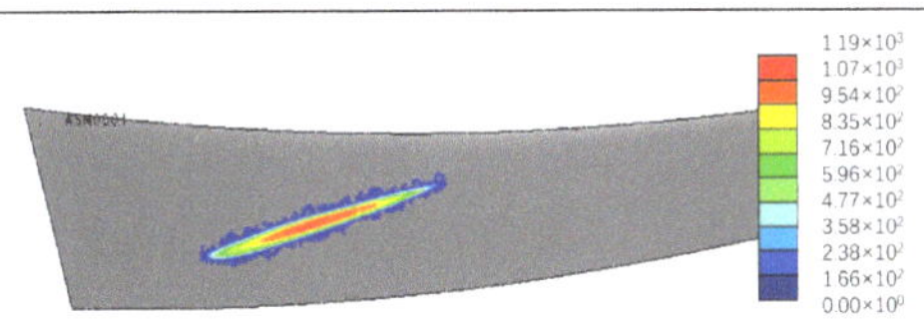

4.4. *Influence of Gear Shaft Spacing Deviation* Δf_{α}

Set $\Delta f_{\alpha} = \pm 0.02$ mm, and take two limit values in the finite element analysis. Adjust the model considering the misalignments and make it no interference between gears. The obtained FEA results are presented in Table 3. It can be noted that the contact pattern is much smaller and the edge contact occurs on the tip of pinion and gear teeth. The contact pattern moves the heel towards the toe of the mating teeth with the increase of Δf_{α}, as shown in Table 3.

Table 3. FEA results with gear shaft spacing deviation.

Error Value	FEA Results (Torque = 100 N·m)
$\Delta f_{\alpha} = -0.02$ mm	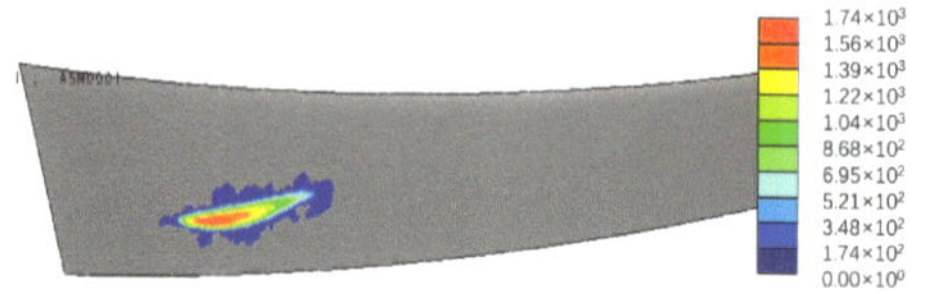
$\Delta f_{\alpha} = 0$ mm	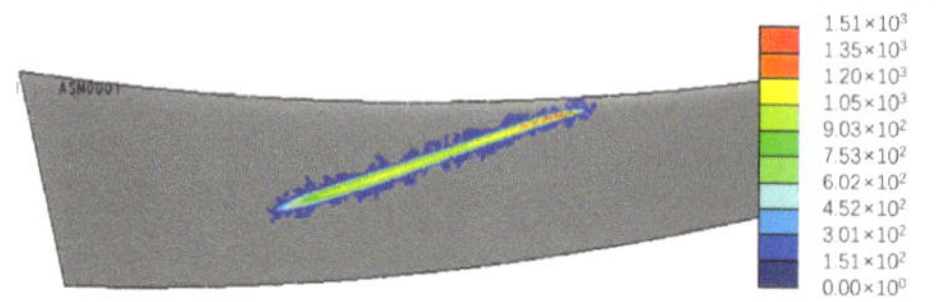
$\Delta f_{\alpha} = +0.02$ mm	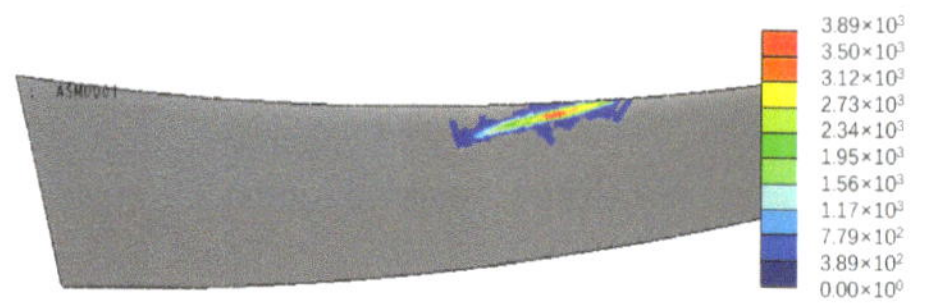

4.5. *Influence of Gear Angular Deviation* ΔE_{Σ}

Set $\Delta E_{\Sigma} = \pm 0.03572^{\circ}$, and take two limit values in the finite element analysis. Adjust the model considering the misalignments and make it no interference between gears. The obtained FEA results are presented in Table 4. It can be seen that the angular misalignment ΔE_{Σ} has a significant influence on contact pattern and by this on the corresponding meshing performance, too. The contact pattern moves the heel towards the toe of the mating teeth with the increase of ΔE_{Σ}, as shown in Table 4.

Table 4. FEA results with gear shaft angle deviation ΔE_{Σ}.

Error Value	FEA Results (Torque = 100 N·m)
$\Delta E_{\Sigma} = -0.03572^{\circ}$	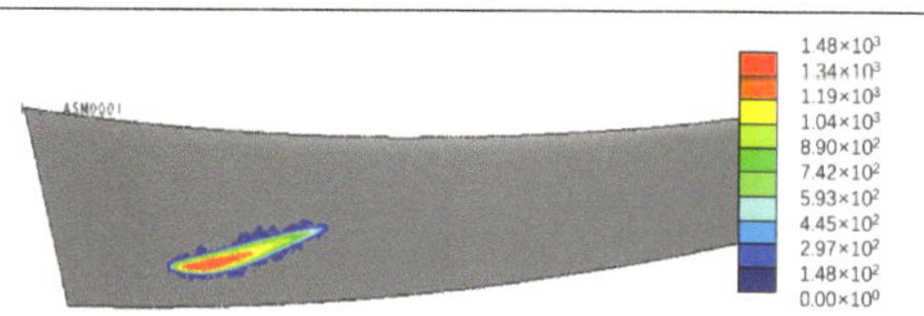
$\Delta E_{\Sigma} = 0^{\circ}$	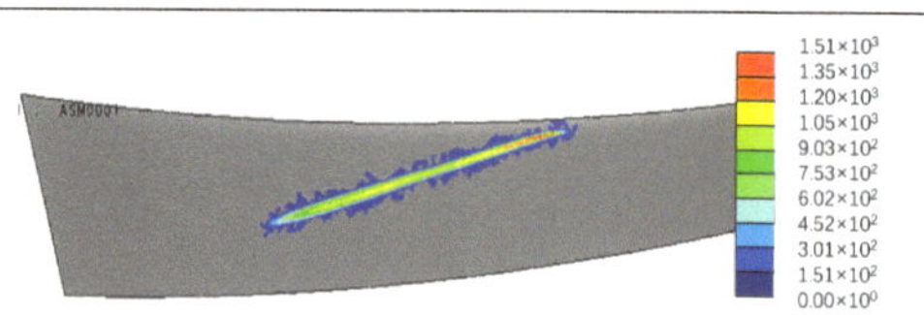
$\Delta E_{\Sigma} = +0.03572^{\circ}$	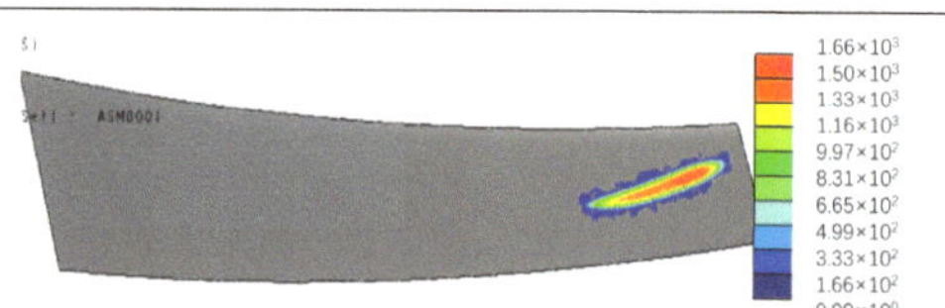

4.6. Comparison with LTCA Software

We also make twelve instantaneous contact patterns under different meshing angles within one tooth surface graph. So that we can get the approximate contact pattern figure of the gear tooth surface, as is shown in Figure 11. Compared with the results of LTCA software and measurement, as shown in Figure 12, this method and LTCA software results are very similar in contact pattern. The LTCA results of spiral bevel gears, based on high-precision virtual machining for misalignments, show high accuracy.

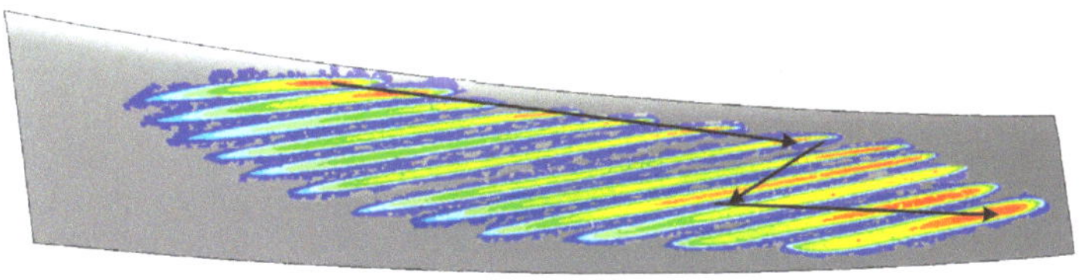

Figure 11. Illustration of contact pattern and path of contact of misalignments are $\Delta A_1 = -0.30$ mm, $\Delta V = 0.20$ mm (convex) (torque = 100 N·m).

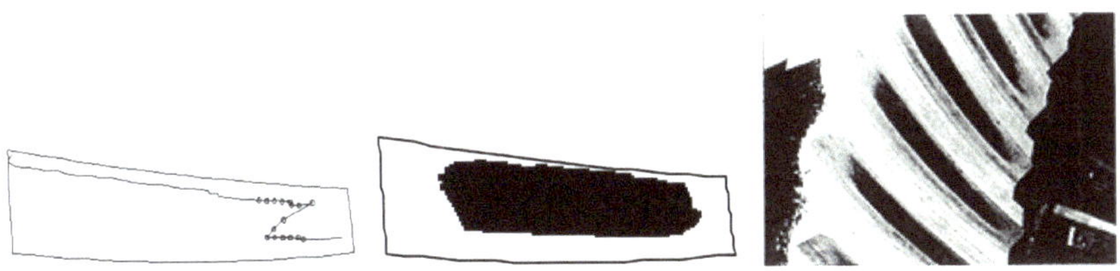

(**a**) (**b**) (**c**)

Figure 12. Results of LTCA software simulation and measurement (torque = 100 N·m). (**a**) Results of path of contact simulation. (**b**) Results of contact pattern simulation. (**c**) Results of contact pattern measurement.

In this section, the geometric models of the spiral bevel gears are precisely generated using the face milling method, resulting in a maximal error of only 0.08094 μm between the theoretical tooth surfaces and the virtual generating tooth surfaces. This high level of precision provides a reliable research platform for the dynamic performance analysis of spiral bevel gears. Building on this, an advanced LTCA method has been developed that incorporates the virtual generating gear with various misalignments. As illustrated in Figure 11, the FEA results demonstrate the apparent influence of these misalignments on the path of contact and the contact pattern, providing deeper insights into the meshing behavior and performance of the gears.

5. Conclusions

In this study, we have explored a comprehensive methodology for LTCA of spiral bevel gears by integrating ease-off surface computation with high-precision virtual generating tooth surfaces. The main objectives were to enhance the accuracy and reliability of LTCA by combining advanced computational methods and validating the approach through practical simulations and theoretical analysis. The results demonstrate significant improvements in predicting and analyzing the performance of spiral bevel gears under various misalignment conditions. We have found that the semi-analytical models significantly improve the prediction of loaded contact patterns by utilizing ease-off surfaces derived from the UGM. This approach effectively combines error-sensitivity analysis with Hertzian contact theory to provide a robust framework for analyzing the contact properties of spiral bevel gears. The ease-off surfaces, optimized through sophisticated corrections in the machining process, ensure high accuracy in the generated tooth surfaces, leading to better performance and durability of the gears. The mathematical kinematical model is directly linked to the machine tool settings of a spiral bevel generator, serving as the foundation for microgeometry optimization and evaluation of spiral bevel gear contact characteristics. This model can be applied to ease-off processes using the generated method. Error-sensitivity analysis delves into the effects of misalignments on the positioning, dimensions, and configuration of the contact pattern. Based on the findings of this analysis, the subsequent conclusions can be drawn:

1. The UGM describes the microgeometry and contact properties of ease-off surfaces, and the error-variation equation is developed to reflect the inherent relationships between errors and variations in machine tool settings.
2. Advanced semi-analytical LTCA method is developed based on the ease-off spiral bevel gears with misalignments. The results of semi-analytical LTCA illustrate an investigation of the influence of misalignments on the path of contact and contact pattern.
3. Through the application of the semi-analytical LTCA method, it is feasible to introduce misalignments during the ease-off process, to relocate the center of the contact pattern to a specific, pre-determined location on the tooth surface.
   Using the semi-analytical LTCA method, misalignments could be introduced during the ease-off process to move the contact pattern center to a certain predetermined position on the tooth surface.

Complementing this, the finite element analysis (FEA) based LTCA model we have also developed considers various misalignment errors, such as pinion offset and adjustment errors along the pinion and gear axis. The validation of this approach through comparisons with LTCA software analysis results confirms its feasibility and effectiveness in predicting the impact of misalignments on contact patterns and load distribution. The integration of findings reveals that the proposed methodologies not only improve the accuracy of LTCA but also offer a comprehensive and reliable tool for analyzing the performance of spiral bevel gears. By incorporating error-sensitivity analysis and virtual generating tooth surfaces, the integrated approach provides a detailed understanding of the contact properties and the impact of misalignments on gear performance. This synthesis highlights the importance of precise error modeling and the role of advanced computational methods in optimizing the design and analysis of spiral bevel gears.

The proposed methodologies also demonstrate the significant impact of ease-off surfaces and high-precision virtual generating surfaces on the contact properties of gears. The semi-analytical models and the FEA-based LTCA models together ensure that the generated tooth surfaces are highly accurate, leading to better noise reduction, improved contact pressure distribution, and enhanced mechanical efficiency. The validation through practical simulations and comparisons with LTCA software analysis results further confirm the robustness of the proposed approach.

Author Contributions: Conceptualization, Y.L. and D.S.; methodology, Y.L.; software, Y.L.; validation, Y.L. and D.S.; formal analysis, Y.L.; investigation, L.C.; resources, L.C.; data curation, Y.L.; writing—original draft preparation, Y.L.; writing—review and editing, X.M.; visualization, L.C.; supervision, L.C.; project administration, L.C. All authors have read and agreed to the published version of the manuscript.

Funding: This research was funded by the project: High-Performance Geometric Constraint Solving Engine, the Project Number is 2023YFB3309003.

Institutional Review Board Statement: Not applicable.

Informed Consent Statement: Not applicable.

Data Availability Statement: Data are contained within the article.

Acknowledgments: The authors gratefully acknowledge the financial support from Huazhong University of Science and Technology for the research and development of intelligent design and basic key technologies of CNC under Grant No. 2021AAB001.

Conflicts of Interest: The authors declare no conflicts of interest.

References

1. Li, G.; Zhu, W. A Review on Up-to-Date Gearbox Technologies and Maintenance of Tidal Current Energy Converters. *Energies* **2022**, *15*, 9236. [CrossRef]
2. Li, G.; Zhu, W. Design and power loss evaluation of a noncircular gear pair for an infinitely variable transmission. *Mech. Mach. Theory* **2021**, *156*, 104137. [CrossRef]
3. Li, G.; Wang, Z.; Zhu, W.; Kubo, A. A function-oriented active form-grinding method for cylindrical gears based on error sensitivity. *Int. J. Adv. Manuf. Technol.* **2017**, *92*, 3019–3031. [CrossRef]
4. He, D.; Ding, H.; Tang, J. A new analytical identification approach to the tooth contact points considering misalignments for spiral bevel or hypoid gears. *Mech. Mach. Theory* **2018**, *121*, 785–803. [CrossRef]
5. Chen, P.; Wang, S.; Zou, H. Efficient semi-analytic method for single tooth contact analysis of loaded spiral bevel gears. *Appl. Math. Model.* **2024**, *129*, 754–779. [CrossRef]
6. Temirkhan, M.; Amrin, A.; Spitas, V.; Spitas, C. Convergence and accuracy problems of the conventional TCA model–Critical analysis and novel solution for crowned spur gears. *Proc. Inst. Mech. Eng. Part C J. Mech. Eng. Sci.* **2024**, *238*, 724–736. [CrossRef]
7. Marciniec, A.; Pacana, J.; Pisula, J.M.; Fudali, P. Comparative analysis of numerical methods for the determination of contact pattern of spiral bevel gears. *Aircr. Eng. Aerosp. Technol.* **2018**, *90*, 359–367. [CrossRef]
8. Li, G.; Wang, Z.; Kubo, A. The modeling approach of digital real tooth surfaces of hypoid gears based on non-geometric-feature segmentation and interpolation algorithm. *Int. J. Precis. Eng. Manuf.* **2016**, *17*, 281–292. [CrossRef]
9. Simon, V.V. Multi-objective optimization of hypoid gears to improve operating characteristics. *Mech. Mach. Theory* **2020**, *146*, 103727. [CrossRef]
10. Moslem, M.; Zippo, A.; Iarriccio, G.; Bergamini, L.; Pellicano, F. Loaded and unloaded tooth contact analysis of spiral bevel gears in consideration of misalignments. *Insight-Non-Test. Cond. Monit.* **2022**, *64*, 442–446. [CrossRef]
11. Li, G.; Wang, Z.; Zhu, W. Prediction of surface wear of involute gears based on a modified fractal method. *ASME J. Tribol.* **2019**, *141*, 031603. [CrossRef]
12. Jiang, J.; Luo, Q.; Wang, L.; Qiao, L.; Li, M. Review on logarithmic spiral bevel gear. *J. Braz. Soc. Mech. Sci. Eng.* **2020**, *42*, 400. [CrossRef]
13. Ghionea, A.; Constantin, G.; Predincea, N.; Sandu, C. Adjustemnt Aspects of Gear Hobbing Machines in Diagonal Milling. In Proceedings of the 16th International Conference on Manufacturing Systems—ICMaS, Bucharest, Romania, 7–8 October 2007; pp. 73–76.
14. Simon, V.V. Improvements in the mixed elastohydrodynamic lubrication and in the efficiency of hypoid gears. *Proc. Inst. Mech. Eng. Part J J. Eng. Tribol.* **2020**, *234*, 795–810. [CrossRef]
15. Liu, Z.; Li, F.; Xu, Z.; He, Q. Semi-analytical loaded tooth contact analysis method for spiral bevel gears. *Int. J. Mech. Sci.* **2023**, *253*, 108329. [CrossRef]
16. Li, F.; Wang, S.; Chen, P.; Li, Z.; Li, L.; Peng, Q. Multi-objective variable parameter optimization algorithm for spiral bevel gear tooth surface considering meshing efficiency and contact performance. *J. Mech. Sci. Technol.* **2022**, *36*, 3547–3559. [CrossRef]
17. Ding, H.; Zhou, Y.; Tang, J.; Zhong, J.; Zhou, Z.; Wan, G. A novel operation approach to determine initial contact point for tooth contact analysis with errors of spiral bevel and hypoid gears. *Mech. Mach. Theory* **2017**, *109*, 155–170. [CrossRef]
18. Fu, Y.; Zhuo, Y.; Zhou, X.; Wan, B.; Lv, H.; Wang, Z. Theoretical and experimental study on contact characteristics of spiral bevel gears under quasi-static and large loading conditions. *Appl. Sci.* **2020**, *10*, 5109. [CrossRef]
19. Ding, H.; Li, H.; Shao, W.; Tang, J. Prediction and control for local bearing contact-based collaborative grinding of non-orthogonal aerospace spiral bevel gears. *Mech. Syst. Signal Process.* **2021**, *160*, 107841. [CrossRef]
20. Kolivand, M.; Kahraman, A. A load distribution model for hypoid gears using ease-off topography and shell theory. *Mech. Mach. Theory* **2009**, *44*, 1848–1865. [CrossRef]
21. Simon, V.V. Multi-objective optimization of the manufacture of face-milled hypoid gears on numerical controlled machine tool. *Proc. Inst. Mech. Eng. Part B J. Eng. Manuf.* **2021**, *235*, 1120–1130. [CrossRef]
22. Stanasel, I.; Ghionea, A.; Ghionea, I.; Ungur, P.; Stanasel, O. The Mathematical Model for the Generation of the Cylindrical Gear With Curved Cycloidal Teeth. In Proceedings of the International Design Engineering Technical Conferences and Computers and Information in Engineering Conference, Las Vegas, NV, USA, 4–7 September 2007; Volume 48086, pp. 279–287.
23. Mathijs, V. Advanced Techniques for Numerical Contact Analysis in Spiral Bevel Gears. Ph.D. Thesis, Università della Calabria, Heverlee, Italy, 2019.
24. Mathur, T.D.; Smith, E.C.; Bill, R.C. A novel loaded tooth contact analysis procedure with application to internal–external straight bevel gear mesh in a pericyclic drive. *J. Tribol.* **2021**, *143*, 101201. [CrossRef]
25. Li, G.; Zhu, W. An Active Ease-Off Topography Modification Approach for Hypoid Pinions Based on a Modified Error Sensitivity Analysis Method. *ASME J. Mech. Des.* **2019**, *141*, 093302. [CrossRef]
26. Gabiccini, M. A twist exponential approach to gear generation with general spatial motions. *Mech. Mach. Theory* **2009**, *44*, 382–400. [CrossRef]
27. Simon, V.V. Influence of tooth modifications on tooth contact in face-hobbed spiral bevel gears. *Mech. Mach. Theory* **2011**, *46*, 1980–1998. [CrossRef]
28. Simon, V.V. Manufacture of optimized face-hobbed spiral bevel gears on computer numerical control hypoid generator. *J. Manuf. Sci. Eng.* **2014**, *136*, 031008. [CrossRef]

29. Doe, J. Simulation of hypoid gears dynamic performance. *J. Mech. Eng.* **2020**, *45*, 123–135.
30. *GB 11365-89*; Terms for Resistance Strain Gauges. China Standards Press: Beijing, China, 1989.

machines

MDPI

Article

Analysis of Nonlinear Dynamics of a Gear Transmission System Considering Effects of the Extended Tooth Contact

Fulin Liao [1], Xingyuan Zheng [2], Jianliang Huang [1,*] and Weidong Zhu [3]

1 Department of Applied Mechanics and Engineering, Sun Yat-sen University, Guangzhou 510275, China; liaoflin@mail2.sysu.edu.cn
2 College of Mechanical and Vehicle Engineering, Chongqing University, Chongqing 400044, China; zhengxingyuan@cqu.edu.cn
3 Department of Mechanical Engineering, University of Maryland Baltimore County, Baltimore, MD 21250, USA; wzhu@umbc.edu
* Correspondence: huangjl@mail.sysu.edu.cn

Abstract: Considering the elasticity of gear solid bodies, the load applied to gear teeth will force theoretically separated gear teeth to get into engaging state in advance. This phenomenon is named as the extended tooth contact (ETC). Effects of the ETC directly influence the time-varying mesh stiffness of gear pairs and subsequently alter nonlinear dynamic characteristics of gear transmission systems. Time-vary mesh stiffness, considering effects of the ETC, is thus introduced into the dynamic model of the gear transmission system. Periodic motions of a gear transmission system are discussed in detail in this work. The analytical model of time-varying mesh stiffness with effects of the ETC is proposed, and the effectiveness of the analytical model is demonstrated in comparison with finite element (FE) results. The gear transmission system is simplified as a single degree-of-freedom (DOF) model system by employing the lumped mass method. The correctness of the dynamic model is verified in comparison with experimental results. An incremental harmonic balance (IHB) method is modified to obtain periodic responses of the gear transmission system. The improved Floquet theory is employed to determine the stability and bifurcation of the periodic responses of the gear transmission system. Some interesting phenomena exist in the periodic responses consisting of "softening-spring" behaviors, jump phenomena, primary resonances (PRs), and super-harmonic resonances (SP-HRs), and saddle-node bifurcations are observed. Especially, effects of loads on unstable regions, amplitudes, and positions of bifurcation points of frequency response curves are revealed. Analytical results obtained by the IHB method match very well with those from numerical integration.

Keywords: gear transmission system; time-varying mesh stiffness; periodic response; incremental harmonic balance method

1. Introduction

As a typical power transmission device, gear transmission systems are extensively employed in industries such as aviation, vehicle engineering, and wind power generation. Due to effects of some factors consisting of material defects, overloads, tooth failures, and installation errors, gear transmission systems are often subjected to complex excitations during operation. Time-varying mesh stiffness, as the main source of internal excitations in gear transmission systems, is usually regarded as a key input of gear vibration and dynamic modeling [1–3]. Thus, developing an effective analytical model of time-varying mesh stiffness is the basis of the dynamic modeling and dynamic analysis of gear transmission systems.

Up to now, some methods for calculating time-varying mesh stiffness are being proposed by many researchers, including experimental methods, FE methods, analytical methods, and mixed methods. For the calculation of time-varying mesh stiffness of gear transmission systems, experimental methods are direct and effective strategies. Raghuwanshi et al. [4] proposed the photoelasticity technique to calculate time-varying mesh stiffness of a gear pair, and a comparison between time-varying mesh stiffness obtained by the photoelasticity technique and that from the analytical model demonstrated the effectiveness of the photoelasticity technique. Pandya et al. [5] also employed the photoelasticity technique to obtain time-varying mesh stiffness of a gear pair. Different from experimental methods that require precise and expensive experimental equipments, FE methods can accurately calculate time-varying mesh stiffness of gear transmission systems by using only FE software. Wang et al. [6] obtained the time-varying mesh stiffness of the gear transmission system with high contact ratio by using an FE method. Zhan et al. [7] proposed a new FE method based on the NX 8.5 software, the ANSYS Workbench 14.5 software, and a quasi-static algorithm to determine the time-varying mesh stiffness. Effects of the tip radius and misalignment of the gear pair on the time-varying mesh stiffness are discussed by using the new FE method. Liang et al. [8] proposed three evaluation models of time-varying mesh stiffness by employing a FE method. Compared with experimental methods and FE methods, analytical methods can more flexibly reveal the relationship between gear parameters and time-varying mesh stiffness. In analytical methods, deformations of elastic gears are regarded as a superposition of the elastic component generated by gear teeth and the fillet foundation of gears. Yang et al. [9] developed a potential energy method to calculate time-varying mesh stiffness caused by deformations of teeth; the potential energy method considered tooth deformations under Hertzian contact, bending, and axial compression. Liang et al. [10] developed an improved analytical model of time-varying mesh stiffness by taking the deformation of the fillet foundation and geometric characteristics of the tooth profile into account. Meng et al. [11] proposed an analytical model of time-varying mesh stiffness; time-varying mesh stiffness of gear teeth with different crack lengths can be obtained using the proposed analytical model. Additionally, effects of spalling with different widths, lengths, and locations on time-varying mesh stiffness are discussed. Liu [12] developed an analytical model to predict time-varying mesh stiffness of a single-stage planetary gear train with different degrees of surface wear. Considering the effects of tooth profile modification on time-varying mesh stiffness, Liu et al. [13] established an analytical model of time-varying mesh stiffness. Meng et al. [14] proposed an anisotropic 3D fractal rough tooth surface model using the fractal geometry theory, and developed a revised analytical model of time-varying mesh stiffness based on the anisotropic 3D fractal rough tooth surface model. Tian et al. [15] established a time-varying mesh stiffness model considering elastohydrodynamic lubrication to calculate mesh stiffness of a planetary gear set. Zhou et al. [16] developed an improved analytical model of time-varying mesh stiffness of a gear pair considering root crack. Zhang et al. [17] proposed an improved time-varying mesh stiffness model of a double-helical planetary gear transmission system considering coupling effects of temperature, fluid, and structure. Considering the centrifugal force in a high-speed working condition, Zheng et al. [18] developed an analytical-FEM framework to integrate the centrifugal field into mesh stiffness of high-speed spur gears, and time-varying mesh stiffness of high-speed spur gears can be accurately predicted using the developed model. The extended tooth contact (ETC) effect stemming from the elasticity of a solid body is not considered in an analytical model of time-varying mesh stiffness in the above-mentioned literature. The ETC effect forces theoretically separated teeth get into engagement in advance and increases the actual contact ratio, which will influence time-varying mesh stiffness and further change dynamic characteristics of gear transmission systems. Ma et al. [19] proposed an analytical model to determine time-varying mesh stiffness of a gear pair with tip relief, where the ETC effect, nonlinear contact stiffness, and revised fillet-foundation stiffness were considered. Ma et al. [20] developed an improved analytical model for the calculation of time-varying mesh stiffness of a cracked spur gear; effects of the reduction of fillet-foundation stiffness and the ETC effect were considered.

Dynamic responses of gear transmission systems can be mainly determined by using numerical methods, analytical methods, and semi-numerical and semi-analytical methods based on a dynamic model of the systems. For the calculation of dynamic responses of gear transmission systems, numerical methods are the most simple. Generally, numerical methods for obtaining dynamic responses of gear transmission systems consist of the fourth-order Runge–Kutta (RK) method, the Newmark-β method, and the Gill numerical integration. Jiang et al. [21] obtained dynamic responses of a gear transmission system considering multi-frequency excitation by employing the numerical integration, and discussed the primary resonance of the system. Wan et al. [22] determined dynamic responses of a gear transmission system considering translational and torsional vibrations by using the Newmark-β method, and analyzed effects of geometric transmission error, bearing stiffness, and mesh stiffness on dynamic responses, and concluded that dynamic responses of the gear transmission system are periodic signals because of the periodic variation of time-varying mesh stiffness. Liu et al. [23] investigated the nonlinear vibration of a multi-meshing gear system considering tooth profile modification by using the numerical integration, and studied effects of tooth profile modification on the nonlinear vibration of the system. Pan et al. [24] calculated dynamic responses of a gear-shaft-bearing transmission system considering some nonlinear factors consisting of time-varying mesh stiffness, fractional backlash, and static transmission error by applying the RK method, and discussed effects of contact temperature, fractional backlash, and random load on dynamic characteristics of the system. However, the above-mentioned investigations on dynamic responses of gear transmission systems using numerical methods only discussed local dynamic characteristics of the systems.

In order to investigate dynamic characteristics of gear transmission systems in detail, global dynamic responses of the systems need to be obtained. Analytical methods, including the multiple-scale (MS) method, are usually employed to calculate global dynamic responses of gear transmission systems. Moradi et al. [25] obtained forced vibration responses of a gear transmission system with backlash nonlinearity by using the MS method, and analyzed primary resonance, super-harmonic resonance, and sub-harmonic resonance based on the responses of the system. Mo et al. [26] investigated primary resonance characteristic and stability of a gear-rolling-bearing system considering time-varying mesh stiffness, transmission error, and backlash by employing the MS method, and discussed effects of damping, time-varying mesh stiffness, and load on the primary resonance of the system. In the case of determining dynamic responses, the harmonic balance (HB) method and the IHB method belong to the semi-numerical and semi-analytical method, and the HB method and the IHB method are applicable to both strong nonlinear systems and weak nonlinear systems. Bruzzone and Rosso [27] proposed two models to analyze gear dynamics: dynamic overloads caused by meshing can be estimated by the first model with a one-dimensional approach, and the second model includes flexibilities of shafts, bearings, and the gearbox housing. Bruzzone et al. [28] analyzed the dynamic response of a spur gear pair using a refined finite element model. Nonlinear phenomena such as hysteretic jumps, sub-harmonic resonance, and super-harmonic resonance are predicted by the proposed model. Bruzzone et al. [29] estimated deflections, load sharing attributes, and contact conditions of meshing teeth in a spur gear transmission system based on a three-dimensional model. Oliveri et al. [30] predicted the dynamic effect of a parametric model of gear engagement using the HB method, and investigated contributions of the actual static transmission error and the contact ratio on gear dynamics. Alshyyab et al. [31] studied the sub-harmonic resonance of a multi-meshing gear transmission system by using the HB method, and predicted period-doubling motions that cause the system get into chaotic motions. Hou et al. [32] calculated nonlinear dynamic responses of a spur gear pair with fractional-order backlash under the combined action of internal and external excitations by applying the IHB method, and investigated effects of parameters on vibration characteristics including amplitudes of resonance and frequencies of resonance.

In this work, the analytical model of time-varying mesh stiffness is formulated to bridge the gap between the extended tooth contact (ETC) and nonlinear dynamics of the gear transmission system, and verification of its effectiveness is performed. The dynamic model of the gear transmission system considering time-varying mesh stiffness is demonstrated in comparison with experimental results. Global dynamic responses of the gear transmission system are accurately determined by using the IHB method. The stability and bifurcation of nonlinear dynamic responses are analyzed by the improved Floquet theory. In particular, effects of loads on amplitudes, unstable regions, and positions of bifurcation points of frequency response curves are revealed in detail.

Primary goals of this work are to investigate the nonlinear vibration of the gear transmission system considering effects of the ETC and analyzed effects of loads on amplitudes, unstable regions, and positions of bifurcation points of frequency response curves. The rest of this paper is organized as follows: The analytical model of time-varying mesh stiffness is proposed, and ordinary differential equations governing the motion of the gear transmission system are formulated using Newton's second law in Section 2. The IHB method is modified to determine dynamic responses of the gear transmission system in Section 3. The stability and bifurcation of dynamic responses are examined by employing the improved Floquet theory in Section 4. Jump phenomena, "softening-spring" characteristics, bifurcation behaviors, and resonances of the gear transmission system are discussed in Section 5, and effects of loads on amplitudes, unstable regions, and positions of bifurcation points of frequency response curves are revealed. Primary conclusions of this work are presented in Section 6.

2. Modeling of the Gear Transmission System and Time-Varying Mesh Stiffness

Time-varying mesh stiffness is one of primary parametric excitations of gear transmission systems; the analytical model of time-varying mesh stiffness is proposed. Dynamic models of gear transmission systems are the basis of analyzing nonlinear dynamic behaviors of the systems. The single-DOF dynamic model of the gear transmission system with time-varying mesh stiffness is developed by using the lumped mass method.

2.1. Analytical Model of Time-Varying Mesh Stiffness

As the main source of high-frequency internal excitations of gear transmission systems, the derivation of an accurate model of time-varying mesh stiffness is an important procedure in the process of modeling. In a number of published studies, the time-varying mesh stiffness $K(t)$ between two engaging gears is considered to have no relation to effects of the ETC. Yang et al. [33] and Liu et al. [34] proposed an analytical model of time-varying mesh stiffness that is not related to effects of the ETC.

In the process of engaging gear pairs, deformations of tooth pairs will cause tooth pairs get into engaging state in advance and result in tooth pairs getting out engaging state in delay, which extend the contact duration of gear pairs. This phenomenon is named as the extended tooth contact (ETC) [19] and will influence time-varying mesh stiffness. Deformations of tooth pairs are dominated by loads acting on engaging gear pairs. Effects of the ETC on time-varying mesh stiffness are thus considered. Our previous work [35] proposed the analytical model of time-varying mesh stiffness considering effects of the ETC. In order to clarify, the analytical model of time-varying mesh stiffness with effects of the ETC is briefly introduced on the basis of the literature [35]. It should be noted that the proposed analytical model of time-varying mesh stiffness can be implemented with a fast numerical method, which removed restrictions of classical fillet-foundation formulas and covered both thick-walled and thin-walled gears.

The flexibility of a gear pair primarily consists of beam actions of teeth, the elasticity of the fillet foundation, and contact compliances of tooth pairs. The time-varying mesh stiffness of tooth pairs is thus decomposed as the tooth stiffness, the fillet foundation stiffness, and the tooth pair contact stiffness.

A tooth is usually considered an non-uniform cantilever beam. Along the line of action, the reaction of tooth pairs to the local mesh force is denoted as F_m, which is regarded as the superposition of beam actions considering axial compression, bending, and shearing. On the basis of the potential energy method [9], the tooth stiffness K_t is expressed as

$$\frac{1}{K_t} = \frac{1}{K_a} + \frac{1}{K_b} + \frac{1}{K_s}, \tag{1}$$

where K_a, K_b, and K_s indicate the axial compressive stiffness, the bending stiffness, and the shearing stiffness, respectively, and K_a, K_b, and K_s can be denoted as

$$\begin{cases} \frac{1}{K_a} = \int_{-\alpha_F}^{\theta_b} \frac{(\theta_b - \theta)\cos\theta \sin^2\alpha_F}{2EL[\sin\theta + (\theta_b - \theta)\cos\theta]} d\theta \\ \frac{1}{K_b} = \int_{-\alpha_F}^{\theta_b} \frac{3\{1 + \cos\alpha_F[(\theta_b - \theta)\sin\theta - \cos\theta]\}^2 (\theta_b - \theta)\cos\theta}{2EL[\sin\theta + (\theta_b - \theta)\cos\theta]^3} d\theta \\ \frac{1}{K_s} = \int_{-\alpha_F}^{\theta_b} \frac{1.2(1+\nu)(\theta_b - \theta)\cos\theta\cos^2\alpha_F}{EL[\sin\theta + (\theta_b - \theta)\cos\theta} d\theta \end{cases}, \tag{2}$$

respectively. α_F, θ_b, E, ν, and L are the action angle of the local mesh force F_m, the half tooth angle at the base circle, Young's modulus, Poisson's ratio, and the tooth width, respectively.

In the process of engaging, the contact compliance of tooth pairs is generated. In order to consider the effects of the load dependency and nonlinearity of tooth contact, the following semi-empirical formula is formulated:

$$K_h = \frac{F_m}{\delta_h} = \frac{E^{0.9} L^{0.8} F_m^{0.1}}{1.275}, \tag{3}$$

where K_h and δ_h denote the Hertzian contact stiffness of a tooth pair and the corresponding contact compliance, respectively.

Considering effects of the deformation response of a tooth and the flexibility of the fillet foundation on tooth deflection, fillet foundation stiffness components are obtained as follows:

$$\begin{cases} \frac{1}{K_f} = \frac{\cos^2\alpha_F}{EL} \sum_{s=0}^{2} [(\frac{u_F}{S_f})^s \Phi_s] \\ \frac{1}{K_{f21}} = \frac{\cos\alpha_{F1}\ \cos\alpha_{F2}}{EL} (\Theta_0^{2,1} + \Theta_1^{2,1}\frac{u_{F1}}{S_f} + \Theta_2^{2,1}\frac{u_{F2}}{S_f} + \Theta_3^{2,1}\frac{u_{F1}u_{F2}}{S_f^2}) \\ \frac{1}{K_{f12}} = \frac{\cos\alpha_{F1}\ \cos\alpha_{F2}}{EL} (\Theta_0^{1,2} + \Theta_1^{1,2}\frac{u_{F2}}{S_f} + \Theta_2^{1,2}\frac{u_{F1}}{S_f} + \Theta_3^{1,2}\frac{u_{F1}u_{F2}}{S_f^2}) \end{cases}, \tag{4}$$

where K_f denotes the stiffness corresponding to the local mesh force F_m; K_{f21} indicates the stiffness corresponding to the local mesh force F_{m1}; K_{f12} indicates the stiffness corresponding to the local mesh force F_{m2}; F_m, F_{m1}, and F_{m2} are mesh forces of tooth pairs under different mesh situations; and u_F, u_{F1}, and u_{F2} are external moment arms corresponding to local mesh forces F_m, F_{m1}, and F_{m2}, respectively. Additionally, S_f is the junction arc length between the fillet foundation and the tooth, and Φ_s, $\Theta_s^{1,2}$, and $\Theta_s^{2,1}$ $(s = 0, 1, 2)$ are load independency infinite series that are calculated in detail in [35].

For an engaging gear pair, the driving gear is called pinion (marked as "p"), and the driven gear is named gear (marked as "g"). In the process of meshing, deflections of tooth pairs vary with the variation of engaging positions. Recalling Equations (1), (3), and (4), the load–deflection constitutive relation can be obtained as follows:

$$\begin{cases} \delta_1(\alpha, F_1, F_2) = F_1 \sum_{j=p,g}(\frac{1}{K_{t1,j}} + \frac{1}{K_{f11,j}} + \frac{1}{2K_{h1}}) + F_2 \sum_{j=p,g} \frac{1}{K_{f12,j}} \\ \delta_2(\alpha, F_1, F_2) = F_2 \sum_{j=p,g}(\frac{1}{K_{t2,j}} + \frac{1}{K_{f22,j}} + \frac{1}{2K_{h2}}) + F_1 \sum_{j=p,g} \frac{1}{K_{f21,j}} \end{cases}, \tag{5}$$

where $\delta(\alpha, F_1, F_2)$ is the total elastic deformation of a certain tooth pair, and subscripts $\cdot_1$ and $\cdot_2$ indicate quantities corresponding to tooth pair 1 and tooth pair 2, respectively, which are defined in [35]. Similarly, subscripts $\cdot_p$ and $\cdot_g$ denote quantities corresponding to the pinion and the gear, respectively, and K_{f11} and K_{f22} are local fillet foundation stiffnesses corresponding to tooth pair 1 and tooth pair 2, respectively.

On the basis of the small deformation assumption, the starting angle φ_1 and the ending angle φ_2 of the extended tooth contact within one mesh cycle are explicitly expressed as

$$\begin{cases} \beta_2 - \varphi_1 \approx \frac{T}{r_p^2} \sum_{j=p,g}[\frac{1}{K_{t2,j}(\beta_2)} + \frac{1}{K_{f22,j}(\beta_2)} + \frac{1}{2K_{h2}(T)} - \frac{1}{K_{f12,j}(\beta_2)}] + \frac{\epsilon}{r_p} \\ \varphi_2 - \beta_3 \approx \frac{T}{r_p^2} \sum_{j=p,g}[\frac{1}{K_{t1,j}(\beta_3)} + \frac{1}{K_{f11,j}(\beta_3)} + \frac{1}{2K_{h1}(T)} - \frac{1}{K_{f21,j}(\beta_3)}] + \frac{\epsilon}{r_p} \end{cases}, \tag{6}$$

where β_2 and β_3 indicate the starting angle and the ending angle of double teeth contact over one mesh cycle, respectively; T is the torque acting on gear pairs; r_p is the base circle radius of the pinion; and ϵ is the non-loaded transmission error. Additionally, the Fourier form of time-varying mesh stiffness considering effects of the ETC by using the fast Fourier transformation (FFT) can be obtained as follows:

$$K(t, T) = K_0(T) + \sum_{i=1}^{\infty} K_i(\epsilon, T)\cos(2i\pi f_m t - \psi_i), \tag{7}$$

where $K_0(T)$, $K_i(\epsilon, T)$, f_m, ε, and ψ_i denote the average mesh stiffness, the Fourier coefficient of the i-th-order harmonic term, the mesh frequency, the contact ratio, and the phase angle of the i-th-order harmonic term, respectively.

2.2. Dynamic Model of the Gear Transmission System

The single-DOF dynamic model of the gear transmission system considering time-varying mesh stiffness is proposed by the lumped mass method, as shown in Figure 1. The driving gear with the inertia I_p, the mass m_p, and the base circle radius r_p and the driven gear with the inertia I_g, the mass m_g, and the base circle radius r_g are modeled as two disks represented by the pinion and the gear, respectively. The mesh between the pinion and the gear is described by the nonlinear displacement function $f(x)$, the viscous damping c, and the time-varying mesh stiffness with effects of the ETC $K(t, T)$. α_p and α_g are torsional displacements of the pinion and the gear, respectively. T_p and T_g denote torques acting on the pinion and the gear, respectively. Additionally, $2b$ is the gear backlash. The translational vibration along the line of action and the friction force between engaging teeth are assumed to be negligible [33]

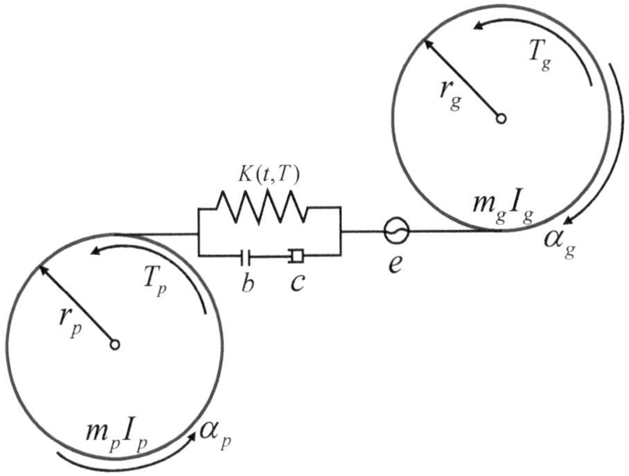

Figure 1. The single-DOF dynamic model of the gear transmission system considering time-varying mesh stiffness.

By employing Newton's second law, ordinary differential equations governing motions of the gear transmission system considering time-varying mesh stiffness are formulated as

$$I_p\ddot{\alpha}_p + r_p F_m = T_p \tag{8}$$
$$I_g\ddot{\alpha}_g - r_g F_m = -T_g, \tag{9}$$

where F_m is the dynamic mesh force, which can be described as

$$F_m = c(r_p\dot{\alpha}_p - r_g\dot{\alpha}_g) + K(t,T)f(r_p\alpha_p - r_g\alpha_g). \tag{10}$$

By defining the dynamic transmission error (DTE) $x = r_p\alpha_p - r_g\alpha_g$, Equations (8)–(10) can be rewritten as

$$m\ddot{x} + c\dot{x} + K(t,T)f(x) = F, \tag{11}$$

where the nonlinear displacement function $f(x)$ can be expressed as

$$f(x) = \begin{cases} x - b, & x > b \\ 0, & -b < x < b \\ x + b, & x < -b \end{cases} \tag{12}$$

and

$$m = \frac{I_p I_g}{r_p^2 I_g + r_g^2 I_p}, \tag{13}$$
$$F = \frac{T_p}{r_p} = \frac{T_g}{r_g}. \tag{14}$$

A dimensionless form of Equation (11) can be derived by introducing $\bar{x} = \frac{x}{b_c}$, $\omega_n = \sqrt{\frac{K_0(t,T)}{m}}$, $\xi = \frac{c}{2m\omega_n}$, $\bar{F} = \frac{F}{mb_c\omega_n^2}$, and $\bar{t} = \omega_n t$, where b_c is the characteristic length [36].

$$\ddot{\bar{x}} + 2\xi\dot{\bar{x}} + \bar{K}(\bar{t},T)f(\bar{x}) = \bar{F}. \tag{15}$$

3. IHB Method for Periodic Responses of the Gear Transmission System

Steady-state periodic responses of the gear transmission system considering time-varying mesh stiffness and a nonlinear displacement function are predicted by utilizing the IHB method. The IHB method is an efficient method for determining periodic or period-doubling responses of smooth systems with general nonlinearities or non-smooth systems with piecewise linearities [37,38], and has been extended to obtain periodic responses of multi-DOF nonlinear systems, such as a two-DOF vehicle system [39] and a three-DOF unit-cell system [40].

By defining a new time variable,

$$\tau = \omega\bar{t}, \tag{16}$$

Equation (15) becomes

$$\omega^2\bar{x}'' + 2\omega\xi\bar{x}' + \bar{K}(\tau,T)f(\bar{x}) = \bar{F}, \tag{17}$$

where ω is an unknown fundamental frequency, and a prime denotes differentiation with respect to τ.

The procedure of the IHB method for obtaining periodic responses mainly consists of the incremental process that is the Newton–Raphson iterative procedure to linearize the differential Equation (17) and the process of harmonic balance that is also called the Galerkin procedure [41,42]. In the process of increment, $\bar{x}_0$ and ω_0 denote the initial state of vibration; the neighboring state can be obtained by adding increments $\Delta\bar{x}$ and $\Delta\omega$ to $\bar{x}_0$ and ω_0, respectively.

$$\bar{x} = \bar{x}_0 + \Delta\bar{x}, \quad \omega = \omega_0 + \Delta\omega. \tag{18}$$

Substituting Equation (18) into Equation (17), and neglecting high-order terms of increments, the linearized incremental equation can be formulated as

$$\omega_0^2\Delta\bar{x}'' + 2\omega_0\xi\Delta\bar{x}' + \bar{K}(\tau,T)\frac{\mathrm{d}f}{\mathrm{d}\bar{x}}|_{\bar{x}=\bar{x}_0}\Delta\bar{x} = \bar{R} - 2(\omega_0\bar{x}_0'' + \xi\bar{x}_0')\Delta\omega, \tag{19}$$

where

$$\bar{R} = \bar{F} - [\omega_0^2\bar{x}_0'' + 2\omega_0\xi\bar{x}_0' + \bar{K}(\tau,T)f(\bar{x}_0)] \tag{20}$$

is a correction term to prevent the incrementation process from drifting away from actual solutions [43].

The second step of the IHB method is the process of harmonic balance. Under parametric excitation and forced excitation, periodic responses of the gear transmission system can be predicted by expanding $\bar{x}_0$ in truncated multiple Fourier series and employing the Galerkin procedure [44]. Hence, $\bar{x}_0$ and its increment $\Delta\bar{x}$ can be expressed as

$$\bar{x}_0 = \sum_{n=1}^{n_c} a_{(n-1)/p}\cos\frac{(n-1)\tau}{p} + \sum_{n=1}^{n_s} b_{n/p}\sin\frac{n\tau}{p} = \mathbf{C}_s\mathbf{A}, \tag{21}$$

$$\Delta\bar{x} = \sum_{n=1}^{n_c} \Delta a_{(n-1)/p}\cos\frac{(n-1)\tau}{p} + \sum_{n=1}^{n_s} \Delta b_{n/p}\sin\frac{n\tau}{p} = \mathbf{C}_s\Delta\mathbf{A}, \tag{22}$$

where

$$\mathbf{C}_s = \begin{bmatrix} 1 & \cos\frac{\tau}{p} & \cos\frac{2\tau}{p} \cdots & \cos\frac{(n_c-1)\tau}{p} & \sin\frac{\tau}{p} & \sin\frac{2\tau}{p} & \cdots & \sin\frac{n_s\tau}{p} \end{bmatrix}, \tag{23}$$

$$\mathbf{A} = \begin{bmatrix} a_0 & a_{1/p} & a_{2/p} & \cdots & a_{(n_c-1)/p} & b_{1/p} & b_{2/p} & \cdots & b_{n_s/p} \end{bmatrix}^T, \tag{24}$$

$$\Delta\mathbf{A} = \begin{bmatrix} \Delta a_0 & \Delta a_{1/p} & \Delta a_{2/p} & \cdots & \Delta a_{(n_c-1)/p} & \Delta b_{1/p} & \Delta b_{2/p} & \cdots & \Delta b_{n_s/p} \end{bmatrix}^T, \tag{25}$$

in which $a_{(n-1)/p}$, $b_{n/p}$, $\Delta a_{(n-1)/p}$, and $\Delta b_{n/p}$ are Fourier coefficients; n_c and n_s are numbers of cosine and sine harmonic terms retained, respectively; and p is a positive integer that is introduced to predict possible period-doubling responses [45].

Differentiating Equations (21) and (22) yields

$$\begin{aligned} \bar{x}_0' &= \mathbf{C}_s'\mathbf{A}, \quad \Delta\bar{x}_0' = \mathbf{C}_s'\Delta\mathbf{A}, \\ \bar{x}_0'' &= \mathbf{C}_s''\mathbf{A}, \quad \Delta\bar{x}_0'' = \mathbf{C}_s''\Delta\mathbf{A}. \end{aligned} \tag{26}$$

Substituting Equations (21), (22) and (26) into Equation (19) and employing the process of harmonic balance yields

$$\begin{aligned} &\int_0^{2p\pi} \delta(\Delta\bar{x})^T \left[\omega_0^2\Delta\bar{x}'' + 2\omega_0\xi\Delta\bar{x}' + \bar{K}(\tau,T)\frac{\mathrm{d}f}{\mathrm{d}\bar{x}}|_{\bar{x}=\bar{x}_0}\Delta\bar{x}\right]\mathrm{d}\tau \\ &\qquad = \int_0^{2p\pi} \delta(\Delta\bar{x})^T\left[\bar{R} - 2(\omega_0\bar{x}_0'' + \xi\bar{x}_0')\Delta\omega\right]\mathrm{d}\tau. \end{aligned} \tag{27}$$

Then a set of linearized equations expressed in terms of $\Delta\mathbf{A}$ and $\Delta\omega$ is obtained as follows:

$$\mathbf{K}_{mc}\Delta\mathbf{A} = \mathbf{R} - \mathbf{R}_\omega\Delta\omega, \tag{28}$$

where

$$\mathbf{K}_{mc} = \omega_0^2\mathbf{M} + 2\omega_0\xi\mathbf{C} + \mathbf{K}', \tag{29}$$

$$\mathbf{R} = \mathbf{F} - (\omega_0^2\mathbf{MA} + 2\omega_0\xi\mathbf{CA} + \mathbf{R}^K), \tag{30}$$

$$\mathbf{R}_\omega = 2(\omega_0\mathbf{MA} + \xi\mathbf{CA}), \tag{31}$$

in which

$$\begin{aligned} &\mathbf{M} = \int_0^{2p\pi} \mathbf{C}_s^T\mathbf{C}_s''\mathrm{d}\tau, \quad \mathbf{C} = \int_0^{2p\pi} \mathbf{C}_s^T\mathbf{C}_s'\mathrm{d}\tau, \quad \mathbf{F} = \int_0^{2p\pi} \mathbf{C}_s^T\bar{F}\mathrm{d}\tau, \\ &\mathbf{R}^K = \int_0^{2p\pi} \mathbf{C}_s^T\bar{K}(\tau,T)f(\bar{x}_0)\mathrm{d}\tau, \quad \mathbf{K}' = \int_0^{2p\pi} \mathbf{C}_s^T\bar{K}(\tau,T)\frac{\mathrm{d}f}{\mathrm{d}\bar{x}}|_{\bar{x}=\bar{x}_0}\mathbf{C}_s\mathrm{d}\tau. \end{aligned} \tag{32}$$

Since the time-varying mesh stiffness with effects of the ETC $K(\tau, T)$ is expressed as

$$\bar{K}(\tau,T) = 1 + \sum_{i=1}^{n_1} c_i\cos(i\tau) + \sum_{i=1}^{n_1} d_i\sin(i\tau), \tag{33}$$

the matrix $\mathbf{K}'$ and the vector $\mathbf{R}^K$ can be divided into following three parts, respectively:

$$\mathbf{K}' = \mathbf{K}^{1\prime} + \mathbf{K}^{2\prime} + \mathbf{K}^{3\prime}, \tag{34}$$

$$\mathbf{R}^K = \mathbf{R}^{1K} + \mathbf{R}^{2K} + \mathbf{R}^{3K}, \tag{35}$$

where

$$\mathbf{K}^{1\prime} = \int_0^{2p\pi} \mathbf{C}_s^T \frac{\mathrm{d}f}{\mathrm{d}\bar{x}}|_{\bar{x}=\bar{x}_0} \mathbf{C}_s \mathrm{d}\tau, \quad \mathbf{K}^{2\prime} = \int_0^{2p\pi} \sum_{i=1}^{n_1} \mathbf{C}_s^T c_i \cos(i\tau) \frac{\mathrm{d}f}{\mathrm{d}\bar{x}}|_{\bar{x}=\bar{x}_0} \mathbf{C}_s \mathrm{d}\tau,$$
$$\mathbf{K}^{3\prime} = \int_0^{2p\pi} \sum_{i=1}^{n_1} \mathbf{C}_s^T d_i \sin(i\tau) \frac{\mathrm{d}f}{\mathrm{d}\bar{x}}|_{\bar{x}=\bar{x}_0} \mathbf{C}_s \mathrm{d}\tau, \quad \mathbf{R}^{1K} = \int_0^{2p\pi} \mathbf{C}_s^T f(\bar{x}_0) \mathrm{d}\tau,$$
$$\mathbf{R}^{2K} = \int_0^{2p\pi} \sum_{i=1}^{n_1} \mathbf{C}_s^T c_i \cos(i\tau) f(\bar{x}_0) \mathrm{d}\tau, \quad \mathbf{R}^{3K} = \int_0^{2p\pi} \sum_{i=1}^{n_1} \mathbf{C}_s^T d_i \sin(i\tau) f(\bar{x}_0) \mathrm{d}\tau. \quad (36)$$

Calculations of the matrix $\mathbf{K}'$ and the vector $\mathbf{R}^K$ are different with the mass matrix $\mathbf{M}$ or the damping matrix $\mathbf{C}$ due to the nonlinear displacement function $f(\bar{x}_0)$ contained in $\mathbf{K}'$ and $\mathbf{R}^K$. The detailed formulation process of $\mathbf{K}^{j\prime}$ and $\mathbf{R}^{jK}$ $(j = 1, 2, 3)$ is provided in Appendix A.

The solution process of Equation (28) begins with a guessed initial solution that need not be accurate. Frequency response curves can be traced point by point as long as the iterative procedure is repeated until the norm of $\mathbf{R}$ is less than 1.0×10^{-10}. In the increment process, three incremental methods containing the frequency increment, the amplitude increment, and the arc-length increment are employed to trace frequency response curves. In general, three incremental methods need to be used alternately. The computer program of the IHB method is written using Matlab 9.13, which can be downloaded using the site at https://ww2.mathworks.cn/ (accessed on 13 February 2025). A logic diagram of the IHB method is presented in Figure 2.

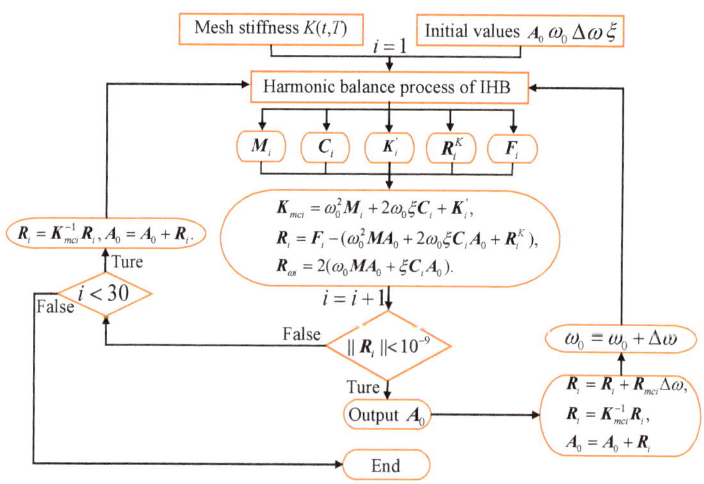

Figure 2. A logic diagram the IHB method.

4. Analyzing of Stability and Bifurcation Analysis of Periodic Responses

The stability and bifurcation types of periodic responses of the gear transmission system considering time-varying mesh stiffness and a nonlinear displacement function are determined by employing the improved Floquet theory. After steady-state solutions of a nonlinear system are obtained by using the IHB method, the stability and bifurcation types of periodic solutions can be examined by adding a small perturbation $\Delta\bar{x}$ on $\bar{x}_0$, as follows:

$$\bar{x} = \bar{x}_0 + \Delta\bar{x}. \quad (37)$$

Substituting Equation (37) into Equation (17) and noting that $\bar{x}_0$ satisfies Equation (17), the linearized equation, by taking $\Delta\bar{x}$ as unknown, is obtained by neglecting high-order terms as follows:

$$\omega^2\Delta\bar{x}'' + 2\omega\xi\Delta\bar{x}' + \bar{K}(\tau,T)\frac{\mathrm{d}f}{\mathrm{d}\bar{x}}|_{\bar{x}=\bar{x}_0}\Delta\bar{x} = 0. \tag{38}$$

Equation (38) is called the perturbed equation that perturbed from the equilibrium state $\bar{x}_0$ [46]. The stability of steady-state periodic solutions depends on the stability of solutions of Equation (38), which is a linearized ordinary differential equation with a four-order Fourier series $\bar{K}(\tau,T)$. The stability of the solutions of Equation (38) can be determined by applying the Floquet theory in the period $2p\pi$. Let

$$\mathbf{q} = [\Delta\bar{x} \quad \Delta\bar{x}']^T, \tag{39}$$

Equation (38) can be described in state space form as

$$\mathbf{q}' = \mathbf{Q}(\tau)\mathbf{q}, \tag{40}$$

where

$$\mathbf{Q} = \begin{bmatrix} 0 & 1 \\ Q_{21}(\tau) & Q_{22} \end{bmatrix}, \tag{41}$$

in which

$$Q_{21}(\tau) = \frac{-\bar{K}(\tau,T)\frac{\mathrm{d}f}{\mathrm{d}\bar{x}}|_{\bar{x}=\bar{x}_0}}{\omega^2}, \quad Q_{22} = \frac{-2\xi}{\omega}. \tag{42}$$

Since $\bar{x}_0$ is a periodic function of τ with the period $\mathrm{T} = 2p\pi$, $Q_{21}(\tau)$ is a periodic function with the same period T as well.

For Equation (40), there exists a fundamental set of solutions as follows:

$$\mathbf{y}_i = \begin{bmatrix} y_{1i} & y_{2i} & \cdots & y_{Ni} \end{bmatrix}^T, \quad i = 1,2,\cdots,N, \tag{43}$$

where $N = 2\mathrm{n}$, and n is number of DOFs of the gear transmission system. This fundamental set can be described in matrix form as follows:

$$\mathbf{Y} = \begin{bmatrix} y_{11} & y_{12} & \cdots & y_{1N} \\ y_{21} & y_{22} & \cdots & y_{2N} \\ \vdots & \vdots & & \vdots \\ y_{N1} & y_{N2} & \cdots & y_{NN} \end{bmatrix}. \tag{44}$$

Equation (44) satisfies the matrix equation

$$\mathbf{Y}' = \mathbf{Q}(\tau)\mathbf{Y}, \tag{45}$$

where $\mathbf{Q}(\tau)$ is a periodic function with the period T; i.e., $\mathbf{Q}(\tau) = \mathbf{Q}(\tau + \mathrm{T})$, $\mathbf{Y}(\tau)$ is a fundamental matrix solution. Hence, $\mathbf{Y}(\tau)$ can be described by

$$\mathbf{Y}(\tau) = \mathbf{P}\mathbf{Y}(\tau + \mathrm{T}), \tag{46}$$

where **P** is an non-singular constant matrix that is called the transition matrix.

The Floquet theory states that the stability criteria of nonlinear systems depends on eigenvalues of the transition matrix **P**. If all moduli of the eigenvalues of the transition matrix $\mathbf{P}$ are less than 1, motions of nonlinear systems are bounded and the periodic solutions of the systems are stable; otherwise, the motions of nonlinear systems are unbounded and the periodic solutions of the systems are unstable. Key to determining the stability of nonlinear systems is to obtain the eigenvalues of the transition matrix $\mathbf{P}$ by using an efficient and accurate method. Friedmann et al. [47] stated that the method for approximating the transition matrix developed by Hsu et al. [48] is the most efficient one. The proposed method divided the period T into a number of equal parts, and the integration of the transition matrix $\mathbf{P}$ over each equal part was performed. In this work, the period T is divided into I equal parts denoted by $[\tau_1 \quad \tau_2], [\tau_2 \quad \tau_3], \cdots, [\tau_{I+1} \quad \tau_{I+2}]$, where $\tau_1 = 0, \tau_{I+2} = 2p\pi$. The magnitude of the i-th equal part is

$$\Delta_i = \tau_{i+1} - \tau_i, \tag{47}$$

where $i = 1, 2, 3, \cdots, I+1$.

Define the step function S as

$$S(i) = \begin{cases} 1, & l_i \geq 0 \\ 0, & l_i < 0 \end{cases}, \tag{48}$$

where $l_1, l_2, l_3, \cdots, l_i, \cdots, l_{I+1}$ denote values of the sign function $|\bar{x}| - \dfrac{b}{b_c} = 0$ in equal parts $[\tau_1 \quad \tau_2], [\tau_2 \quad \tau_3], \cdots, [\tau_i \quad \tau_{i+1}], \cdots, [\tau_{I+1} \quad \tau_{I+2}]$, respectively.

In the i-th equal part, the periodic coefficient matrix $\mathbf{Q}(\tau)$ is replaced by the constant matrix as follows:

$$\mathbf{Q}_i(\tau) = \begin{bmatrix} 0 & 1 \\ [Q_{21}]_i & [Q_{22}]_i \end{bmatrix}, \tag{49}$$

where

$$[Q_{21}]_i = -S(i)\frac{1}{\Delta_i}\int_{\tau_{i-1}}^{\tau_i} \frac{\bar{K}(\tau, T)}{\omega^2} \mathrm{d}\tau, \ [Q_{22}]_i = -2\frac{1}{\Delta_i}\int_{\tau_{i-1}}^{\tau_i} \frac{2\xi}{\omega} \mathrm{d}\tau. \tag{50}$$

Finally, the transition matrix $\mathbf{P}$ can be expressed as

$$\mathbf{P} = \prod_{i=1}^{I} \left[\mathbf{I} + \sum_{l=1}^{L} \frac{(\Delta_i \mathbf{Q}_i)^l}{l!} \right], \tag{51}$$

where L is the number of retained terms in the Taylor expansion.

5. Numerical Results and Discussion

5.1. Verification of Analytical Model of Time-Varying Mesh Stiffness

The correctness of an analytical model of time-varying mesh stiffness is verified by performing comparisons between analytical results and FE results of time-varying mesh stiffness, and effects of loads on time-varying mesh stiffness are discussed.

Comparisons between analytical results and FE results of time-varying mesh stiffness for loads of 100 Nm and 180 Nm are displayed in Figure 3, where dash lines represent analytical results and solid lines represent FE results. It can be seen that time-varying mesh stiffness from the analytical model developed in our previous work is a little larger than that obtained by the FE model.

Figure 3a,c present comparisons between analytical results and FE results of time-varying mesh stiffness of the gear transmission system neglecting effects of the ETC for loads of 100 Nm and 180 Nm, respectively. It can be observed that curves of time-varying mesh stiffness from the analytical model coincide with those obtained by the FE model over a mesh period, regions of double teeth contact of time-varying mesh stiffness from the analytical model are smaller than those from the FE model, and regions of single tooth contact of time-varying mesh stiffness from the analytical model are larger than those from the FE model.

Figure 3b,d show comparisons between analytical results and FE results of time-varying mesh stiffness of the gear transmission system considering effects of the ETC for loads of 100 Nm and 180 Nm, respectively. It can be observed that curves of time-varying mesh stiffness from the analytical model agree very well with those obtained by the FE model over a mesh period. Thus, it can be concluded that the analytical model of time-varying mesh stiffness with effects of the ETC is effective, which can accurately express variations of time-varying mesh stiffness. Time-varying mesh stiffness of the gear transmission system can thus be accurately obtained using the proposed analytical model for any loads unless failures of teeth are generated.

Curves of time-varying mesh stiffness under different loads are displayed in Figure 4. The increment amplitude between two loads is 40 Nm. It can be seen that time-varying mesh stiffness increases with the increase in loads, and when loads $T > 100$ Nm, the incremental amplitude of time-varying mesh stiffness decreases with the increase in loads. Curves of time-varying mesh stiffness of the gear transmission system neglecting effects of the ETC under different loads are displayed in Figure 4a. It can be observed that regions of double teeth contact and regions of single tooth contact are invariant with the increase in loads. Curves of time-varying mesh stiffness of the gear transmission system considering effects of the ETC under different loads are shown in Figure 4b. It can be seen that region of double teeth contact increases with the increase in loads, and regions of single tooth contact decreases with the increase in loads. It can be concluded that the ETC has important influences on regions of double teeth contact and regions of single tooth contact of time-varying mesh stiffness, which can subsequently affect dynamic behaviors of the gear transmission system considering effects of the ETC.

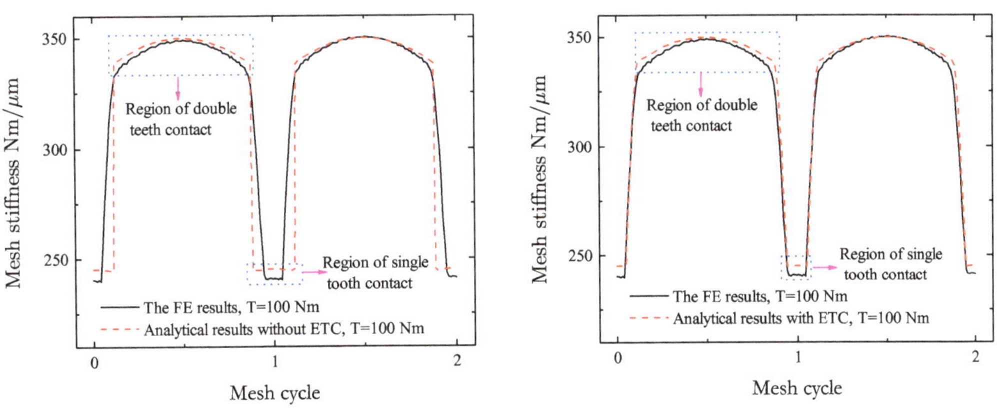

(a) (b)

Figure 3. *Cont.*

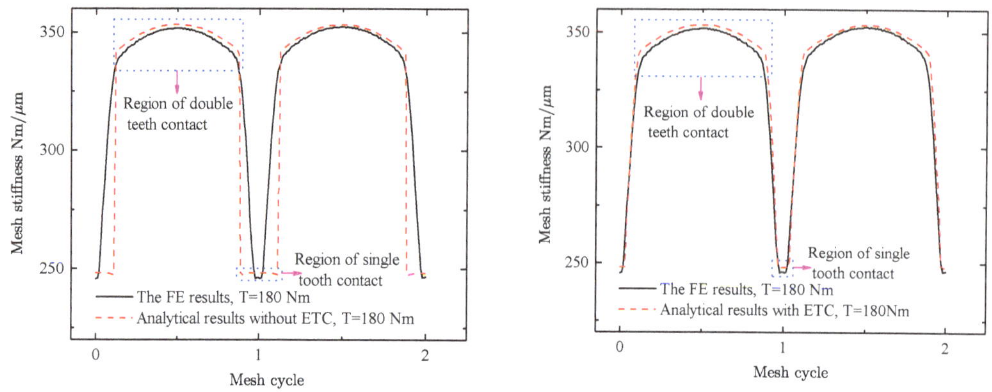

(c) (d)

Figure 3. Comparisons between analytical results and FE results of the time-varying mesh stiffness of the gear transmission system for loads of 100 Nm and 180 Nm: (**a**) the gear transmission system neglecting effects of the ETC for a load of 100 Nm, (**b**) the gear transmission system considering effects of the ETC for a load of 100 Nm, (**c**) the gear transmission system neglecting effects of the ETC for a load of 180 Nm, and (**d**) the gear transmission system considering effects of the ETC for a load of 180 Nm.

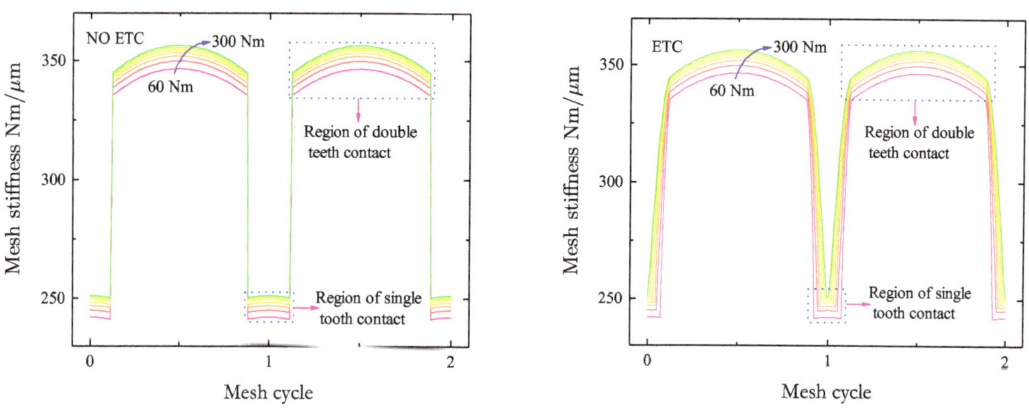

(a) (b)

Figure 4. Effects of loads on the time-varying mesh stiffness of the gear transmission system: (**a**) the gear transmission system neglecting effects of the ETC and (**b**) the gear transmission system considering effects of the ETC.

5.2. Verification of Dynamic Model and Analyzing of Nonlinear Phenomena of the Gear Transmission System

In this subsection, PR, SP-HR, nonlinear "softening-spring" characteristics, and bifurcations of the gear transmission system in consideration of time-varying mesh stiffness are discussed. The number of cosine terms is n_c, and the number of sine terms is n_s. Equation (21) can be written as

$$\bar{x}_0 = a_0 + \sum_{n=1}^{n_s} A_{n/p} \cos((n/p)\tau + \phi_{n/p}), \quad (52)$$

where

$$A_{n/p} = \sqrt{a_{n/p}^2 + b_{n/p}^2}, \quad \phi_{n/p} = \arctan\left(-b_{n/p}, a_{n/p}\right). \tag{53}$$

Results from the IHB method are verified in comparison with experimental results measured by Kahraman and Blankenship [49] based on a gear test rig. Experimental settings used in the process of collecting experimental data are presented as below: (a) a pair of spur test gears, (b) spherical roller bearings, (c) rigid split-line bearing pedestals, (d) test gear shafts, (e) reaction gear box, (f) compliant elastomer couplings, (g) flywheels, (h) a reaction gear set, (i) split hub couplings, (j) locking pins, (k) a 12 kW motor, (l) four miniature piezoelectric accelerometers, (m) slip rings, (n) optical encoders, (o) belt drive, and (p) polymer granite base. The sample size was set as a gear rotational speed range of 600 rpm to 4000 rpm with an increment of 50 rpm. A flowchart of collecting and analyzing experimental data is shown in Figure 5.

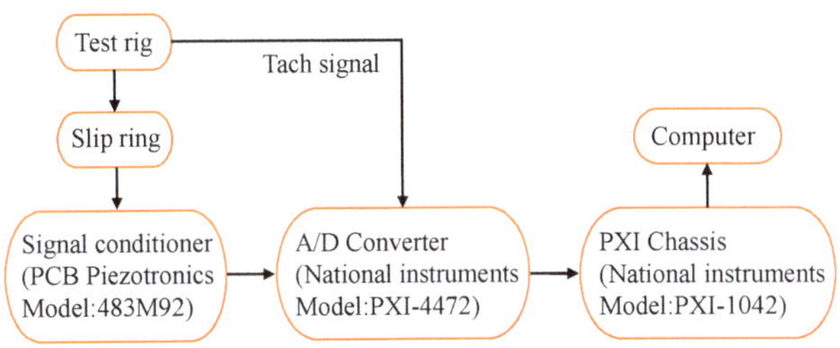

Figure 5. Flowchart of collecting and analyzing experimental data.

In the published literature [49], the DTE was measured by attaching four miniature piezoelectric accelerometers tangentially to the pinion and the gear. An analog signal representing the DTE was processed by employing a programmable spectrum analyzer in real time. Frequency spectra of the DTE were obtained by performing the FFT, and the r-th mesh harmonic amplitude A_r of the DTE were generated from the resulting frequency spectra. At last, an equivalent root-mean-square (rms) amplitude A_{rms} of an alternating component of the DTE is proposed to represent nonlinear responses of the gear transmission system, and A_{rms} can be expressed as

$$A_{rms} = \sqrt{\sum_{r=1}^{3} A_r^2}. \tag{54}$$

In order to obtain the equivalent root-mean-square (rms) amplitude A_{rms} of results from the IHB method to describe nonlinear responses of the gear transmission system, A_{rms} of results from the IHB method is obtained as follows:

$$A_{rms} = \sqrt{\sum_{n=1}^{3} A_{n/p}^2}. \tag{55}$$

The amplitude A_{rms} of dynamic responses from experimental results [49] (represented by squares), numerical results [35] (represented by triangles), and the IHB method results (represented by line) are compared in Figure 6. In the process of calculating the IHB method results, time-varying mesh stiffness of the gear transmission system considering effects of the ETC for a load of 170 Nm is considered. It can be observed that the frequency response curve represented by the IHB method results exhibits similar trends as compared with those from experimental results and numerical results, and these frequency response curves display three SP-HR, one PR, and nonlinear "softening-spring" behaviors represented by frequency response curves bent to the left in PR regions, caused by separations of engaging teeth. Compared with frequency response curves from experimental results and numerical results, an unstable region between the point "S_1" and the point "S_2" is observed in the frequency response curve from the IHB method results. Bifurcations and the jump phenomenon of the frequency response curve represented by the IHB method results are displayed in Figure 6. It can be seen that there are two saddle-node bifurcation points (marked as "S_1" and "S_2") in Figure 6. At saddle-node bifurcation points, periodic solutions become unstable and cause a jump phenomenon from a stable solution to another one, as evidenced by a real Floquet multiplier leaving the unit circle through the +1 direction along the real axis (Table 1). When sweeping the frequency from the left to the right, the stable periodic solution loses its stability at the saddle-node bifurcation point S_1 and jumps upward to the point B. When sweeping the frequency from the right to the left, the stable periodic solution becomes unstable at the saddle-node bifurcation point S_2 and jumps downward to the point A.

Table 1. Floquet multipliers for different frequencies near bifurcation points "S_1" and "S_2" in the frequency response curve in Figure 6.

	ω	First and Second Floquet Multipliers λ_1, λ_2	Moduli of λ_1, λ_2
Near point S_1	0.89821	0.52104, 0.99442	0.52104, 0.99442
	0.89823	0.51303, 1.00993	0.51303, 1.00993
Near point S_2	0.76731	0.46205, 1.00236	0.46205, 1.00236
	0.76734	0.46349, 0.99925	0.46349, 0.99925

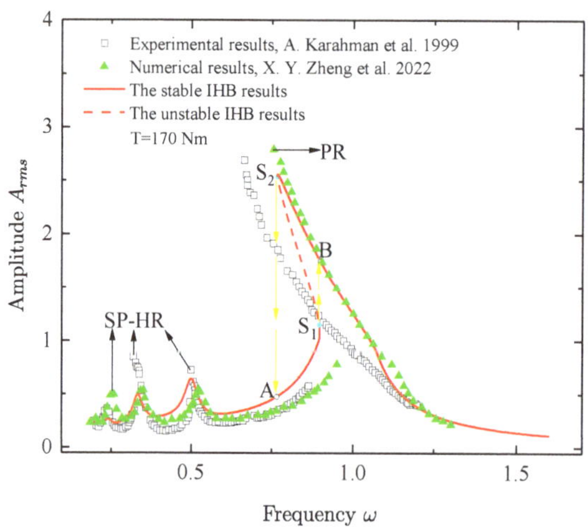

Figure 6. Equivalent root-mean-square (rms) amplitude A_{rms} from experimental results [49], numerical results [35], and the IHB method results.

Figure 7 exhibits a time history map and a phase plane diagram of the periodic response from the IHB method results presented in Figure 6 when $\omega = 0.5990$. The results obtained by the IHB method (represented by dots) match very well with that calculated by the fourth-order RK method (represented by a solid line) in Figure 7a, which indicate that the IHB method has sufficient accuracy to determine periodic responses of the gear transmission system in consideration of time-varying mesh stiffness and piecewise linearity. It can be seen that a phase plane diagram from the IHB method and the fourth-order RK method is one circle in Figure 7b, which indicate that the periodic response is period-1.

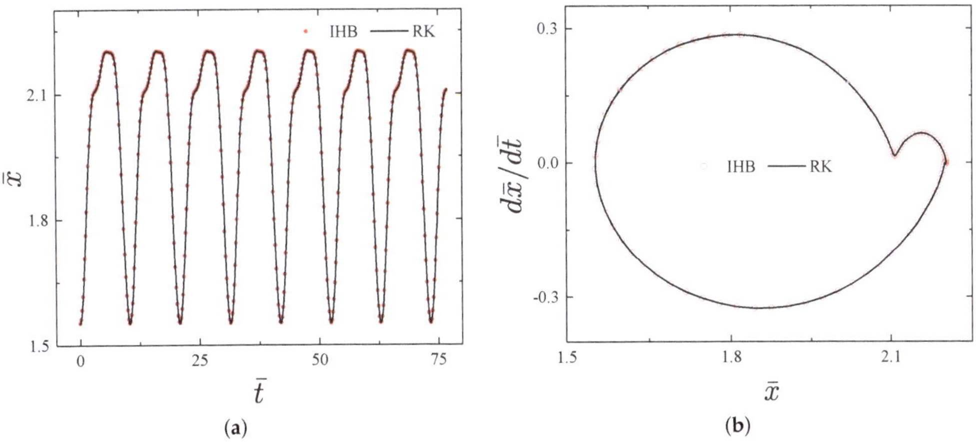

Figure 7. Time history map and phase plane diagram of the periodic response when $\omega = 0.5990$: (**a**) time history map and (**b**) phase plane diagram.

5.3. Effects of Loads on Nonlinear Dynamic Responses of the Gear Transmission System

In this work, time-varying mesh stiffness with effects of the ETC is considered. Effects of loads on nonlinear dynamic responses of the gear transmission system considering effects of the ETC and neglecting effects of the ETC are discussed in this subsection.

Figure 8 exhibits frequency response curves of the gear transmission system considering effects of the ETC and neglecting effects of the ETC under different loads, and it can be seen that amplitudes of frequency response curves of the gear transmission system increase with the increase in loads, and frequency response curves display nonlinear "softening-spring" behaviors, which caused by the separation between two engaging teeth. Additionally, it can be observed that nonlinear "softening-spring" behaviors represented by frequency response curves bent to the left gradually weaken with the increase in loads in Figure 8, which induced by the degree of separation between two engaging teeth decreases with the increase in loads. Under same loads, nonlinear "softening-spring" behaviors of frequency response curves of the gear transmission system considering effects of the ETC are more obvious than that of the system neglecting effects of the ETC, which caused by time-varying mesh stiffness of the system considering effects of the ETC, which is larger than that of the system neglecting effects of the ETC. Effects of loads on amplitudes of frequency response curves of the gear transmission system neglecting effects of the ETC are more obvious than that of loads on amplitudes of frequency response curves of the system considering effects of the ETC. Amplitudes of frequency response curves of the gear transmission system neglecting effects of the ETC are larger than that of frequency response curves of the system considering effects of the ETC.

The frequency response curves of $\omega \sim A_{rms}$ exist in two SP-HRs around $\omega = 0.331$ and $\omega = 0.5$ and one PR near $\omega = 1$ in Figure 8a,b, and it can be seen that the frequency response curves display nonlinear "softening-spring" behaviors characterized by frequency response curves bent to the left at PR and SP-HR around $\omega = 0.5$, which caused by the separation of engaging teeth. Additionally, amplitudes at PR are larger than that at SP-HR of the frequency response curves.

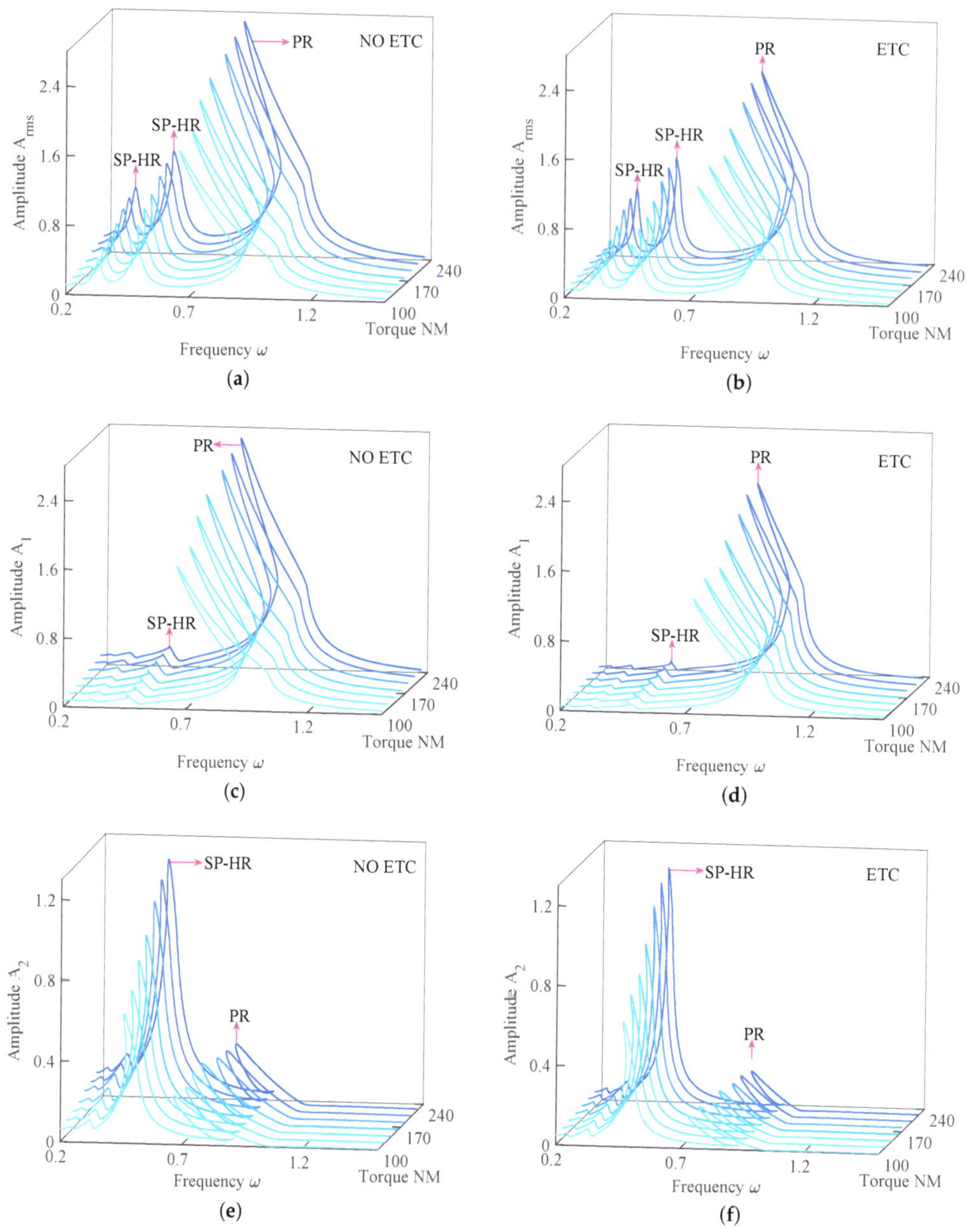

Figure 8. Nonlinear dynamic responses of the gear transmission system under different loads: (**a**) the equivalent root-mean-square amplitude A_{rms} of the gear transmission system neglecting effects of the ETC and (**b**) the equivalent root-mean-square amplitude A_{rms} of the gear transmission system considering effects of the ETC; (**c**) the first-order harmonic amplitude A_1 of the gear transmission system neglecting effects of the ETC and (**d**) the first-order harmonic amplitude A_1 of the gear transmission system considering effects of the ETC; and (**e**) the second-order harmonic amplitude A_2 of the gear transmission system neglecting effects of the ETC and (**f**) the second-order harmonic amplitude A_2 of the gear transmission system considering effects of the ETC.

In Figure 8c,d, the frequency response curves of $\omega \sim A_1$ exhibit obvious PR phenomena. SP-HR phenomena in the frequency response curves of $\omega \sim A_1$ are not obvious, and amplitudes at PR are larger than that at SP-HR of the frequency response curves. The reason is that PR phenomena are mainly dominated by the first-order harmonic amplitude. Additionally, the frequency response curves show obvious nonlinear "softening-spring" behaviors at PR.

In Figure 8e,f, the frequency response curves of $\omega \sim A_2$ display SP-HR phenomena around $\omega = 0.5$ and PR phenomena near $\omega = 1.0$. Amplitudes at SP-HR around $\omega = 0.3$ are very small. Amplitudes at SP-HR around $\omega = 0.5$ are obviously larger than that at PR near $\omega = 1.0$ in the frequency response curves. The reason is that SP-HR phenomena around $\omega = 0.5$ in the frequency response curves are primarily dominated by the second-order harmonic amplitude. Additionally, the frequency response curves show nonlinear "softening-spring" behaviors at SP-HR around $\omega = 0.5$ and PR near $\omega = 1.0$.

5.4. Effects of Loads on Bifurcation Characteristics of Frequency Response Curves

Besides effects of loads on amplitudes, resonance phenomena, and nonlinear "softening-spring" behaviors of frequency response curves, loads can also influence bifurcation characteristics of frequency response curves. Effects of loads on bifurcation characteristics of frequency response curves are investigated in detail in this subsection.

The unstable phenomena and bifurcation behaviors of period responses of the gear transmission system considering time-varying mesh stiffness are exhibited in Figure 9. Solid lines and dash lines denote stable period responses and unstable period responses, respectively, in Figure 9, and saddle-node bifurcation points in frequency response curves are denoted by black points that are marked as "S". One can see that there exist some saddle-node bifurcation points at regions of PR and regions of SP-HR in Figure 9, and frequency response curves generate jump phenomena from one stable period solution to another one at saddle-node bifurcation points, and saddle-node bifurcation points move upward in the vertical direction with increasing loads, which lead to unstable regions in frequency response curves decreasing with increasing loads (marked by red arrows).

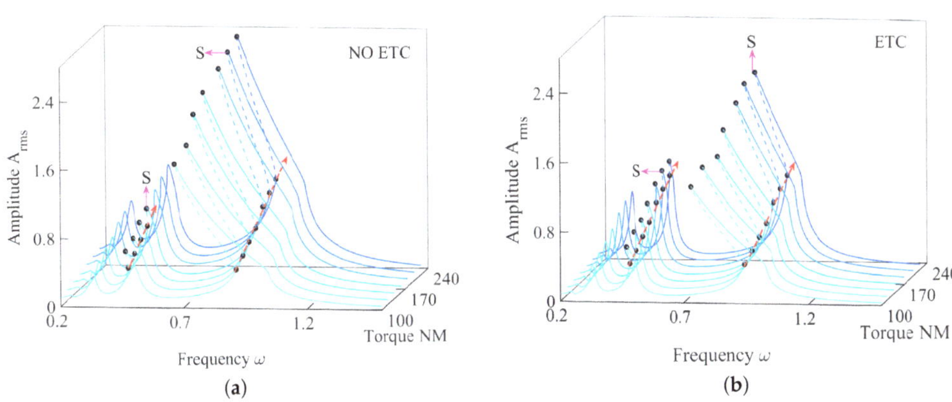

Figure 9. Effects of loads on the bifurcation characteristics of the gear transmission system: (**a**) bifurcation characteristics of the gear transmission system neglecting the effects of the ETC and (**b**) bifurcation characteristics of the gear transmission system considering the effects of the ETC.

Frequency response curves of the gear transmission system neglecting effects of the ETC under different loads are shown in Figure 9a. It can be observed that unstable regions of frequency response curves at SP-HR and PR gradually decrease with the increase in loads, and saddle-node bifurcation points of frequency response curves at SP-HR gradually disappear with the increase in loads. Frequency response curves of the gear transmission system considering effects of the ETC under different loads are exhibited in Figure 9b. It can be seen that the decrease in unstable regions at SP-HR and PR of frequency response curves with the increase in loads is not discovered, and the disappearance of saddle-node bifurcation points at SP-HR and PR of frequency response curves with the increase in loads is not observed.

6. Conclusions

Nonlinear dynamic responses of a single-DOF model of the gear transmission system considering time-varying mesh stiffness are investigated in detail in this work. An analytical model of time-varying mesh stiffness with effects of the ETC is developed, and the correctness of the analytical model is demonstrated in comparison with results from the FE method. Governing equations of the single-DOF model of the gear transmission system are proposed by using the Newton's second law, and the effectiveness of the single-DOF model is verified as compared with experimental results. The IHB method is improved to determine periodic responses of the single-DOF model of the gear transmission system. The Floquet theory combined with piecewise linearity is extended to analyze the stability and bifurcation characteristics of periodic responses. Results from the IHB method are in agreement with those obtained by numerical integration, which is verified by a time history diagram and a phase plane map.

Effects of loads on time-varying mesh stiffness are discussed, and it can be concluded that time-varying mesh stiffness increases with the increase in loads. Periodic responses of the single-DOF model of the gear transmission system considering time-varying mesh stiffness are analyzed in detail. In the periodic responses of the single-DOF model of the gear transmission system, nonlinear "softening-spring" behaviors characterized by frequency response curves bent to the left, jump phenomena from one stable periodic solution to another one, saddle-node bifurcation, PR, and SP-HR are observed. Effects of loads on the periodic responses of the single-DOF model of the gear transmission system are discussed, amplitudes of frequency response curves increase with the increase in loads, and nonlinear "softening-spring" behaviors of frequency response curves gradually weaken with the increase in loads. Unstable regions of frequency response curves of the gear transmission system neglecting effects of the ETC gradually decrease with the increase in loads, and saddle-node bifurcation points of frequency response curves of the gear transmission system considering effects of the ETC gradually disappear with the increase in loads. Additionally, this work provides a theoretical foundation to analyze nonlinear behaviors of gear transmission systems.

Author Contributions: Conceptualization, J.H. and W.Z.; Methodology, J.H.; Software, F.L.; Formal analysis, F.L. and X.Z.; Data curation, F.L.; Writing—original draft, F.L.; Writing—review & editing, J.H. and W.Z.; Supervision, J.H.; Project administration, J.H.; Funding acquisition, J.H. and W.Z. All authors have read and agreed to the published version of the manuscript.

Funding: Financial support from the National Natural Science Foundation of China (Grant No. 12372028), the National Science Foundation (Grant No. 2329791), and the Guangdong Basic and Applied Basic Research Foundation (Grant No. 2022A1515011809) is gratefully acknowledged.

Data Availability Statement: The authors attest that all data for this study are included in the paper.

Conflicts of Interest: There are no conflicts of interest.

Appendix A. The Detailed Formulation Process of the Matrix $\mathbf{K}^{j'}$ and the Vector $\mathbf{R}^{jK}$

The detailed formulation process of the matrix $\mathbf{K}^{j'}$ and the vector $\mathbf{R}^{jK}$ is presented as follows:

$$\mathbf{R}^{jK} = \mathbf{K}^{j'}\mathbf{A} + \mathbf{R}^{jNL}b, \quad (j = 1,2,3), \tag{A1}$$

$$\mathbf{K}^{j'} = \begin{bmatrix} \begin{bmatrix} \mathbf{K}_{11}^{j'} \\ \mathbf{K}_{21}^{j'} \end{bmatrix} & \begin{bmatrix} \mathbf{K}_{12}^{j'} \\ \mathbf{K}_{22}^{j'} \end{bmatrix} \end{bmatrix}, \mathbf{R}^{jNL} = \begin{bmatrix} \begin{bmatrix} \mathbf{R}_{1}^{jNL} \\ \mathbf{R}_{2}^{jNL} \end{bmatrix} \end{bmatrix}, \tag{A2}$$

where

$$\begin{aligned}
\left[\mathbf{K}_{11}^{1'}\right]_{u,v} =& \frac{1}{2}\sum_{s=0}^{1+n} H_m(s)p\left[\frac{\sin((u+v-2)/p)\theta(s+1) - \sin((u+v-2)/p)\theta(s)}{u+v-2}\right. \\
&+ \left.\frac{\sin((u-v)/p)\theta(s+1) - \sin((u-v)/p)\theta(s)}{u-v}\right], \\
&(u = 1,2,\cdots,n_c, v = 1,2,\cdots,n_c), \\
\left[\mathbf{K}_{21}^{1'}\right]_{u,v} =& -\frac{1}{2}\sum_{s=0}^{1+n} H_m(s)p\left[\frac{\cos((u+v-1)/p)\theta(s+1) - \cos((u+v-1)/p)\theta(s)}{v-u+1}\right. \\
&- \left.\frac{\cos((v-u+1)/p)\theta(s+1) - \cos((v-u+1)/p)\theta(s)}{v-u+1}\right], \\
&(u = 1,2,\cdots,n_c, v = 1,2,\cdots,n_s), \\
&\left[\mathbf{K}_{21}^{1'}\right] = \left[\mathbf{K}_{12}^{1'}\right]^T, \\
\left[\mathbf{K}_{22}^{1'}\right]_{u,v} =& -\frac{1}{2}\sum_{s=0}^{1+n} H_m(s)p\left[\frac{\sin((u+v)/p)\theta(s+1) - \sin((u+v)/p)\theta(s)}{u+v}\right. \\
&+ \left.\frac{\sin((u-v)/p)\theta(s) - \sin((u-v)/p)\theta(s+1)}{u-v}\right], \\
&(u = 1,2,\cdots,n_s, v = 1,2,\cdots,n_s), \\
\left[\mathbf{R}_{1}^{1NL}\right]_{u} =& \sum_{s=0}^{1+n} H_m(s)S_m(s)p\left[\frac{\sin((u-1)/p)\theta(s+1) - \sin((u-1)/p)\theta(s)}{u-1}\right], \\
&(u = 1,2,\cdots,n_c), \\
\left[\mathbf{R}_{2}^{1NL}\right]_{u} =& \sum_{s=0}^{1+n} H_m(s)S_m(s)p\left[\frac{\cos(u/p)\theta(s) - \cos(u/p)\theta(s+1)}{u}\right], \\
&(u = 1,2,\cdots,n_s),
\end{aligned} \tag{A3}$$

$$
\begin{aligned}
\left[\mathbf{K}_{11}^{2\prime}\right]_{u,v} =& \frac{1}{4}\sum_{i=1}^{n_1}\sum_{s=0}^{1+n} c_i H_m(s) p \left[\frac{\sin((u+v-2+pi)/p)\theta(s+1) - \sin((u+v-2+pi)/p)\theta(s)}{u+v-2+pi}\right. \\
&+ \frac{\sin((u+v-2-pi)/p)\theta(s+1) - \sin((u+v-2-pi)/p)\theta(s)}{u+v-2-pi} \\
&+ \frac{\sin((u-v+pi)/p)\theta(s+1) - \sin((u-v+pi)/p)\theta(s)}{u-v+pi} \\
&\left.+ \frac{\sin((u-v-pi)/p)\theta(s+1) - \sin((u-v-pi)/p)\theta(s)}{u-v-pi}\right], \\
&(u = 1,2,\cdots,n_c, v = 1,2,\cdots,n_c), \\
\left[\mathbf{K}_{12}^{2\prime}\right]_{u,v} =& -\frac{1}{4}\sum_{i=1}^{n_1}\sum_{v=0}^{1+n} c_i H_m(s) p \left[\frac{\cos((v+u-1+pi)/p)\theta(s+1) - \cos((v+u-1+pi)/p)\theta(s)}{v+u-1+pi}\right. \\
&+ \frac{\cos((v+u-1-pi)/p)\theta(s+1) - \cos((v+u-1-pi)/p)\theta(s)}{v+u-1-pi} \\
&+ \frac{\cos((v-u+1+pi)/p)\theta(s+1) - \cos((v-u+1+pi)/p)\theta(s)}{v-u+1+pi} \\
&\left.+ \frac{\cos((v-u+1-pi)/p)\theta(s+1) - \cos((v-u+1-pi)/p)\theta(s)}{v-u+1-pi}\right], \\
&(u = 1,2,\cdots,n_c, v = 1,2,\cdots,n_s), \\
&\left[\mathbf{K}_{21}^{2\prime}\right] = \left[\mathbf{K}_{12}^{2\prime}\right]^T, \\
\left[\mathbf{K}_{22}^{2\prime}\right]_{u,v} =& \frac{1}{4}\sum_{i=1}^{n_1}\sum_{v=0}^{1+n} c_i H_m(s) p \left[\frac{\sin((u-v+pi)/p)\theta(s+1) - \sin((u-v+pi)/p)\theta(s)}{u-v+pi}\right. \\
&+ \frac{\sin((u-v-pi)/p)\theta(s+1) - \sin((u-v-pi)/p)\theta(s)}{u-v-pi} \\
&- \frac{\sin((u+v+pi)/p)\theta(s+1) + \sin((u+v+pi)/p)\theta(s)}{u+v+pi} \\
&\left.- \frac{\sin((u+v-pi)/p)\theta(s+1) + \sin((u+v-pi)/p)\theta(s)}{u+v-pi}\right], \\
&(u = 1,2,\cdots,n_s, v = 1,2,\cdots,n_s), \\
\left[\mathbf{R}_1^{2NL}\right]_u =& \frac{1}{2}\sum_{i=1}^{n_1}\sum_{v=0}^{1+n} c_i H_m(s) S_m(s) p \left[\frac{\sin((pi+u-1)/p)\theta(s+1) - \sin((pi+u-1)/p)\theta(s)}{pi+u-1}\right. \\
&\left.+ \frac{\sin((pi-u+1)/p)\theta(s+1) - \sin((pi-u+1)/p)\theta(s)}{pi-u+1}\right], (u = 1,2,\cdots,n_c), \\
\left[\mathbf{R}_2^{2NL}\right]_u =& \frac{-1}{2}\sum_{i=1}^{n_1}\sum_{s=0}^{1+n} c_i H_m(s) S_m(s) p \left[\frac{\cos((u+pi)/p)\theta(s+1) - \cos((u+pi)/p)\theta(s)}{u+pi}\right. \\
&\left.+ \frac{\cos((u-pi)/p)\theta(s+1) - \cos((u-pi)/p)\theta(s)}{u-pi}\right], (u = 1,2,\cdots,n_s),
\end{aligned} \tag{A4}
$$

and

$$\begin{aligned}
\left[\mathbf{K}_{11}^{3\prime}\right]_{u,v} =& \frac{1}{4}\sum_{i=1}^{n_1}\sum_{s=0}^{1+n} d_i H_m(s) p\Bigg[\frac{-\cos((u+v-2+pi)/p)\theta(s+1)+\cos((u+v-2+pi)/p)\theta(s)}{u+v-2+pi}\\
&+\frac{\cos((u+v-2-pi)/p)\theta(s+1)-\cos((u+v-2-pi)/p)\theta(s)}{u+v-2-pi}\\
&+\frac{-\cos((u-v+pi)/p)\theta(s+1)+\cos((u-v+pi)/p)\theta(s)}{u-v+pi}\\
&+\frac{\cos((u-v-pi)/p)\theta(s+1)-\cos((u-v-pi)/p)\theta(s)}{u-v-pi}\Bigg],\\
&(u=1,2,\cdots,n_c, v=1,2,\cdots,n_c),\\
\left[\mathbf{K}_{12}^{3\prime}\right]_{u,v} =& -\frac{1}{4}\sum_{i=1}^{n_1}\sum_{v=0}^{1+n} d_i H_m(s) p\Bigg[\frac{-\sin((v+u-1+pi)/p)\theta(s+1)+\sin((v+u-1+pi)/p)\theta(s)}{v+u-1+pi}\\
&+\frac{\sin((v+u-1-pi)/p)\theta(s+1)-\sin((v+u-1-pi)/p)\theta(s)}{v+u-1-pi}\\
&+\frac{\sin((u-v-1+pi)/p)\theta(s+1)-\sin((u-v-1+pi)/p)\theta(s)}{u-v-1+pi}\\
&+\frac{-\sin((u-v-1-pi)/p)\theta(s+1)+\sin((u-v-1-pi)/p)\theta(s)}{u-v-1-pi}\Bigg],\\
&(u=1,2,\cdots,n_c, v=1,2,\cdots,n_s),\\
&\left[\mathbf{K}_{21}^{3\prime}\right] = \left[\mathbf{K}_{12}^{3\prime}\right]^T,\\
\left[\mathbf{K}_{22}^{3\prime}\right]_{u,v} =& \frac{1}{4}\sum_{i=1}^{n_1}\sum_{v=0}^{1+n} d_i H_m(s) p\Bigg[\frac{\cos((u-v+pi)/p)\theta(s+1)-\cos((u-v+pi)/p)\theta(s)}{u-v+pi}\\
&+\frac{-\cos((u-v-pi)/p)\theta(s+1)+\cos((u-v-pi)/p)\theta(s)}{u-v-pi}\\
&+\frac{-\cos((u+v+pi)/p)\theta(s+1)+\cos((u+v+pi)/p)\theta(s)}{u+v+pi}\\
&+\frac{\cos((u+v-pi)/p)\theta(s+1)-\cos((u+v-pi)/p)\theta(s)}{u+v-pi}\Bigg],\\
&(u=1,2,\cdots,n_s, v=1,2,\cdots,n_s),\\
\left[\mathbf{R}_{1}^{3NL}\right]_{u} =& \frac{1}{2}\sum_{i=1}^{n_1}\sum_{v=0}^{1+n} d_i H_m(s) S_m(s) p\Bigg[\frac{-\cos((pi+u-1)/p)\theta(s+1)+\cos((pi+u-1)/p)\theta(s)}{pi+u-1}\\
&+\frac{\cos((-pi+u-1)/p)\theta(s+1)-\cos((-pi+u-1)/p)\theta(s)}{-pi+u-1}\Bigg], (u=1,2,\cdots,n_c),\\
\left[\mathbf{R}_{2}^{3NL}\right]_{u} =& \frac{-1}{2}\sum_{i=1}^{n_1}\sum_{s=0}^{1+n} d_i H_m(s) S_m(s) p\Bigg[\frac{-\sin((u+pi)/p)\theta(s+1)+\sin((u+pi)/p)\theta(s)}{u+pi}\\
&+\frac{\sin((u-pi)/p)\theta(s+1)-\sin((u-pi)/p)\theta(s)}{u-pi}\Bigg], (u=1,2,\cdots,n_s),
\end{aligned} \tag{A5}$$

in which n denotes the number of zeros of the equation $|\bar{x}_0| - \dfrac{b}{b_c} = 0$ within the interval $[0 \quad 2p\pi]$; $\theta(1), \theta(2), \cdots, \theta(n)$ $(\theta(1) < \theta(2) < \cdots < \theta(n))$ represent zeros of the equation $|\bar{x}_0| - \dfrac{b}{b_c} = 0$, as presented in Figure A1; $\theta(0) = 0$, $\theta(1+n) = 2p\pi$; and H_m and S_m are step functions that depend on the sign functions $|\bar{x}_0| - \frac{b}{b_c} = 0$ and $\bar{x}_0 - \frac{b}{b_c} = 0$, respectively, and can be described as

$$H_m(s) = \begin{cases} 1, & v_s \geq 0 \\ 0, & v_s < 0, \end{cases} \tag{A6}$$

$$S_m(s) = \begin{cases} -1, & v_s' \geq 0 \\ 1, & v_s' < 0, \end{cases} \tag{A7}$$

respectively. Here, $s = 0, 1, 2, \cdots, n$, and $v_0, v_1, v_2, \cdots, v_s$ and $v_0', v_1', v_2', \cdots, v_s'$ denote the values of sign functions $|\bar{x}_0| - \frac{b}{b_c} = 0$ and $\bar{x}_0 - \frac{b}{b_c} = 0$ in the subintervals $[\theta(0) \quad \theta(1)], [\theta(1) \quad \theta(2)], \cdots, [\theta(n) \quad \theta(1+n)]$, respectively. It should be noted that in the process of calculating Equations (A3)–(A5), the values of $\sin(0\theta)/(0\theta)$ and $\cos(0\theta)/(0\theta)$ should be 1 and 0 [50], respectively.

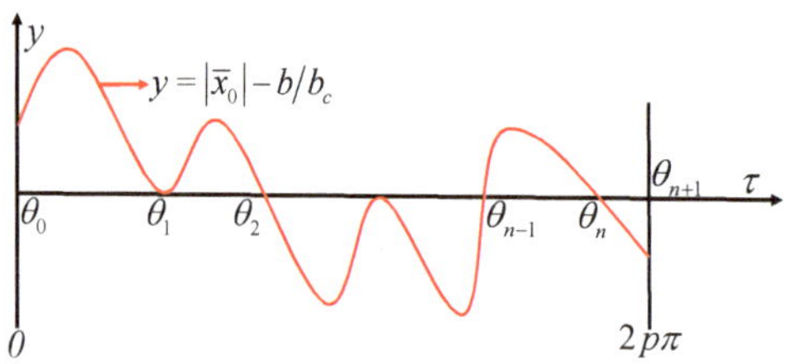

Figure A1. Zeros of the equation $|\bar{x}_0| = b/b_c$.

References

1. Lin, T.J.; Ou, H.G.; Li, R.F. A finite element method for 3D static and dynamic contact/impact analysis of gear drives. *Comput. Methods Appl. Mech. Eng.* **2007**, *196*, 1716–1728. [CrossRef]
2. Eritenel, T.; Parker, R.G. An investigation of tooth mesh nonlinearity and partial contact loss in gear pairs using a lumped-parameter model. *Mech. Mach. Theory* **2012**, *56*, 28–51. [CrossRef]
3. Fang, Y.N.; Liang, X.H.; Zuo, M.J. Effects of friction and stochastic load on transient characteristics of a spur gear pair. *Nonlinear Dyn.* **2018**, *93*, 599–609. [CrossRef]
4. Raghuwanshi, N.K.; Parey, A. Mesh stiffness measurement of cracked spur gear by photoelasticity technique. *Measurement* **2015**, *73*, 439–452. [CrossRef]
5. Yogesh, P.; Parey, A. Crack behavior in a high contact ratio spur gear tooth and its effect on mesh stiffness. *Eng. Fail. Anal.* **2013**, *34*, 69–78.
6. Wang, J.D.; Howard, I. Finite element analysis of high contact ratio spur gears in mesh. *J. Tribol.* **2005**, *127*, 469–483. [CrossRef]
7. Zhan, J.X.; Fard, M.; Jazar, R. A CAD-FEM-QSA integration technique for determining the time-varying meshing stiffness of gear pairs. *Measurement* **2017**, *100*, 139–149. [CrossRef]
8. Liang, X.H.; Zhang, H.S.; Zuo, M.J.; Qin, Y. Three new models for evaluation of standard involute spur gear mesh stiffness. *Mech. Syst. Signal Process.* **2018**, *101*, 424–434. [CrossRef]
9. Yang, D.C.H.; Lin, J.Y. Hertzian damping, tooth friction and bending elasticity in gear impact dynamics. *J. Mech. Transm. Autom. Des.* **1987**, *109*, 189–196. [CrossRef]
10. Liang, X.H.; Zuo, M.J.; Pandey, M. Analytically evaluating the influence of crack on the mesh stiffness of a planetary gear set. *Mech. Mach. Theory* **2014**, *76*, 20–38. [CrossRef]
11. Meng, Z.; Shi, G.X.; Wang, F.L. Vibration response and fault characteristics analysis of gear based on time-varying mesh stiffness. *Mech. Mach. Theory* **2020**, *148*, 103786. [CrossRef]
12. Liu, X.Z. Vibration modelling and fault evolution symptom analysis of a planetary gear train for sun gear wear status assessment. *Mech. Syst. Signal Process.* **2022**, *166*, 108403. [CrossRef]
13. Liu, H.; Zhang, C.; Xiang, C.L.; Wang, C. Tooth profile modification based on lateral-torsional-rocking coupled nonlinear dynamic model of gear system. *Mech. Mach. Theory* **2016**, *105*, 606–619. [CrossRef]
14. Meng, F.S.; Xia, H.; Zhang, X.; Wang, J.; Jin, Y. Study on nonlinear dynamic characteristics of gear system with 3D anisotropic rough tooth surface based on fractal theory. *Int. J. Non-Linear Mech.* **2023**, *150*, 104366. [CrossRef]

15. Tian, H.X.; Zhao, X.J.; Huang, W.K.; Ma, H. A stiffness model for EHL contact on smooth/rough surfaces and its application in mesh stiffness calculation of the planetary gear set. *Tribol. Int.* **2024**, *196*, 109720. [CrossRef]
16. Zhou, W.G.; Zhu, R.P.; Li, Z.W.; Liu, W.; Wang, J. An improved dynamic model of spur gears considering delayed meshing out of the contact point due to root crack. *Mech. Syst. Signal Process.* **2024**, *212*, 111325. [CrossRef]
17. Zhang, T.; Lin, T.J.; Fu, L.Y. Analytical and experimental study on acoustic-vibration characteristics of double-helical planetary gear transmission systems with multi-field coupling effect. *Mech. Syst. Signal Process.* **2025**, *224*, 112143. [CrossRef]
18. Zheng, X.Y.; Luo, W.J.; Hu, Y.M.; He, Z.; Wang, S. Study on the mesh stiffness and nonlinear dynamics accounting for centrifugal effect of high-speed spur gears. *Mech. Mach. Theory* **2022**, *170*, 104686. [CrossRef]
19. Ma, H.; Zeng, J.; Feng, R.J.; Pang, X.; Wen, B. An improved analytical method for mesh stiffness calculation of spur gears with tip relief. *Mech. Mach. Theory* **2016**, *98*, 64–80. [CrossRef]
20. Ma, H.; Pang, X.; Feng, R.J.; Zeng, J.; Wen, B. Improved time-varying mesh stiffness model of cracked spur gears. *Eng. Fail. Anal.* **2015**, *55*, 271–287. [CrossRef]
21. Jiang, Y.; Zhu, H.; Li, Z.; Peng, Z. The nonlinear dynamics response of cracked gear system in a coal cutter taking environmental multi-frequency excitation forces into consideration. *Nonlinear Dyn.* **2016**, *84*, 203–222. [CrossRef]
22. Wan, Z.G.; Cao, H.R.; Zi, Y.Y.; He, W.; He, Z. An improved time-varying mesh stiffness algorithm and dynamic modeling of gear-rotor system with tooth root crack. *Eng. Fail. Anal.* **2014**, *42*, 157–177. [CrossRef]
23. Liu, G.; Parker, R.G. Dynamic modeling and analysis of tooth profile modification for multimesh gear vibration. *J. Mech. Des.* **2008**, *130*, 1121402. [CrossRef]
24. Pan, W.J.; Li, X.P.; Wang, L.L. Nonlinear response analysis of gear-shaft-bearing system considering tooth contact temperature and random excitations. *Appl. Math. Model.* **2019**, *68*, 113–136. [CrossRef]
25. Moradi, H.; Salarieh, H. Analysis of nonlinear oscillations in spur gear pairs with approximated modelling of backlash nonlinearity. *Mech. Mach. Theory* **2012**, *51*, 14–316. [CrossRef]
26. Mo, S.; Zhang, Y.X.; Song, Y.L. Nonlinear vibration and primary resonance analysis of non-orthogonal face gear-rotor-bearing system. *Nonlinear Dyn.* **2022**, *108*, 3367–3389.
27. Bruzzone, F.; Rosso, C. Comparison of two possible dynamic models for gear dynamic analysis. In *Topics in Modal Analysis & Parameter Identification, Proceedings of the 41st IMAC, A Conference and Exposition on Structural Dynamics*; Conference Proceedings of the Society for Experimental Mechanics Series; Springer: Cham, Switzerland, 2023; Volume 9, pp. 183–192.
28. Bruzzone, F.; Rosso, C.; Theodossiades, S. Dynamic reduction technique for nonlinear analysis of spur gear pairs. *Nonlinear Dyn.* **2024**, *112*, 15797–15811. [CrossRef]
29. Bruzzone, F.; Maggi, T.; Marcellini, C.; Rosso, C. Gear teeth deflection model for spur gears: proposal of a 3D nonlinear and non-hertzian approach. *Machines* **2021**, *9*, 223. [CrossRef]
30. Oliveri, L.; Rosso, C.; Zucca, S. Influence of actual static transmission error and contact ratio on gear engagement dynamics. In *Nonlinear Dynamics, Proceedings of the 35th IMAC, A Conference and Exposition on Structural Dynamics*; Conference Proceedings of the Society for Experimental Mechanics Series; Springer: Cham, Switzerland, 2017; Volume 1, pp. 143–154.
31. Al-Shyyab, A.; Kahraman, A. Non-linear dynamic analysis of a multi-mesh gear train using multi-term harmonic balance method: Sub-harmonic motions. *J. Sound Vib.* **2005**, *279*, 417–451. [CrossRef]
32. Hou, J.Y.; Yang, S.P.; Li, Q.; Liu, Yo. Analysis of dynamic characteristics of a fractional-order spur gear pair with internal and external excitations. *J. Comput. Nonlinear Dyn.* **2022**, *17*, 021004. [CrossRef]
33. Yang, Y.; Xia, W.; Han, J.; Song, Y.; Wang, J.; Dai, Y. Vibration analysis for tooth crack detection in a spur gear system with clearance nonlinearity. *Int. J. Mech. Sci.* **2019**, *157*, 648–661. [CrossRef]
34. Liu, G.H.; Hong, J.; Parker, R.G. Influence of simultaneous time-varying bearing and tooth mesh stiffness fluctuations on spur gear pair vibration. *Nonlinear Dyn.* **2019**, *97*, 1403–1424. [CrossRef]
35. Zheng, X.Y.; Hu, Y.M.; He, Z.; Xiao, Y.; Zhang, X. On the extended tooth contact and nonlinear dynamics for spur gears—An analytical model. *Mech. Mach. Theory* **2022**, *175*, 104958. [CrossRef]
36. Kahraman, A.; Singh, R. Non-linear dynamics of a geared rotor-bearing system with multiple clearances. *J. Sound Vib.* **1991**, *144*, 469–506. [CrossRef]
37. Lau, S.L.; Zhang, W.S. Nonlinear vibrations of piecewise-linear systems by incremental harmonic balance method. *J. Appl. Mech.* **1992**, *59*, 153–160. [CrossRef]
38. Choi, Y.S.; Noah, S.T. Forced periodic vibration of unsymmetric piecewise-linear systems. *J. Sound Vib.* **1988**, *121*, 117–126. [CrossRef]
39. Zhou, S.H.; Song, G.Q.; Li, Y.M.; Huang, Z.; Ren, Z. Dynamic and steady analysis of a 2-DOF vehicle system by modified incremental harmonic balance method. *Nonlinear Dyn.* **2019**, *98*, 75–94. [CrossRef]
40. Vorotnikov, K.; Starosvetsky, Y. Bifurcation structure of the special class of nonstationary regimes emerging in the 2D inertially coupled, unit-cell model: Analytical study. *J. Sound Vib.* **2016**, *377*, 226–242. [CrossRef]

41. Huang, J.L.; Zhang, B.X.; Zhu, W.D. Quasi-periodic solutions of a damped nonlinear quasi-periodic mathieu equation by the incremental harmonic balance method with two time scales. *J. Appl. Mech.* **2022**, *89*, 091009. [CrossRef]
42. Liao, F.L.; Huang, J.L.; Zhu, W.D. Nonlinear Vibration of a Multi-Degree-of-Freedom Gear Transmission System With Multipiece-wise Linear Functions. *J. Comput. Nonlinear Dyn.* **2023**, *18*, 041003. [CrossRef]
43. Huang, J.L.; Wang, T.; Zhu, W.D. An incremental harmonic balance method with two time-scales for quasi-periodic responses of a Van der Pol–Mathieu equation. *Int. J. Non-Linear Mech.* **2021**, *135*, 103767. [CrossRef]
44. Ju, R.; Fan, W.; Zhu, W.D. Comparison between the incremental harmonic balance method and alternating frequency/time-domain method. *J. Vib. Acoust.* **2021**, *143*, 024501. [CrossRef]
45. Huang, J.L.; Su, K.L.R.; Lee, Y.Y.R.; Chen, S. Various bifurcation phenomena in a nonlinear curved beam subjected to base harmonic excitation. *Int. J. Bifurc. Chaos* **2018**, *28*, 1830023. [CrossRef]
46. Huang, J.L.; Su, K.L.R.; Chen, S.H. Precise Hsu's method for analyzing the stability of periodic solutions of multi-degrees-of-freedom systems with cubic nonlinearity. *Comput. Struct.* **2009**, *87*, 1624–1630. [CrossRef]
47. Friedmann, P.; Hammond, C.E.; Woo, T.H. Efficient numerical treatment of periodic systems with application to stability problems. *Int. J. Numer. Methods Eng.* **1977**, *11*, 1117–1136. [CrossRef]
48. Hsu, C.S.; Cheng, W.H. Applications of the theory of impulsive parametric excitation and new treatments of general parametric excitation problems. *J. Appl. Mech.* **1973**, *40*, 78–86. [CrossRef]
49. Kahraman, A.; Blankenship, G.W. Effect of involute contact ratio on spur gear dynamics. *J. Mech. Des.* **1999**, *121*, 112–118. [CrossRef]
50. Shen, Y.J.; Yang, S.P.; Liu, X.D. Nonlinear dynamics of a spur gear pair with time-varying stiffness and backlash based on incremental harmonic balance method. *Int. J. Mech. Sci.* **2006**, *48*, 1256–1263. [CrossRef]

machines

Article

Study on the Potential of New Load-Carrying Capacity Descriptions for the Service Life Calculations of Gears

Daniel Vietze *, Josef Pellkofer and Karsten Stahl *

Gear Research Center (FZG), TUM School of Engineering and Design, Technical University of Munich, Boltzmannstraße 15, D-85748 Garching near Munich, Germany
* Correspondence: daniel.vietze@tum.de (D.V.); karsten.stahl@tum.de (K.S.)

Abstract: Calculating the service life of gears under variable loads requires a description of the load-carrying capacity. The current standard for this is the use of the S/N curve. International standards such as ISO 6336 stipulate the use of this approach for the calculation of the service of gears under variable loads. In this paper, five new approaches are developed and evaluated to describe the load-carrying capacity of gears in the load range of finite life. Four methods are based on machine learning, and one uses mathematical regression. To validate the new approaches, the results of an experimental study investigating the service life of gears under variable loads are presented. These results form the basis for the conducted study, which compares the five new methods with the existing approach. The comparison focuses on the ability of the load-carrying capacity descriptions to provide an accurate calculation of the service life and to reduce scattering as much as possible. The results of the study show significant potential for the new methods, especially the one based on a neural network.

Keywords: fatigue life analysis; gears; damage accumulation; machine learning; machine elements

1. Introduction

Gearboxes are an essential part of many machines, such as electric cars, robots, ships, or aircraft. At the heart of the gearbox are, in most cases, the gears. Gears are one of the most common machine elements used to transmit and convert rotational movement [1]. In order to dimension these machine elements according to the load occurring during operation, a description of the load-carrying capacity is essential. The load-carrying capacity of gears depends on the load and must, therefore, be described as a function of the occurring loads. The accepted mathematical approach for this is based on a publication by Basquin [2] dating back to 1910. For instance, ISO 6336 [3,4] specifies this concept for the load-carrying capacity of gears. The approach to describing the load-carrying capacity of gears based on the concept of Basquin allows for a relatively simple procedure for testing and calculation and has proven to be applicable and reliable for gears.

Since the publication of this approach back in 1910, the possibilities for data analysis have changed fundamentally. In particular, the advent of computers opened up new potential. In addition, recent developments in the field of machine learning have created new options for analyzing and describing data. In many fields of engineering, the application of machine learning enabled significant advancements, for example, in the field of condition monitoring [5–7]. The scope of this paper is to evaluate these methods for the description of the load-carrying capacity of gears. More precisely, this paper investigates various alternatives for the description of the load-carrying capacity in the load range of finite service life. The main question of this study is whether it is possible to reduce the scattering of the calculation results for the service life of gears under variable loads by applying one of the developed approaches compared to the results based on the approach, according to Basquin.

First, a brief summary of the state of knowledge regarding the load-carrying capacity of gears is presented. In this context, the above-mentioned approach, according to Basquin, is introduced. This is followed by a description of the new approaches investigated to describe the load-carrying capacity of gears. These approaches are based on mathematical data analysis and machine learning. To validate the suitability of the presented approaches, an experimental study was conducted at FZG. The results are presented in this paper. Finally, the paper compares the newly presented methods with the common approach based on Basquin's publication. The scope of this comparison is the assessment of the accuracy of the calculation of the service life of gears under variable loads.

2. Load-Carrying Capacity of Gears

The load-carrying capacity of gears is usually described using the S/N curve [8,9]. The S/N curve was developed by the German engineer August Wöhler [10] and is, therefore, also known as the "Woehler" curve. A schematic S/N curve is presented in Figure 1. The S/N curve divides the loads into three ranges:

- Loads greater than the static strength cannot be tolerated for a relevant number of load cycles (red area in Figure 1).
- Loads less than the endurance limit can be tolerated for an infinite number of load cycles (green area in Figure 1).
- Loads between these two characteristic values can be tolerated for a limited number of load cycles (yellow area in Figure 1).

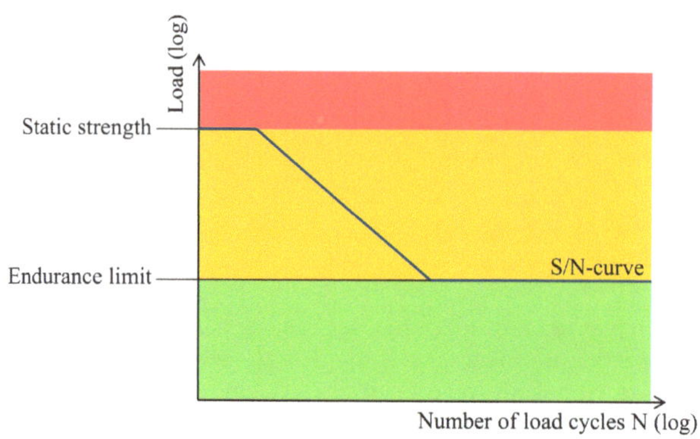

Figure 1. Schematic presentation of a S/N curve.

In the load range of finite life, the correlation between the load L and the number of endurable load cycles N is non-linear and based on the Basquin equation [2]. Due to the logarithmic scale of both axes of the S/N diagram, the correlation appears linear in Figure 1. The Basquin equation [2] is shown in Equation (1).

$$N = C \times L^{-k} \tag{1}$$

The Basquin equation uses only the slope k and the constant C to determine the S/N curve. Therefore, the ability to adapt the equation to describe the load-carrying capacity of a particular machine element or gear is limited. On the other hand, this significantly reduces the amount of data required to determine the S/N curve. Figure 2 shows three different examples of S/N curves. On the left-hand side, the curves according to the Basquin equation are shown with logarithmic scaling of both axes. On the right-hand side, both axes are scaled linearly. The slope k and constant C can be used to adjust the S/N curve, but the general shape and mathematical formula are fixed by the Basquin equation. The three exemplary S/N curves shown in Figure 2 are based on unique combinations of the slope k and constant C. This example is intended to show that the general shape cannot be influenced by choosing different values for the parameters, and, therefore, the possibility of adaption is limited.

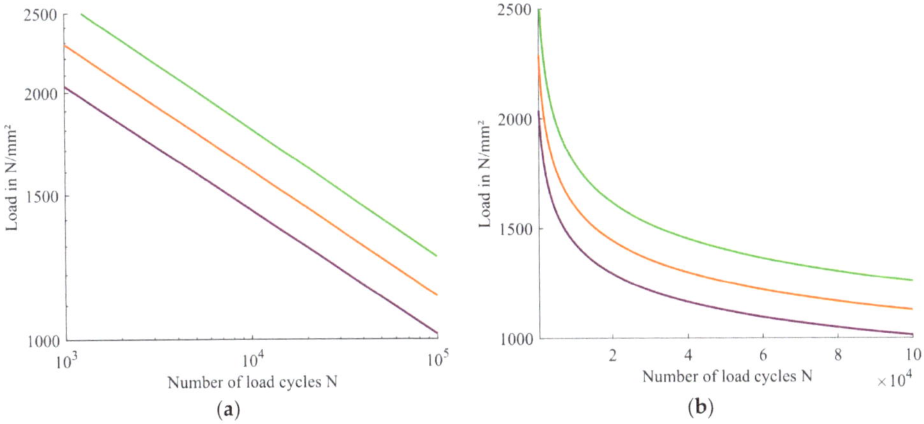

Figure 2. Comparison of different S/N curves in logarithmic (**a**) and linear (**b**) scale.

The application of the Basquin equation to describe the load-carrying capacity is the state of knowledge for gears [8] and is stipulated by standards [9,11].

The procedure for the experimental determination of the S/N curve for gears is described in the FVA Guideline 563/I [9]. It is divided into two parts:

- The determination of the endurance limit of the gear;
- The specification of the S/N curve in the load range of limited service life.

The data for the S/N curve are generated using single load tests only. Gears are tested under a constant load until either the gears fail or the number of load cycles exceeds a certain threshold. If the latter occurs, the test run is classified as a runout. The two parts of the test procedure are separate and usually do not share any data points. To determine the endurance limit, both failures and runouts are required within the set of evaluated data points. Failures are the only valid basis for determining the S/N curve in the load range of limited service life.

The endurance limit can be determined using one of two approaches:

- The horizon method uses a certain number of stress levels, which must be determined by the user. At every load level, a certain number of test points are conducted. This basis is used to calculate the endurance limit.
- The staircase procedure includes a certain number of test runs. The load level of each test run is determined by the result of the previous test run. If a failure occurs, the next test will be performed at a lower load level. On the other hand, if a runout is observed, the next test run applies a higher load. The endurance limit is usually calculated using the method according to Hück [12]. Additional information about this method and its accuracy can be found in [13].

To define the parameters of the Basquin equation, which describes the number of endurable load cycles for loads above the endurance limit, two data points on the S/N curve must be determined. These two points are a combination of a load and the corresponding number of endurable load cycles. Since gear failure is subject to scattering, it is common to perform several test runs at both load levels and use the average number of load cycles $N_{50\%,n}$ according to Equation (2) [9]. Where N_i is the respective number of load cycles at failure for each one of the n test runs.

$$\log_{10} N_{50\%,n} = \frac{1}{n}\sum_{i=1}^{n} \log_{10} N_i \tag{2}$$

Typically, the total number of test runs conducted for gears is in the range of 20 to 40 [9]. Finally, based on the results, it is possible to determine the number of load cycles at the knee point of the S/N curve N_D. This point is the intersection of the sloped part of the S/N curve and the horizontal line representing the endurance limit.

3. Basics of Machine Learning and Application in the Field of Fatigue Life Analysis

Machine learning methods enable computers to solve problems without being specifically programmed for their solution. These methods rely on data to learn the correlation between the input and the output of a problem and, therefore, do not require a ridged algorithm specified by the user [14]. Especially in very complex cases with a lot of data, machine learning may provide a solution for a problem that otherwise may not have been solvable using traditional approaches. Machine learning can be classified into supervised and unsupervised learning. In the case of unsupervised learning, the algorithm only receives a set of input data and tries to find the best possible correlation between the data points. In the case of supervised learning, the output data are also provided, and therefore, the learning process can rely on these data to find the best possible solution for the problem. Later, the trained algorithm can then be used to provide the answer for new input data.

This paper investigates the potential of replacing the Basquin equation as a description of the load-carrying capacity with a machine learning-based model. For the implementation of this approach, a machine learning model solving a regression problem first has to be trained; for this, different methods of machine learning can be applied. The selection of the methods considered within this paper is based on the results and findings of publications in the field of fatigue life analysis. Three commonly used machine learning methods in the field of fatigue life analysis are neural network, support vector machine, and random forest. He et al. [15], for instance, applied all three methods for the estimation of the S/N curve of three different steels. Zhan and Li [16] utilized the neural network and the random forest approach to estimate the service life of aluminum parts.

Mudabbir and Mosavi [17] presented a review of the use of machine learning for the modeling of service life. They include 29 publications and conclude that neural network, support vector machine, and random forest are most commonly used within this field.

Another often-used method is the Gaussian process regression. In [18,19], this machine learning method was applied to estimate the state of damage of an off-shore wind turbine. In [20], the service life under multiaxial loading was predicted based on the Gaussian process regression.

Based on the presented results and publications, the following machine learning methods are applied within this paper:

- Neural network (NN);
- Support vector machine/support vector regression (SVR);
- Random forest (RF);
- Gaussian process regression (GPR).

In the following, the four machine learning methods applied in this paper are introduced in a compact manner. This paper only uses supervised learning. Additional information about machine learning can be found in the relevant literature [14].

3.1. Support Vector Machine

The support vector machine (SVM) is a machine learning method that was originally designed for classification. Classification is a kind of problem where input data have to be divided into two or more classes, for instance, if a picture shows a human or an animal. Further developments to the method enabled the application for regression [14]. Regression is the prediction of a floating output value based on the input, for instance, which air pressure can be expected in a certain area based on the current weather data. The SVM is based on the so-called kernel trick that uses a transformation of the data into higher mathematical dimensions to solve the problem, which could not have been solved in its original dimension. The SVM requires little input data and computational power compared to other methods of machine learning [21].

3.2. Random Forest

The random forest approach is, per se, not an original machine learning method. It can be allocated to the field of ensemble learning. Ensemble learning is the approach to combining several methods or several trained models of one method to solve a problem [14]. The random forest approach combines several decision trees, which are, per se, a machine learning method. Software tools like MATLAB (version 2023b) [22] can be used to automatically train and combine decision trees to build a random forest and solve the problem based on this approach. In the cases of the software MATLAB [22], 100 decision trees are combined for the solution of the regression problem [23].

3.3. Neural Network

The neural network is a machine learning method that uses a structure in the style of the human brain. The network is composed of layers consisting of single neurons. These neurons are linked to each other and, therefore, enable the transfer of data from the input to the output side of the network. During the training process, the weighting of the single neurons and the functions activating each one is determined to fit the data best [14]. Neural networks are part of the field of deep learning, which enabled many of the remarkable results achieved with machine learning in the past years. Neural networks can be scaled to solve very complex problems but, in general, require more input data and computational power compared to other methods of machine learning like the SVM [21].

Within this paper, a feedforward, fully connected neural network designed for regression is applied. Additional information about the used approach can be found in [24].

3.4. Gaussian Process Regression

The Gaussian process regression is commonly used to solve regression problems [25]. The process enables inter- and extrapolation based on data. The approach is capable of performing well with a relatively small database. Additional information about the Gaussian process regression can be found in [25].

4. Experimental Data for the Determination of the S/N Curve

This paper uses data derived from tooth root breakage tests of spur gears. The tests are a continuation of the results presented by the authors in [26]. The data on the gears used are presented in Table 1. The gear geometry was developed at FZG, especially for the application as test gear regarding the tooth root carrying capacity at a pulsator test rig, and was used before in serval research projects at FZG [27,28]. The pulsator used for the experiments is shown in Figure 3, and a detailed description of this test rig is presented in [27,29]. The data generated using a pulsator test rig cannot be directly applied to the real operating conditions of gears; instead, an additional conversion has to be conducted. This is due to differences regarding loading conditions and statistics. Additional information on this topic can be found in [30]. This does not affect the validity of the results presented in this paper. All compared data and results originate from the same experimental setup, and therefore, a direct application of the test data is possible.

Table 1. Basic test gear data.

Gear	Spur Gear
Normal module m_n	5 mm
Number of teeth	24
Face width b	20 mm
Machining	Milling and grinding
Material	Steel 18CrNiMo7-6
Heat treatment	Case-hardened

Figure 3. Pulsator test rig at FZG.

The results of the tests are presented in Figure 4, together with the resulting S/N curve. The endurance limit is determined using the staircase procedure, according to Hück [12]. The sloped section of the S/N curve is based on the results of the test using the load levels corresponding to 70 and 80 kN of pulsator force, respectively. All data points marked in blue are included in the determination of the S/N curve. To create an advanced database, additional test runs were conducted. These test runs would not have been required to determine the S/N curve but are used as an extended basis for the new approaches presented in this paper. The results of the additional runs are marked in orange and are not included in the presented S/N curve.

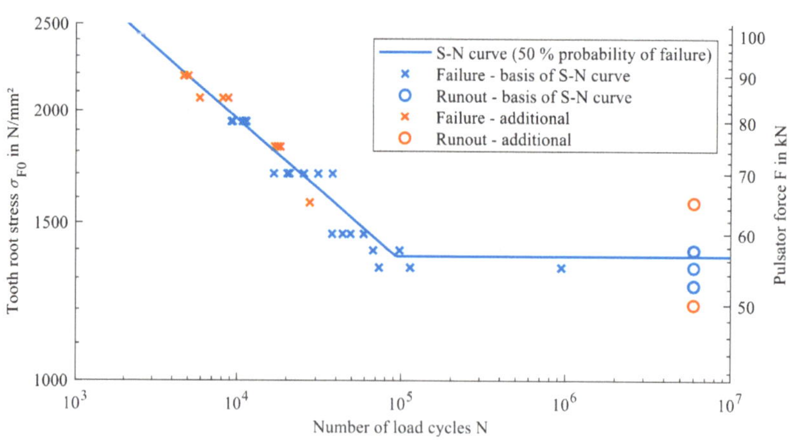

Figure 4. S/N curve of the gears, including all test points with constant load.

The data presented within this paper refer to the nominal tooth root stress σ_{F0}. The nominal tooth root stress caused by the applied pulsator force F is calculated according to ISO 6336-3 [31] standard and is based on Equation (3).

$$\sigma_{F0} = \frac{F}{b * m_n} \times Y_s \times Y_F \quad (3)$$

The stress correction factor Y_s and the form factor Y_F are determined based on geometry data of the actual tooth root of the gears. Therefore, the tooth root curve is measured at FZG using a CNC-controlled precision measuring center by manufacturer Klingelnberg. These data are then used to determine the factors by applying an iterative calculation process.

The numerical values of the S/N curve are listed in Table 2.

Table 2. Data of the S/N curve.

Endurance limit	1375.81 N/mm^2
Slope k	6.33
Constant C	6.90×10^{24}

5. Methods for the Advanced Description of the Load-Carrying Capacity of Gears

The purpose of this paper is to investigate the possibility of using alternative methods to describe the load-carrying capacity of gears. Instead of using the Basquin equation, machine learning and regression using polynomials are applied. All data points representing a failure are used as the database. Therefore, runouts are not included. In total, the database consists of 31 data points. Within this paper, only the sloped section of the S/N curve is described when alternative methods are applied. The endurance limit is set to the value corresponding to the original S/N curve. Five different approaches are investigated as possible alternatives to the Basquin equation. Four approaches are based on machine learning, and one is a polynomial regression. As mentioned, runouts are not included as data for the presented methods. In this paper, the scope is limited to the failures as a starting point for the investigation of new approaches. The information provided by the runouts yields potential for further investigations following this publication.

5.1. *Polynomial Regression*

The general shape of the curve of the Basquin equation is similar to a third-degree polynomial. As mentioned before, the Basquin equation has two coefficients to be adapted to the test data. A third-degree polynomial, on the other hand, has four coefficients c_0–c_3 to be adjusted for. The general formula for this polynomial is shown in Equation (4).

$$L\,(N) = c_3 * N^3 + c_2 * N^2 + c_1 * N + c_0 \quad (4)$$

MATLAB (version 2023b) [22] software was used to determine the polynomial from the test data presented. The coefficients c_0–c_3 were optimized to represent the test points best. During the optimization an additional condition was added. To ensure good compatibility with the endurance limit, the polynomial was optimized to intersect the endurance limit at the same number of load cycles as the Basquin equation. Therefore, the same number of load cycles is set at the knee point. The results for the coefficients c_0–c_3 are presented in Table 3. The resulting new sections of the S/N curve are presented in Figure 5.

Table 3. Coefficients c_0–c_3 of the third-degree polynomial.

c_0	2274.519
c_1	−0.033
c_2	4.067×10^{-7}
c_3	-1.697×10^{-12}

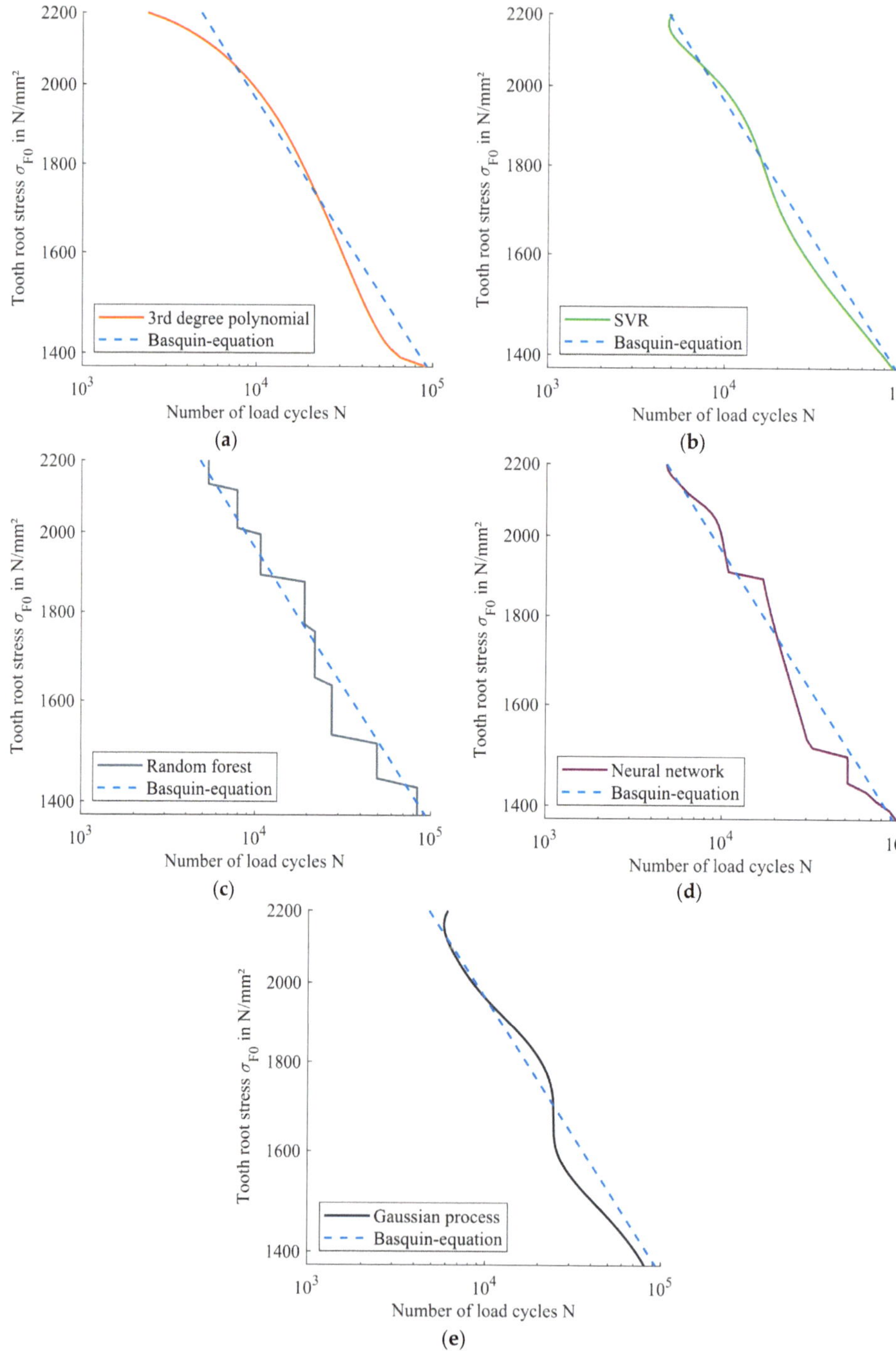

Figure 5. Alternative descriptions of the load-carrying capacity, based on (**a**) a third-degree polynomial, (**b**) a SVR, (**c**) a random forrest, (**d**) a neural network and (**e**) a gaussian process regression.

5.2. Approaches Based on Machine Learning

Machine learning allows the user to evaluate data without any additional input other than the data. For instance, there is no need to choose a specific mathematical approach to describe the correlations within the data. Therefore, this approach has the advantage of being highly adaptable to the input data. For a general introduction to the field of machine learning, this paper refers to the relevant literature [14].

The model based on Gaussian process regression is built using the software MATLAB (version 2023b) [22]. The same dataset applied to the polynomial regression serves as a basis. To ensure plausible results, one data point is removed for the Gaussian process regression. The failure, which occurred at nearly one million load cycles, caused the result to shift far to the right within the S/N diagram; therefore, this point is not considered for this method.

For the training of the three remaining machine learning methods, the input data are divided into 70% training data and 30% test data. The training of the methods is also conducted in the software MATLAB (version 2023b) [22] and uses the internal optimization of this tool. Additional information about machine learning with the software MATLAB can be found in [32]. A specific documentation of the used models and approaches can be found in [23] for the random forest, in [33] for the Gaussian process regression, in [34] for the support vector regression, and in [24] for the neural network.

5.3. Overview

The results of all five approaches are presented in Figure 5. Additionally, the S/N curve based on the Basquin equation is added to each chart for reference.

All five new descriptions of the load-carrying capacity are positioned near the one based on the Basquin equation within the S/N diagram. In general, the new approaches exhibit a more complex shape compared to the Basquin equation. The validation of the new approaches will be conducted within the following chapters using experimental data. At this point, it is important to mention that it is reasonable to argue about the physical logic of the shape of the new load-carrying descriptions. Some of them may seem unintuitive to some. Especially the load-carrying description based on the random forest approach shows an abnormal shape. Within the scope of this study, the suitability will only be validated by the results of the following calculations. This is performed with the intention that it may be possible to increase the calculation accuracy with an approach that seems unintuitive regarding its shape.

6. Experimental Data for Validation

The focus of this publication is to evaluate the accuracy of the service life calculation under variable loads based on the different approaches for the load-carrying capacity. Therefore, data about the service life of gears under variable loads are required. These data are generated using the same experimental setup as described in Section 4. The test rig offers the possibility to program load sequences and test the gears until failure.

Four different load sequences are designed. The design of the load sequences is mostly random. The loads are chosen in a reasonable range, and the overall design is chosen in a way that ensures usable results. For example, high loads are applied for shorter periods of time to avoid early failures. Each load sequence is used for three runs, resulting in a total of twelve data points. The results of the experimental study under variable loads are shown in Figures 6–9. More detailed results can be found in the following chapter.

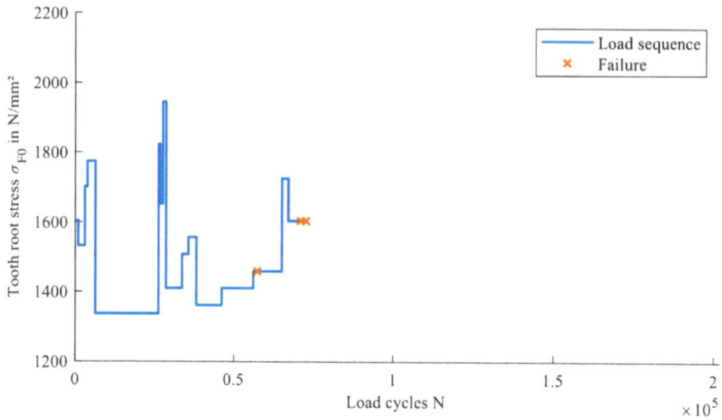

Figure 6. Results of the first test run with variable loads.

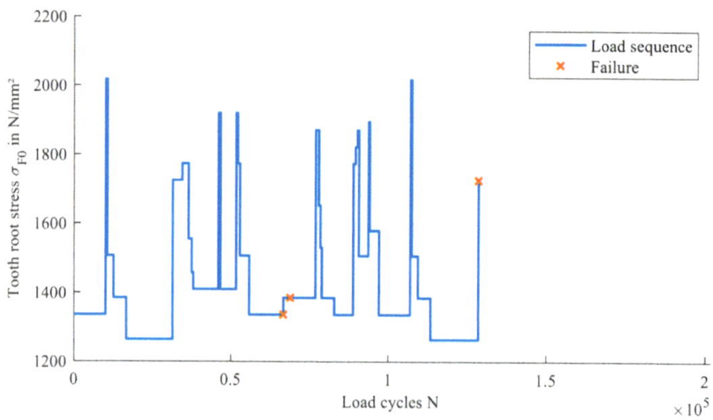

Figure 7. Results of the second test run with variable loads.

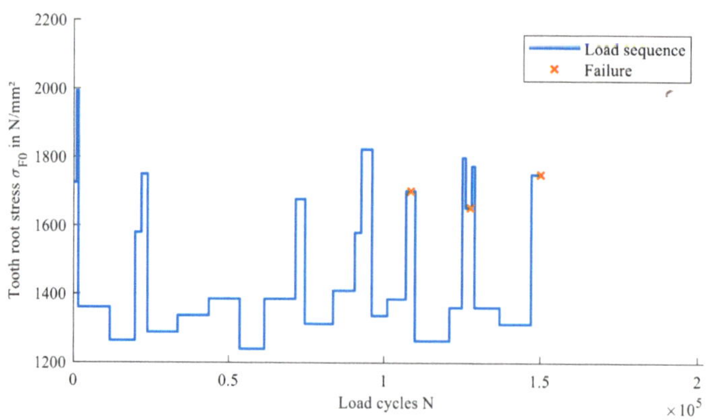

Figure 8. Results of the third test run with variable loads.

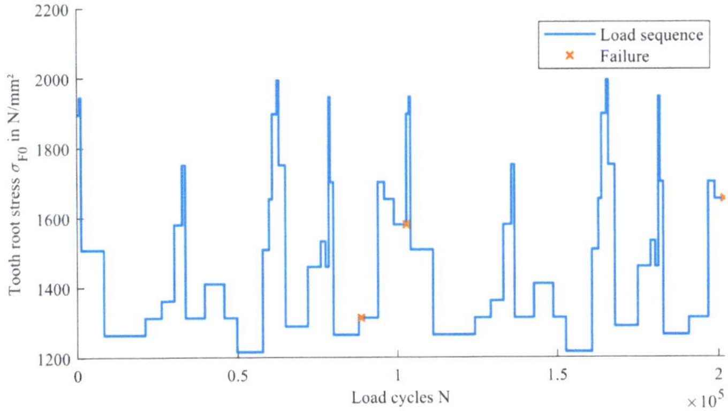

Figure 9. Results of the fourth test run with variable loads.

7. Results of the Validation

To compare the results of the different approaches for the description of the load-carrying capacity, the twelve test runs presented in Section 6 are used to calculate the damage sum D at failure. The damage sum is calculated according to the damage accumulation hypothesis according to Palmgren [35] and Miner [36]. The original form of the hypothesis (DAH Miner original) is used; therefore, loads beneath the endurance limit are not taken into account [37]. The reason for this choice is the design of the new approaches. All five concepts were designed to replace only the part of the S/N curve above the endurance limit. No changes were made to account for lower loads. Therefore, a damage accumulation hypothesis, including loads below the endurance limit, would negatively affect the comparability of the concepts. The use of this damage accumulation hypothesis is also the recommendation of standards [4,38–40] regarding the calculation of the service life of gears under variable loads. The resulting damage sum according to the different load-carrying capacity descriptions is given in Table 4.

Table 4. Damage sums resulting from the test runs with variable load.

Data Point	Damage Sum According to DAH Miner Original Calculated with the Load-Carrying Description Based On					
	Basquin	Polynomial	SVR	RF	NN	GPR
1	0.699	0.806	0.786	0.771	0.789	0.779
2	1.017	1.187	1.173	1.156	1.179	1.161
3	1.063	1.241	1.232	1.216	1.241	1.226
4	0.803	0.893	0.876	0.827	0.847	0.849
5	0.779	0.865	0.847	0.802	0.824	0.821
6	1.462	1.607	1.593	1.513	1.512	1.583
7	1.091	1.172	1.218	1.119	1.111	1.152
8	1.263	1.347	1.412	1.309	1.295	1.325
9	1.473	1.551	1.639	1.510	1.505	1.511
10	0.992	1.092	1.064	1.090	1.062	1.096
11	2.455	2.696	2.702	2.778	2.698	2.751
12	1.279	1.410	1.417	1.462	1.421	1.450

The use of the S/N curve in combination with a damage accumulation hypothesis is common for calculating the service life of gears under variable loads [4]. In theory, failure should occur at a damage sum of $D = 1$. In reality scattering spreads the damage sum at failure over a numerical range [41–43]. This poses a challenge for the design of gears. On the one hand, a certain level of reliability must be ensured [44]. On the other hand, the design of the gear should be as efficient as possible. Therefore, the calculated service life should be as accurate as possible, and the scattering range should be as small as possible. The evaluation of the different approaches considers two ranges of scattering:

- The range of the damage sum at failure of all data points (total scattering);
- The range of the damage sum at failure of the inner 50% of the data points.

For the calculation of the total scattering, one data point is excluded. The corresponding point failed at a damage sum of $D = 2.455$ according to the DAH Miner original based on the Basquin equation. This comparatively high damage sum causes a significant increase in the total scattering and, therefore, causes a less informative result regarding the other data points. Table 5 shows the results of the damage sums calculated using the different approaches. The reference for the results is the calculation using the Basquin equation, which represents the state of knowledge for the service life calculation of gears.

Table 5. Analysis of the results.

	Arithmetic Mean	Median	Scattering (Inner 50%)		Total Scattering	
Basquin	1.20	1.08	0.47	Reference	0.77	Reference
Polynomial	1.32	1.21	0.49	+4.26%	0.80	+3.90%
SVR	1.33	1.23	0.54	+14.89%	0.85	+10.39%
RF	1.30	1.20	0.53	+12.77%	0.74	−3.90%
NN	1.29	1.21	0.51	+8.51%	0.72	−6.49%
GPR	1.31	1.19	0.51	+8.51%	0.80	+3.90%

Overall, the results of all five approaches are comparable, and the differences are within a reasonable range. This corresponds to the expectations since the general shape of all approaches is similar (see Figure 5). The performance regarding the scattering of the inner 50% of the data points is decreased for all of the new approaches compared to the Basquin equation. For the total scattering, the Basquin equation is outperformed by the machine learning approaches using ensemble learning (−3.90%) and a neural network (−6.49%).

All five methods underestimate the average service life of the gears. Both the arithmetic means and the medians of the damage sum at failure are greater than one. This is not an unusual phenomenon. The actual damage sum at failure can deviate from one quite substantially. Therefore, a predicted damage sum at failure or a permissible damage sum has to be determined for the calculations. This damage sum depends, among other things, on the applied damage accumulation hypothesis. Through this approach, deviations within the calculations can be adjusted for. On the other hand, this does not influence the scattering of the damage sum at failure. For a reliable prediction of the service life of gears, the scattering should be as minimal as possible. The focus of the following evaluation is the scattering of the damage sums at failure according to the different approaches for the description of the load-carrying capacity.

For this evaluation, the data are processed to adjust the arithmetic mean of each method to one. Therefore, each data point of one method is divided by the arithmetic mean of all data points of this method. Based on this recalculation, it is possible to assess the ratio between the arithmetic mean of the damage sum at failure and the scattering of the damage sum at failure. The recalculated data are presented in Table 6.

Table 6. Analysis of the relative results.

	Arithmetic Mean	Median	Scattering (Inner 50%)		Total Scattering	
Basquin	1.00	0.90	0.40	Reference	0.65	Reference
Polynomial	1.00	0.92	0.37	−7.50%	0.61	−6.15%
SVR	1.00	0.92	0.40	±0%	0.64	−1.54%
RF	1.00	0.92	0.41	+2.50%	0.57	−12.31%
NN	1.00	0.94	0.39	−2.50%	0.56	−13.85%
GPR	1.00	0.91	0.39	−2.50%	0.61	−6.15%

When analyzing this interpretation of the data, it is interesting to notice that the Basquin equation is outperformed by each of the five new approaches in terms of total scattering. Focusing on the scattering of the inner 50%, the neural network, the third-degree polynomial, and the Gaussian process regression also result in lower scattering compared to the results based on the Basquin equation.

This approach to manipulating service life data is uncommon in service life calculations. It is used in this paper to better understand the differences between the methods. All five new approaches result in an increase in the average damage sum at failure (see Table 5). Therefore, these methods are more conservative than the Basquin equation. The results of the recalculated scattering show an important fact. The new approaches do not simply scale up the calculated damage sums. The new methods fundamentally change the weighting of different loads. Although the average damage sum is increased by 7.50% by using the neural network, the total scattering is decreased by 13.85% in relative terms.

8. Conclusion and Outlook

The scope of this paper was to investigate the potential of new approaches to describe the load-carrying capacity of gears. The Basquin equation was used as a reference, which is the state of knowledge for the description of the load-carrying capacity of gears. Five new methods have been developed within this paper. Four of them are based on machine learning, and one applies polynomial regression. All five methods produce applicable results.

The scattering of the results is one of the major aspects when evaluating a method for the service life calculation. With respect to this aspect, none of the new approaches can outperform the reference set by the Basquin equation. The approaches using support vector regression and a neural network result in a reduction in the total scattering but in an increase in the scattering of the inner 50% of the data points. When examining the relative scattering, the neural network, the third-degree polynomial, and the Gaussian process regression are able to outperform the Basquin equation.

This is quite an impressive result, considering that the database of only 31 points is quite small for the application of machine learning and that the evaluation of the performance was conducted with a different dataset. The training of the method was based on single load tests, and the validation was based on variable load tests. An alternative approach for the description of the load-carrying capacity would be the use of variable load data as training data for machine learning. Due to the increased requirements regarding the amount of data, this approach has not yet been successfully implemented. Another important step in proceeding with the research regarding this will be the application of the methods for real-world data on gears. Currently, the research focuses on test rig data because this type of data is more suitable for the investigation of the general suitability and the potential of the methods. This is due to the reduced number of influences on the service life and load-carrying capacity compared to real-world operating conditions.

The results of this paper show that there is potential for new approaches to describe the load-carrying capacity of gears. The ability of the presented approaches to better fit the experimental data is potentially beneficial for the service life calculation. Even a small increase in accuracy has the potential to increase the efficiency of the dimensioning of gears. This could result in overall cost savings and a reduction in the carbon footprint.

The potential of the presented methods goes far beyond the presented results. Increasing the amount of training data could improve the performance even more. In addition, it is possible to expand the methods to include variable load test data in the determination of an alternative S/N curve.

The potential reduction in the scattering of the calculated damage sum at failure can be especially significant for applications in the field of engineering. One example is the prediction of the remaining service life of gearboxes based on damage accumulation, as suggested by Foulard et al. [45,46]. For this prediction, the damage sum at failure has to be assumed by the engineer. The remaining service life is then calculated based on the ratio of the current damage sum and the assumed damage sum at failure. The scattering of the actual damage sum at failure decreases the accuracy of the prediction significantly. Therefore, the presented approaches create a potential for engineers to increase the accuracy of their service life calculations and predictions without being much more complex compared to the established approach.

The application of approaches based on machine learning and mathematical regression may seem more complex compared to the Basquin equation at first, but it is possible to create an automated program for the training and validation of different methods for the load-carrying capacity description. The user only needs to input the test data. Therefore, it is possible to implement the presented methods with the same or even less effort required by the engineer compared to the traditional approach.

Author Contributions: Conceptualization, D.V.; methodology, D.V.; software, D.V.; validation, D.V.; investigation, D.V.; writing, D.V.; supervision, J.P. and K.S. All authors have read and agreed to the published version of the manuscript.

Funding: The presented results are based on the research project STA 1198/19-1, supported by the German Research Foundation e.V. (DFG). The authors would like to thank the DFG for the sponsorship and support received.

Data Availability Statement: Data are contained within the article.

Conflicts of Interest: The authors declare no conflicts of interest.

Nomenclature

		Unit
b	Face width	mm
C	Constant of the S/N curve	-
c	Coefficients of the polynomial	-
D	Damage sum	-
F	Pulsator force	N
k	Slope of the S/N curve	-
L	Load	N/mm^2 or N
m_n	Normal module	mm
N	Number of endurable load cycles	-
N_D	Number of load cycles at the knee point of the S/N curve	-
Y_F	Form factor	-
Y_S	Stress correction factor	-
σ_{F0}	Tooth root stress	N/mm^2

References

1. Niemann, G.; Winter, H. *Machine Elements—Part 2 (In German: Maschinenelemente—Band 2: Getriebe Allgemein, Zahnradgetriebe—Grundlagen, Stirnradgetriebe)*, 2nd ed.; Springer: Berlin/Heidelberg, Germany, 2003.
2. Basquin, O. The exponential law of endurance tests. In Proceedings of the Thirteenth Annual Meeting, Trenton, NJ, USA, 8 November 1910; pp. 625–630.

3. *ISO 6336-1:2019-11*; Calculation of Load Capacity of Spur and Helical Gears—Part 1: Basic Principles, Introduction and General Influence Factors. ISO. Geneva, Switzerland, 2019.
4. *ISO 6336-6:2019-11*; Calculation of Load Capacity of Spur and Helical Gears—Part 6: Calculation of Service Life under Variable Load. ISO: Geneva, Switzerland, 2019.
5. Dong, Z.; Zhao, D.; Cui, L. An intelligent bearing fault diagnosis framework: One-dimensional improved self-attention-enhanced CNN and empirical wavelet transform. *Nonlinear Dyn.* **2024**, *112*, 6439–6459. [CrossRef]
6. Zhao, D.; Wang, T.; Chu, F. Deep convolutional neural network based planet bearing fault classification. *Comput. Ind.* **2019**, *107*, 59–66. [CrossRef]
7. Dong, Z.; Zhao, D.; Cui, L. Non-negative wavelet matrix factorization-based bearing fault intelligent classification method. *Meas. Sci. Technol.* **2023**, *34*, 115013. [CrossRef]
8. Hein, M. Holistic Approach for Fatigue Life Design and Testing of Gearwheels (In German: Zur Ganzheitlichen Betriebsfesten Auslegung und Prüfung von Getriebezahnrädern). Ph.D. Thesis, Technical University of Munich, München, Germany, 2018.
9. Tobie, T.; Matt, P. *FVA 563 I—FVA Guideline: Recommendations for the Standardisation of Load Capacity Tests on Hardened and Tempered Cylindrical Gears*; Research Association for Drive Technology (FVA): Frankfurt am Main, Germany, 2012.
10. Wöhler, A. Report on the experiments carried out on the Royal Lower Silesian-Mark Railway with devices for measuring the bending and twisting of railroad car axles during travel. (In German: Bericht über die Versuche, welche auf der Königl. Niederschlesisch-Märkischen Eisenbahn mit Apparaten zum Messen der Biegung und Verdrehung von Eisenbahnwagen-Achsen während der Fahrt angestellt wurden). *Z. Bauwes.* **1858**, *8*, 641–678.
11. *DIN 50100:2022-12*; Load Controlled Fatigue Testing—Execution and Evaluation of Cyclic Tests at Constant Load Amplitudes on Metallic Specimens and Components. DIN: Berlin, Germany, 2022.
12. Hück, M. An improved procedure for the evaluation of staircase procedure tests (In German: Ein verbessertes Verfahren zur Auswertung von Treppenstufenversuchen). *Z. Werkst.* **1982**, *14*, 147–176.
13. Müller, C.; Wächter, M.; Masendorf, R.; Esderts, A. Accuracy of fatigue limits estimated by the staircase method using different evaluation techniques. *Int. J. Fatigue* **2017**, *100*, 296–307. [CrossRef]
14. Russell, S.J.; Norvig, P. *Artificial Intelligence: A Modern Approach*, 4th ed.; Pearson: Hoboken, NJ, USA, 2021.
15. He, L.; Wang, Z.; Akebono, H.; Sugeta, A. Machine learning-based predictions of fatigue life and fatigue limit for steels. *J. Mater. Sci. Technol.* **2021**, *90*, 9–19. [CrossRef]
16. Zhan, Z.; Li, H. A novel approach based on the elastoplastic fatigue damage and machine learning models for life prediction of aerospace alloy parts fabricated by additive manufacturing. *Int. J. Fatigue* **2021**, *145*, 106089. [CrossRef]
17. Mudabbir, M.; Mosavi, A. Machine Learning for Modeling Service Life: Comprehensive Review, Bibliometrics Analysis and Taxonomy. In Proceedings of the 2023 IEEE 27th International Conference on Intelligent Engineering Systems (INES), Nairobi, Kenya, 26–28 July 2023; pp. 235–244. [CrossRef]
18. Liu, D.P.; Ferri, G.; Heo, T.; Marino, E.; Manuel, L. On long-term fatigue damage estimation for a floating offshore wind turbine using a surrogate model. *Renew. Energy* **2024**, *225*, 120238. [CrossRef]
19. Wilkie, D.; Galasso, C. Gaussian process regression for fatigue reliability analysis of offshore wind turbines. *Struct. Saf.* **2021**, *88*, 102020. [CrossRef]
20. Karolczuk, A.; Słoński, M. Application of the Gaussian process for fatigue life prediction under multiaxial loading. *Mech. Syst. Signal Process.* **2022**, *167*, 108599. [CrossRef]
21. Döbel, I.; Leis, M.; Vogelsang, M.; Neustroev, D.; Patzka, H.; Rüping, S.; Voss, A.; Wegele, M.; Welz, J. *BMBF Nr. 01IS17019—Machine Learning—Competence, Applications and Research Deamnd (In German: Maschinelles Lernen—Kompetenzen, Anwendungen und Forschungsbedarf—Ergebnisbericht)*; Fraunhofer Society: Munich, Germany, 2018.
22. The MathWorks, Inc. MATLAB Version 2023b. Available online: https://de.mathworks.com/products/matlab.html (accessed on 19 April 2024).
23. The MathWorks, Inc. MATLAB Documentation—Fitrensemble. Available online: https://de.mathworks.com/help/stats/fitrensemble.html (accessed on 22 April 2024).
24. The MathWorks, Inc. MATLAB Help Center—Fitrnet. Available online: https://de.mathworks.com/help/stats/fitrnet.html#mw_f074fe92-8d2f-48a0-b8f9-8c7d7a6ddf54 (accessed on 19 April 2024).
25. Wang, J. An Intuitive Tutorial to Gaussian Process Regression. *Comput. Sci. Eng.* **2023**, *25*, 4–11. [CrossRef]
26. Vietze, D.; Pellkofer, J.; Stahl, K. Non-Linear Analysis of Gear-Fatigue-Damage under Variable Load: 23FTM08. In Proceedings of the AGMA Fall Technical Meeting, FTM 2023, Detroit, MI, USA, 16–18 October 2023.
27. Illenberger, C.; Tobie, T.; Stahl, K. *FVA-Nr. 610/III—Booklet 1328—Gearing steel worldwide Part III (fina report) (In German: Zahnradstähle weltweit III—Abschlussbericht—Referenzwerte zur Zahnradtragfähigkeit moderner Zahnradstähle im Weltmarkt—Teil III)*; Forschungsvereinigung Antriebstechnik e.V.: Frankfurt am Main, Germany, 2019.
28. Saddei, P.; Schurer, S.; Güntner, C.; Steinbacher, M.; Tobie, T.; Zoch, H.-W.; Stahl, K. *IGF-Nr. 17903 N/2—FVA-Nr. 513/III—Heft 1248—Randschichtgefüge—Abschlussbericht—Alternative mehrphasige Randschichtgefüge beim Einsatzhärten zur Steigerung der Festigkeitseigenschaften von verzahnten Getriebebauteilen*; Forschungsvereinigung Antriebstechnik e.V.: Frankfurt am Main, Germany, 2017.
29. Weber, C.; Tobie, T.; Stahl, K. Rapid and Precise Manufacturing of Special Involute Gears for Prototype Testing: 19FTM20. In Proceedings of the AGMA Fall Technical Meeting, FTM 2019, Detroit, MI, USA, 14–16 October 2019.

30. Bonaiti, L.; Rosa, F.; Rao, P.M.; Concli, F.; Gorla, C. Gear root bending strength: Statistical treatment of Single Tooth Bending Fatigue tests results. *Eng. Res.* **2022**, *86*, 251–258. [CrossRef]
31. *ISO 6336-3:2019-11*; Calculation of Load Capacity of Spur and Helical Gears—Part 3: Calculation of Tooth Bending Strength: Corrected Version 2020-11. ISO: Geneva, Switzerland, 2019.
32. The MathWorks, Inc. Machine Learning with MATLAB. 2022. Available online: https://de.mathworks.com/content/dam/mathworks/ebook/gated/machine-learning-ebook-all-chapters.pdf (accessed on 19 April 2024).
33. The MathWorks, Inc. MATLAB Documentation—Fitrgp. Available online: https://de.mathworks.com/help/stats/fitrgp.html (accessed on 22 April 2024).
34. The MathWorks, Inc. MATLAB Documentation—Fitrsvm. Available online: https://de.mathworks.com/help/stats/fitrsvm.html#buxgtub (accessed on 22 April 2024).
35. Pålmgren, A. Service Life of Ball Gearings (In German: Die Lebensdauer von Kugellager). *Z. Vereines Dtsch. Ingenieure* **1924**, *58*, 339–341.
36. Miner, M.A. Cumulative Damage in Fatigue. *J. Appl. Mech.* **1945**, *12*, 159–164. [CrossRef]
37. Haibach, E. *Operational Strength—Methods and Data for Calculation (In German: Betriebsfestigkeit—Verfahren und Daten zur Bauteilberechnung)*, 3rd ed.; Springer: Berlin/Heidelberg, Germany, 2006.
38. *DIN 3990-6:1994-12*; Calculation of Load Capcity of Cylindrical Gears—Part 6: Calculation of Service Strength. DIN: Berlin, Germany, 1994.
39. *DNVGL-CG-0036:2015-12*; Calculation of Gear Rating for Marine Transmissions. DNVGL: Bærum, Norway, 2015.
40. *ANSI/AGMA 2101-D04*; Fundamental Rating Factors and Calculation Methods for Involute Spur and Helical Gear Teeth. AGMA: Alexandria, VA, USA, 2004.
41. Stahl, K.; Michaelis, K.; Höhn, B.-R.; Winter, H. *IGF-Nr. 11154, FVA-Nr. 304—Booklet 580—Service Life Statistics (Final Report) (In German: Lebensdauerstatistik—Statistische Methoden zur Beurteilung von Bauteillebensdauer und Zuverlässigkeit und ihre Beispielhafte Anwendung auf Zahnräder)*; Research Association for Drive Technology (FVA): Frankfurt, Germany, 1999.
42. Geitner, M.; Tobie, T.; Stahl, K. *FVA-Nr. 610 IV—Booklet 1432—Materials 4.0 (Final Report) (In German: Werkstoffe 4.0—Erweiterte Datenanalyse zur Bewertung des Einflusses von Werkstoff und Wärmebehandlungseigenschaften auf die Zahnradtragfähigkeit)*; Research Association for Drive Technology (FVA): Frankfurt, Germany, 2021.
43. Götz, S.; Eulitz, K.-G. *Fatgiue Life Analysis (In German: Betriebsfestigkeit)*, 2nd ed.; Springer: Berlin/Heidelberg, Germany, 2022.
44. Gudehus, H.; Zenner, H. *Guideline for Fatigue Life Analysis (In German: Leitfaden für eine Betriebsfestigkeitsrechnung)*, 4th ed.; Verlag Stahleisen: Düsseldorf, Germany, 1999.
45. Foulard, S.; Rinderknecht, S.; Ichchou, M.; Perret-Liaudet, J. Real-time and online lifetime monitoring of automotive transmissions: Determining of the remaining lifetime of transmission components with the aid of torque measure. In Proceedings of the International Conference on Gears: VDI-Berichte 2199, München, Germany, 7–9 October 2013.
46. Foulard, S.; Rinderknecht, S.; Perret-Liaudet, J.; Ichchou, M. Online and real-time damage calculation in automotive transmissions Application to remaining service life estimation. In Proceedings of the International Gear Conference 2014: Conference Proceedings Volume II, Lyon Villeurbanne, France, 26–28 August 2014; pp. 933–949.

machines

MDPI

Article

Feature Extraction of a Planetary Gearbox Based on the KPCA Dual-Kernel Function Optimized by the Swarm Intelligent Fusion Algorithm

Yan He [1], Linzheng Ye [1,2,*] and Yao Liu [1,2]

1 School of Mechanical Engineering, North University of China, Taiyuan 030051, China; zbhy91@163.com (Y.H.); liuyao@nuc.edu.cn (Y.L.)

2 Shanxi Key Laboratory of Advanced Manufacturing Technology, North University of China, Taiyuan 030051, China

* Correspondence: yelinzheng@nuc.edu.cn

Abstract: The feature extraction problem of coupled vibration signals with multiple fault modes of planetary gears has not been solved effectively. At present, kernel principal component analysis (KPCA) is usually used to solve nonlinear feature extraction problems, but the kernel function selection and its blind parameter setting greatly affect the performance of the algorithm. For the optimization of the kernel parameters, it is very urgent to study the theoretical modeling to improve the performance of kernel principal component analysis. Aiming at the deficiency of kernel principal component analysis using the single-kernel function for the nonlinear mapping of feature extraction, a dual-kernel function based on the flexible linear combination of a radial basis kernel function and polynomial kernel function is proposed. In order to increase the scientificity of setting the kernel parameters and the flexible weight coefficient, a mathematical model for dual-kernel parameter optimization was constructed based on a Fisher criterion discriminant analysis. In addition, this paper puts forward a swarm intelligent fusion algorithm to increase this method's advantages for optimization problems, involving the shuffled frog leaping algorithm combined with particle swarm optimization (SFLA-PSO). The new fusion algorithm was applied to optimize the kernel parameters to improve the performance of KPCA nonlinear mapping. The optimized dual-kernel function KPCA (DKKPCA) was applied to the feature extraction of planetary gear wear damage, and had a good identification effect on the fuzzy damage boundary of the planetary gearbox. The conclusion is that the DKKPCA optimized by the SFLA-PSO swarm intelligent fusion algorithm not only effectively improves the performance of feature extraction, but also enables the adaptive selection of parameters for the dual-kernel function and the adjustment of weights for the basic kernel function through a certain degree of optimization; so, this method has great potential for practical use.

Keywords: swarm intelligence optimization; kernel principal component analysis; dual-kernel function; planetary gearbox; feature extraction

1. Introduction

Feature extraction is an important prerequisite for pattern recognition. It can be divided into the methods based on signal processing and those based on learning [1]. The former was developed earlier, and its typical algorithms include the Fourier transform, Gabor filter, and wavelet transform [2]. The latter has emerged in the last ten years; this method uses the dimension reduction method to map the data from the original space onto a certain low-dimensional space. The data in the low-dimensional space can greatly reflect the essential characteristics of the original space. The typical algorithms include principal component analysis (PCA), kernel principal component analysis (KPCA) [3], linear discriminant analysis (LDA) [4], etc. The feature extraction method based on signal processing involves extracting features from the transform domain and emphasizing the individual information from the samples, while the method based on learning mainly considers the statistical characteristics of all the samples and extracts the typical features that can express the original information.

In recent years, some achievements have been made using the proposed kernel learning method in the research fields of feature extraction, pattern recognition, data mining, and image and signal processing. To a certain extent, it solves the nonlinear problem in an actual system and improves the accuracy of pattern recognition and prediction. For example, a kernel principal component analysis (KPCA) and the observer method were applied to detect faults in a hydraulic system by analyzing the variation in water volume, and achieved results for the detection of faults in complex and nonlinear systems [5]. A rolling bearing fault diagnosis method based on the Volterra series and a kernel principal component analysis (KPCA) has been proposed [6]. However, the kernel learning method has encountered a bottleneck problem that affects the performance of the algorithm, that is, the selection of the kernel function and its parameters. The kernel method is based on the kernel function, which can be divided into the local kernel and global kernel at present. For the so-called local kernel, only when the distance between the data points is very close or the two features are similar will the value of the kernel function be greatly affected; for the global kernel, on the contrary, a large distance between the data points or a significant feature difference can also have a great impact on the value of the kernel function. In general, the radial basis function (RBF) kernel is the typical local kernel, and the polynomial kernel is the global kernel, and their mapping performances are different when their parameters are changed. The different kernel functions and their parameters reflect the differences in their mapping performance for the overall or local data.

In the practical application of the kernel analysis method, the type of kernel function is generally selected according to the data characteristics of practical problems [7]. Each kernel function has its own limitations. In reality, one kernel function mapping can only reflect a certain feature, and often cannot effectively describe the whole feature. Moreover, with the complexity and diversification of engineering problems, there is an increasing number of problems, such as large sample data sets or uneven high-dimensional data, and the defects of the single-kernel function are more obvious. Therefore, researchers began to try to combine the multi-kernel functions. Lei et al. proposed a relevance vector machine prediction method based on an adaptive multi-kernel combination, which was applied to the remaining useful life prediction of machinery [8]. Fu et al. proposed a combined kernel correlation vector machine and a quantum particle swarm optimization algorithm for transformer fault diagnosis [9]. Deng et al. modified kernel principal component analysis using a dual-weighted local outlier factor and applied this method to linear process monitoring [10]. Pan et al. put forward a mean-class kernel principal component fault diagnosis method with a combined kernel function for monitoring the distillation process [11].

In recent years, it has been proven that the performance of nonlinear mapping using multiple kernels is better than that using the single-kernel model or the single-kernel machine combination model. The construction methods for the combinatorial kernel can be divided as follows: the first method is a multi-kernel linear combination synthesis, in which the shape is like $K = \sum_{i=1}^{q} \beta_j K_i, \beta_j \geq 0, \sum_j^q \beta_j = 1$, where q is the number of basic kernel functions, K_i is the basic kernel function, and β is the weight coefficient; the second is a multi-kernel extended synthesis method, which is an attempt to achieve the fusion of the different kernel matrices through the concept of summation "averaging" [12]. For example, Lv et al. derived a hybrid kernel function combining the RBF kernel with the polynomial kernel, and introduced it into an extreme learning machine to improve the accuracy of an intrusion detection system [13]. Nithya et al. proposed a method for kidney disease detection and segmentation using an artificial neural network and a multi-kernel K-means clustering algorithm for ultrasound images [14]. Ouyang et al. proposed a hybrid improved kernel LDA and PNN algorithm for efficient face recognition [15]. Afzal et al. achieved deep multilayer kernel learning in kernel vector machines for classification [16].

For the combinatorial kernel function, the scientific setting of the kernel parameters is the key to improving the performance of the kernel learning algorithm. In past studies, the kernel parameters were set according to the experimental or empirical value with considerable blindness. However, the current theoretical research is mainly based on the experimental correction method, which is inefficient, as well as data-related methods and intelligence based on optimization methods [17]. Applying the swarm intelligence optimization method to the study of the optimization of the kernel function and its parameters represents a new research hotspot for improving the status quo and enhancing the performance of the kernel learning method [18,19].

Particle swarm optimization (PSO) and the shuffled frog leaping algorithm (SFLA) are categorized as swarm intelligence optimization algorithms [20,21]. PSO is a method derived from the behavior characteristics of biological populations, and is used to solve optimization problems. In order to optimize the search, the behavioral characteristics of birds are simulated, and competition among the individuals is achieved. The SFLA simulates the optimization of the group foraging of frogs by introducing a meme evolution mechanism into the group search. Organizational social behavior is used to replace the natural selection mechanisms of evolutionary algorithms, and this method has been widely used in optimization problems, path selection, and fault diagnosis [22].

As complex compound gear transmission systems, planetary gearboxes with a compact structure, large transmission ratio, high transmission efficiency, smooth operation, and other characteristics, have been widely used in automobile gearboxes, wind turbines [23], and aviation engine settings. They play an important role in industries such as coal, energy, advanced manufacturing, and wind power. Due to the harsh working environment, gear and bearing damage and failures often occur and, due to different degrees of damage, diverse fault modes, and complex vibration signal transmission paths, the problems of feature extraction, state recognition, and the fault diagnosis of planetary gear multi-fault-mode coupled vibration signals have not been effectively solved. At present, KPCA is commonly used to solve nonlinear feature extraction problems. This paper focuses on the transmission boxes of wind turbines and establishes a planetary gearbox simulation fault experimental platform for feature extraction and fault diagnosis research.

In this paper, aiming at the deficiency of the nonlinear mapping of the single-kernel function for KPCA feature extraction, a dual-kernel function with a flexible linear combination of the RBF and polynomial function is proposed. In order to increase the scientificity of the setting of the kernel parameters and the flexible weight coefficient, a mathematical model for the optimization of the kernel parameters was constructed, based on Fisher's discriminant thought. Making full use of the advantages of the swarm intelligence fusion algorithm for solving optimization problems, the SFLA-PSO fusion algorithm is presented, and is used to find the optimal solution for kernel parameters and the flexible weight coefficient. Lastly, a KPCA of the dual-kernel function optimized by the SFLA-PSO fusion algorithm was applied to the feature extraction of planetary gear wear.

The paper is organized as follows: First, it introduces the kernel learning method and summarizes the combination kernel function and its kernel parameter's important role in feature extraction with KPCA. The necessity of kernel parameter optimization with the swarm intelligence optimization algorithm is put forward. In Section 2, the SFLA and PSO fusion mechanism is described, and the flow chart for the swarm intelligence algorithm is given. The constitution of the dual-kernel function of the KPCA and its optimization model, as well as the simulation and comparative analysis of the Iris data, are described in Section 3. In addition, the experimental scheme for planetary gearbox failure is described in Section 4. The feature extraction for the planetary gearbox based on the KPCA dual-kernel parameter optimized by the SFLA-PSO is presented, and the simulation results are discussed in detail in Section 5. Finally, we present the conclusion in Section 6.

2. Swarm Intelligence Fusion Algorithm

2.1. The SFLA and PSO Fusion Mechanism

As a new branch of evolutionary algorithms, the swarm intelligence optimization algorithm has many advantages, such as its simplicity, easy implementation, low requirements for the mathematical behavior of the target problem, and high efficiency, so it has attracted the attention of scholars [24]. One form of the swarm intelligence optimization algorithm, PSO, has the advantage of fast convergence speed, but it easily falls into local convergence, while the SFLA has a strong global search and optimization ability. Similarly, as in evolution, it has to go through a series of processes, such as frog population generation, evolution, crossover, and information exchange. The SFLA evolution mechanism is more flexible, but its computational complexity is high, and its convergence speed is slow. Therefore, the fusion strategy of different algorithms aims to utilize the unique advantages of the original algorithms, achieving mutual coordination and complementary advantages, and forming a robust and more efficient algorithm [25]. The fusion of the swarm intelligence algorithm involves the construction of a new fusion algorithm, replacing the original algorithm, which involves the strategies of population size, the population structure relationship, individual learning or interaction, and the population evolution mode (selecting, eliminating, and updating individuals). According to the concept of elite strategy, the SFLA algorithm is combined with the PSO algorithm in this study, the strategy of "two-level optimization and internal and external circulation" is implemented, and the SFLA-PSO swarm intelligent fusion algorithm is formed. Compared to other improved PSO and SFLA methods, the concept of the fusion algorithm is simple, no additional parameters are added, and it is easy to implement.

The SFLA-PSO fusion strategy is as follows: Firstly, the randomly generated particles are divided into n_{pso} subgroups of equal size. The particles in each subgroup evolve independently according to the PSO mechanism to achieve the optimization of the first level. Then, the optimal particles in each subgroup are taken out to form a new population, i.e., the initial frog group. According to the SFLA mechanism, the optimal frog position is that which realizes the optimization of the second level. Finally, the best position in the SFLA subgroup is reflected in the speed of the particle swarm optimization update, which can effectively guide PSO evolution and avoid falling into the local optimum. Using this strategy not only reduces the blindness of the initial SFLA group, but also enables a fine search within the SFLA.

The particle velocity and position update for the PSO optimization of the first level are shown in Equations (1) and (2) [20]:

$$v_{jid}(t+1) = \omega \times v_{jid}(t) + c_1 r_1 \left(p_{jid} - x_{jid}(t)\right) + c_2 r_2 \left(p_{jgd} - x_{jid}(t)\right) \quad (1)$$

$$x_{jid}(t+1) = x_{jid}(t) + v_{jid}(t+1) \quad (2)$$

where $j = 1, 2, \ldots, n_{\text{pso}}$ is the group number of the particle swarm; $i = 1, 2, \ldots, n_{\text{p}}$ denotes the ith particle; p_{jid} and p_{jgd} represent the current optimal position and the global optimal position of the particles in the jth group, respectively; and x_{jid} is the current position of the ith particle in the jth group. v_{jid} is the current velocity of the ith particle in the jth group, $v_{jid} \in [-v_{\text{limit}}, v_{\text{limit}}]$, v_{limit} is the maximum speed limit. C_1 and c_2 are the learning factors, r_1 and r_2 are uniform random numbers within the range [0, 1], and ω is the inertia weight. t is the evolution generation.

In the SFLA optimization of the second stage, the frog's velocity and position update are shown in Equations (3)–(5) [21].

The step length update formula is

$$S_i = r \times (P_{\text{b}} - P_{\text{w}}) \quad (3)$$

$$S_i = r \times (P_g - P_{\text{w}}) \quad (4)$$

where r is a uniform random number within [0, 1], P_b is the optimal frog position in the meme group, P_w is the worst frog position in the meme group, P_g is the best frog position in the whole group, and S_i is the frog step length, $S_{min} \le S_i \le S_{max}$.

The position update formula for the worst-positioned frog is

$$P'_w = P_w + S_i \tag{5}$$

where P'_w is the position of the updated frog, which is within the range $[P_{min}, P_{max}]$. When the position of the worst-positioned individual is not improved through this evolution, P_g, the best of the entire frog group, is used to replace P_b, as shown in Equation (4), and then Formula (5) is used for further calculations.

2.2. Fusion Algorithm Flow

The flow chart for the SFLA-PSO fusion algorithm is shown in Figure 1. The specific optimization process is as follows:

(1) Relevant parameters of the SFLA-PSO fusion algorithm are set.

The parameters for PSO include the total number of particles in the initial particle swarm N, the number of particles per particle swarm n_p, the number of particle groups n_{pso}, the particle dimension d, the total number of iterations T_{max}, the inertia weight ω, learning factors c_1, c_2, and the velocity limit v_{limit}. The parameters of the SFLA include the number of meme groups m, the number of frogs in each meme group n, the total number of iterations of the SFLA G_{max}, the number of local cycles L_{max}, and the maximum step S_{max} of the SFLA, where $N = n_{pso} \times n_p$ and $n_{pso} = n \times m$.

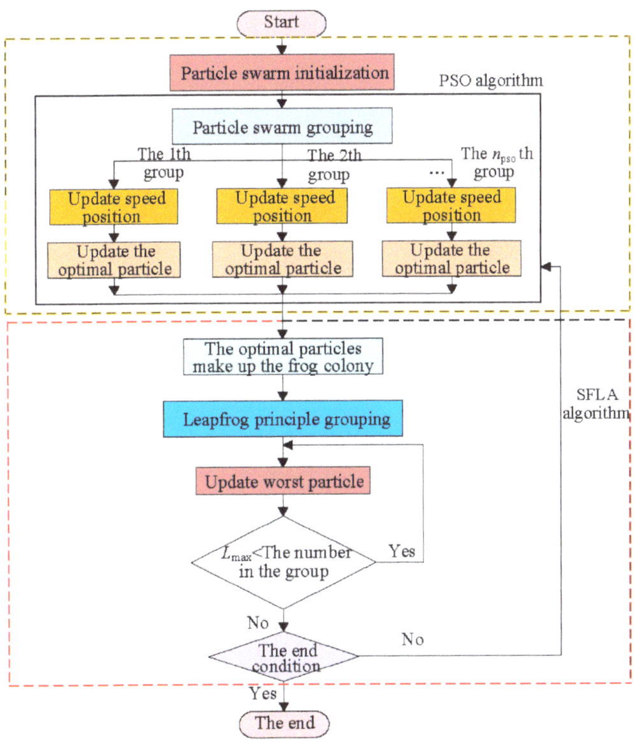

Figure 1. SFLA-PSO fusion algorithm flow chart.

(2) The particle swarm is initialized, and each particle is initialized. After the fitness is sorted, each particle is divided into n_{pso} groups with n_{p} particles in each group.
(3) The PSO mode iterative optimization is carried out for each group, the fitness of each particle is evaluated, and the optimal particle P_{jid} of each group is recorded.
(4) According to Formulas (1) and (2) for PSO, the velocity and position of each group particle are updated to complete the optimization of the first level.
(5) Each group of optimal particles (called elites) optimized by PSO are regarded as all the frogs of the SFLA, and their fitness is evaluated and ranked.
(6) According to the SFLA grouping mechanism, the elite particles are divided into m groups with n particles in each group.
(7) According to the updating Formulas (3)–(5) of the SFLA, the local optimal value and the local worst value for the population are recorded. After all the groups are updated, the global optimal value is updated to find the best X_{gd}, and the second-level optimization is completed.
(8) If the end condition is satisfied, that is, the predetermined accuracy requirement is met or the set maximum iteration number is reached, X_{gd} is the output, and the algorithm is stopped; otherwise, the procedure returns to (4).

3. The Composition and Optimization of the Dual-Kernel Function of KPCA

3.1. The Principle of KPCA Algorithm Feature Extraction

The detection data can be abstracted as vector $\boldsymbol{y}_i = (y_i^1, y_i^2, \cdots, y_i^L)$, where y_i^L is the Lth-dimension data of $\boldsymbol{y}_i$. The KPCA feature extraction steps are as follows [26]:

(1) The data are mapped to the high-dimensional feature space via the nonlinear transformation $\Phi(\boldsymbol{y}_i)$, and the kernel matrix can be calculated. The covariance matrix $\boldsymbol{C}^{\text{F}}$ in the feature space of the training sample after mapping is

$$C^{\text{F}} = \frac{1}{M}\sum_{i=1}^{M} \Phi(\boldsymbol{y}_i)\Phi(\boldsymbol{y}_i)^{\text{T}} \tag{6}$$

where M is the number of training samples. The kernel matrix $\boldsymbol{K}$ is defined as

$$[\boldsymbol{K}]_{ij} = K_{ij} = \left\langle \Phi(\boldsymbol{y}_i), \Phi\left(\boldsymbol{y}_j\right) \right\rangle = K(\boldsymbol{y}_i \boldsymbol{y}_j), i, j = 1, 2, \cdots, M \tag{7}$$

where K is the kernel function, which is defined as a function that represents the inner product $\left\langle \Phi(\boldsymbol{y}_i), \Phi\left(\boldsymbol{y}_j\right) \right\rangle$ of vector $\boldsymbol{y}_i$ and $\boldsymbol{y}_j$ through nonlinear mapping to a feature space, using two vectors in the original space as a function in order to achieve nonlinear mapping.

(2) Data centralization: If

$$\sum_{i=1}^{M} \Phi(\boldsymbol{y}_i) \neq 0$$

then the centralized processing is carried out. $\boldsymbol{K}$ is replaced by $\boldsymbol{K}^* = \boldsymbol{K} - \boldsymbol{AK} - \boldsymbol{KA} + \boldsymbol{AKA}$. In the formula, $A_{ij} = \frac{1}{M}$.

(3) The eigenvalues λ and eigenvectors $\boldsymbol{v}$ of the covariance matrix $\boldsymbol{C}^{\text{F}}$ are calculated, satisfying the following conditions:

$$\lambda \boldsymbol{v} = \boldsymbol{C}^{\text{F}} \boldsymbol{v} \tag{8}$$

By substituting Equation (7) into Equation (8), the following equation can be obtained:

$$\boldsymbol{M}\lambda\boldsymbol{\alpha} = \boldsymbol{K}\boldsymbol{\alpha} \tag{9}$$

The eigenvalues greater than zero, $\lambda_1, \lambda_2, \ldots, \lambda_p$, and the corresponding eigenvectors, $\alpha_1, \alpha_2, \ldots, \alpha_p$, are obtained by solving Equation (9), giving

$$V^k = \sum_{i=1}^{M} \alpha_i^k \Phi(y_i) \tag{10}$$

(4) Extraction of the principal component: the projection $\Phi(y)$ on the feature vector space V^k is calculated:

$$V^k \cdot \Phi(y) = \sum_{i=1}^{M} \alpha_i^k \Phi(y_i)\Phi(y) \tag{11}$$

where $\Phi(y_i)\cdot\Phi(y)$ can be calculated using the kernel technique, which is a type of function that makes $K_{ij} = \Phi(y_i)\cdot\Phi(y_j)$ hold. Let

$$g_k(y) = V^k \cdot \Phi(y) = \sum_{i}^{M} \alpha_i^k K(y_i, y) \tag{12}$$

where K is the kernel function and $g_k(y)$ becomes the kth nonlinear kernel principal component corresponding to $\Phi(y)$, and all projected values are used as the feature vectors of the samples.

$$g_k(y) = [g_1(y_i), g_2(y_i), \cdots, g_p(y_i)\ ] \tag{13}$$

As long as the first r maximum eigenvalues of K satisfy $\sum_{j=1}^{r} \lambda_j / \sum_{j=1}^{p} \lambda_j \geq 85\%$, then r feature vectors can be extracted from the above p feature components; thus, the feature vectors of the sample can be extracted as

$$y_i' = [g_1(y_i), g_2(y_i), \cdots, g_r(y_i)\ ]$$

The common kernel functions are as follows:
Polynomial kernel function (Poly):

$$K(y_i, y_j) = [(y_i \cdot y_j) + 1]^{d_1} \tag{14}$$

Radial basis kernel function (RBF):

$$K(y_i, y_j) = \exp(-\|y_i - y_j\|^2 / 2\sigma^2) \tag{15}$$

Sigmoid kernel function (Sigmoid):

$$K(y_i, y_j) = \tanh[\nu(y_i, y_j) + c] \tag{16}$$

3.2. The Composition of the Dual-Kernel Function of Flexible Weight Linear Combination

At present, the combination kernel method typically employs a new kernel function generated by a linear combination of the simple kernel functions, which satisfies the Mercer condition and has had some successful applications [26,27]. However, there is no theoretical basis for the parameter selection or the combination of kernel functions, and the uneven distribution of samples cannot be solved satisfactorily, which greatly limits the expression ability of the decision function. Therefore, there is an urgent need to establish a mathematical model for combinatorial kernel optimization using theory and to theoretically optimize the weight coefficient and parameters of the combined kernel function using an intelligent algorithm [28].

Due to the different characteristics of different kernel functions, their performance at solving specific problems varies significantly. Therefore, the combined kernel is the best method with which to select the kernel function, and it can avoid the problem of kernel function selection to some extent. The dual kernel can adopt the direct sum kernel, weighted sum kernel, and weighted polynomial extension kernel. In many practical applications, the Gaussian kernel function shows excellent properties. In this study, the Gaussian kernel function was used as the main function, and other kernel functions were selected for linear combination. The Gaussian kernel function is a typical local kernel. In order to balance the local kernel with the global kernel, the typical global kernel–polynomial kernel is used for the combination. In the Gaussian kernel function, the RBF is a typical local kernel. So, the RBF and the polynomial kernel were used in this study as a flexible linear combination. The dual-kernel function expression is shown in Equation (17):

$$K(y_i \cdot y_j) = \gamma \cdot [(y_i \cdot y_j) + 1]^{d_1} + (1-\gamma) \cdot \exp(-\|y_i - y_j\|^2/2\sigma^2) \quad (17)$$

where γ is the flexible weight of the two kernel functions, with $1 \geq \gamma \geq 0$, and σ and d_1 are the parameters of the combined kernel function.

With the help of Fisher's criteria (FC) of minimum within-class distance and maximum class distance [29], a model with which to measure the class discrimination of the feature space data was constructed. For two kinds of problems, the classification can be determined by constructing the discriminant function and criterion [30].

In the feature space, the data samples are $\boldsymbol{y}_1(y_{11}, y_{12}, \cdots, y_{1i}), \boldsymbol{y}_2(y_{21}, y_{22}, \cdots, y_{2j})$ ($i = 1, 2, \ldots, n_1, j = 1, 2, \ldots, n_2$), and the mean vectors of the two groups of samples are

$$\boldsymbol{\mu}_1 = \frac{1}{n_1}\sum_{i=1}^{n_1} \Phi(y_{1i}) \quad (18)$$

$$\boldsymbol{\mu}_2 = \frac{1}{n_2}\sum_{j=1}^{n_2} \Phi(y_{2j}) \quad (19)$$

The square of the distance between the two groups is as follows:

$$\begin{aligned} D_b &= \|\boldsymbol{\mu}_1 - \boldsymbol{\mu}_2\|^2 = (\boldsymbol{\mu}_1 - \boldsymbol{\mu}_2)^T(\boldsymbol{\mu}_1 - \boldsymbol{\mu}_2) \\ &= \frac{1}{n_1^2}\sum_{i=1}^{n_1}\sum_{j=1}^{n_2} k(y_{1i}, y_{1j}) - \frac{2}{n_1 n_2}\sum_{i=1}^{n_1}\sum_{j=1}^{n_2} k(y_{1i}, y_{2j}) + \frac{1}{n_2^2}\sum_{i=1}^{n_1}\sum_{j=1}^{n_2} k(y_{2i}, y_{2j}) \end{aligned} \quad (20)$$

The square of the dispersion in $\boldsymbol{y}_k$ is

$$D_{w1} = \sum_{i=1}^{n_1}\|\Phi(y_{1i}) - \boldsymbol{\mu}_1\|^2 = \sum_{i=1}^{n_1}\Phi(y_{1i})^T\Phi(y_{1i}) - n_1\boldsymbol{\mu}_1^T\boldsymbol{\mu}_1 = \sum_{i=1}^{n_1} k(y_{1i}, y_{1i}) - \frac{1}{n_1}\sum_{i=1}^{n_1}\sum_{j=1}^{n_2} k(y_{1i}, y_{1j}) \quad (21)$$

$$D_{w2} = \sum_{j=1}^{n_2}\|\Phi(y_{2j}) - \boldsymbol{\mu}_2\|^2 = \sum_{j=1}^{n_2}\Phi(y_{2j})^T\Phi(y_{2j}) - n_2\mu_2^T\mu_2 = \sum_{j=1}^{n_2} k(y_{2j}, y_{2j}) - \frac{1}{n_2}\sum_{i=1}^{n_1}\sum_{j=1}^{n_2} k(y_{2i}, y_{2j}) \quad (22)$$

According to Fisher's criterion, the objective function is established; that is, the fitness function of the swarm intelligence optimization is

$$J_{Fisher}(\gamma, d_1, \sigma) = \frac{D_{w1} + D_{w2}}{D_b} \quad (23)$$

Because the mapping process is realized by means of the kernel function, the final transformation is to find γ, d_1, and σ to obtain the minimum value of the J_{Fisher} function. The optimal kernel parameters γ^*, $d_1{}^*$, and σ^* can be found via the SFLA-PSO to achieve the minimum value of J_{Fisher}.

3.3. The Specific Steps of Dual-Kernel Parameter Optimization

The FC discriminant function J_{Fisher} is used as the fitness of the SFLA-PSO algorithm to optimize the parameters γ, d_1, and σ of the dual-kernel function. The specific process is as follows:

(1) According to Fisher's criterion, the samples are input, and the sum of the square of the distance between class D_b and within class D_{w1} and D_{w2} are calculated using Formulas (20)–(22).
(2) The optimized objective function J_{Fisher} is constructed using Equation (23) and is used as the fitness of the swarm intelligence optimization.
(3) The parameters of the SFLA-PSO swarm intelligent fusion algorithm are set, and the particle swarm optimization is initialized.
(4) The initial population is generated randomly, the fitness of the individuals are calculated, and the velocity and position are updated according to Formulas (1) and (2) of the PSO strategy.
(5) The optimal particles optimized in the first layer of PSO are taken as all the initial frogs of the SFLA and are grouped once more.
(6) According to the fitness size, the frog's velocity and position are updated based on the frog leaping update Formulas (3)–(5) to determine the best advantage.
(7) If the objective function J_{Fisher} satisfies the termination condition, the optimal value $J_{\text{Fisher}}(\gamma^*, d_1{}^*, \sigma^*)$ is output, and the algorithm is stopped. Otherwise, it returns to (4).

3.4. Simulation and Comparative Analysis

In order to compare the application effect of the KPCA feature extraction method based on the dual-kernel function proposed in this paper (DKKPCA for short), a comprehensive comparison test was carried out.

3.4.1. Iris Plant Database

In the simulation experiment, the Iris data set was selected, and its characteristics were used as the data source. It is divided into three types of patterns, and there are 50 data sets in each category. Each data point includes four data sets, namely, the length and width of the calyx and the length and width of the petals. They are commonly taken as test sets and training sets in data mining and data classification. The first class is linearly separable from the second and the third classes, while the second and the third classes are nonlinearly separable. The distribution diagram is shown in Figure 2.

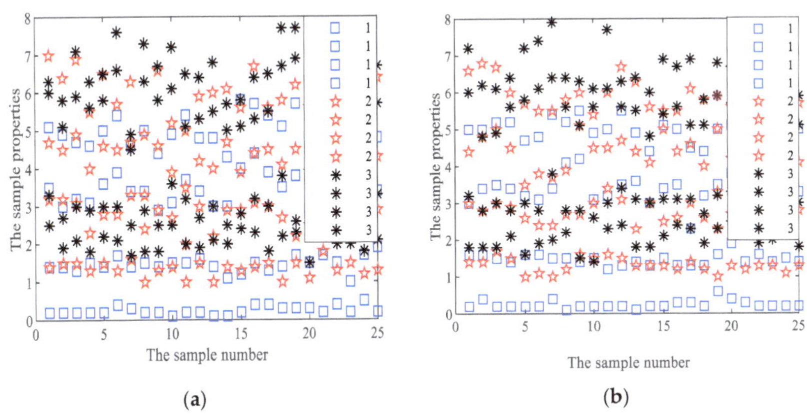

Figure 2. Iris sample distribution: (**a**) the training sample distribution; (**b**) the test sample distribution.

3.4.2. Parameter Optimization of the Iris Data Set with the Dual-Kernel Function Based on the SFLA-PSO

In the simulation analysis, the first 25 samples of the Iris data set were taken as the training samples, and the SFLA-PSO was used to optimize the kernel parameters. The parameters of the SFLA-PSO were as follows: $N = 200$, $n_{PSO} = 20$, $n_p = 10$, $c_1 = c_2 = 2$, the PSO iteration number was 10, $F = 20$, $m = 5$, $n = 4$, $L_{max} = 10$, $T_{max} = 50$, $d = 10$, and $S_{max} = 20$.

The optimization iteration process is shown in Figure 3. It can be seen that the minimum value can be obtained in the 50 evolution algebras, and the three optimized parameters are $\gamma^* = 0.057$, $d_1{}^* = 1$, and $\sigma^* = 3.1$.

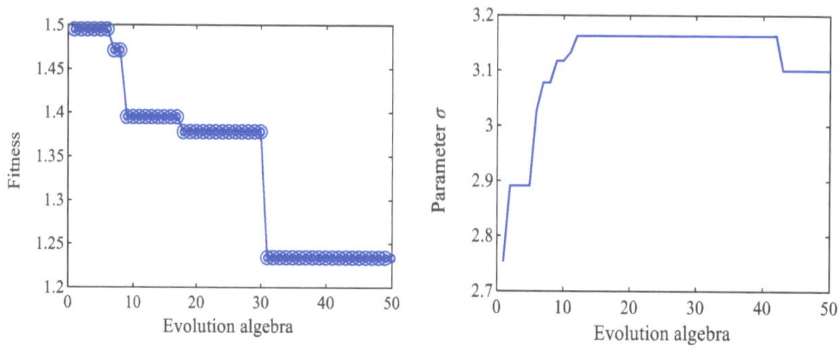

(a) (b)

Figure 3. Evolution process of kernel parameters: (**a**) SFLA-PSO evolution course; (**b**) evolution course of kernel parameters σ.

3.4.3. DKKPCA Feature Extraction of the Iris Data

The optimized parameters γ^*, $d_1{}^*$, and σ^* were used with the Iris data for the DKKPCA analysis, and the results are shown in Figure 4a,b. Figure 4a shows the DKKPCA scatter diagram for the principal component of the Iris data, and Figure 4b shows the histogram for the kernel principal component contribution rate. It can be seen that, due to the combination of the RBF kernel and the polynomial kernel function, the microscopic effect of the local kernel function and the amplification effect of the global kernel function are comprehensively used, the weight γ and kernel parameters d_1 and σ are optimized, and the clustering effect of the Iris samples is very obvious. Three types of Iris patterns are clustered in their respective centers, and the data projection points for categories 1 and 2 are very compact; the distance between classes is large, and the regional boundaries between classes are obvious.

For the comparative analysis, the analysis results of the single-kernel KPCA are shown in Figure 5a,b, which are the polynomial kernel function and the RBF kernel function, respectively. From Figure 5a, it can be found that the three types of data are mixed together in the feature space using KPCA and cannot be distinguished. It can be observed from Figure 5b that class 1 can be obviously separated from class 2 and class 3. However, some of the projection points for the data from classes 2 and 3 are still mixed due to their nonlinear inseparability, and the boundary between class 2 and class 3 is not obvious. The KPCA results are listed in Table 1 for the dual-kernel function (DKKPCA), the polynomial kernel function (KPCA_Poly), and the radial basis kernel (KPCA_RBF). The contribution rate of the first two principal components using KPCA is more than 85%; that is, Iris can completely replace the original four attributes by using two attributes, and the features are reduced by half.

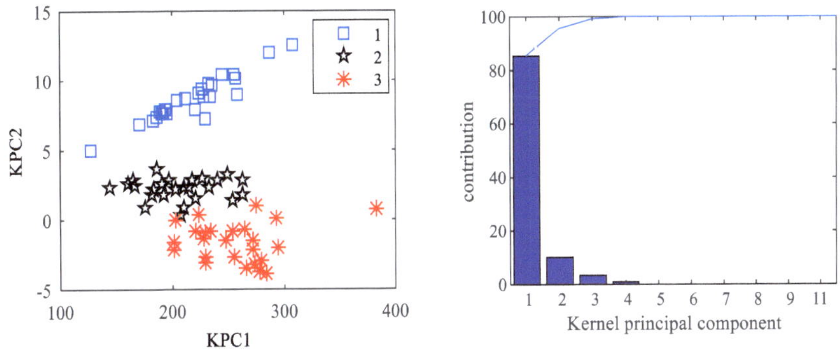

(**a**) (**b**)

Figure 4. KPCA results for Iris data using the optimized dual-kernel parameters (**a**) $\gamma = 0.057$, $d_1 = 1$, and $\sigma = 3.1$; (**b**) histogram of the DKKPCA contribution rate.

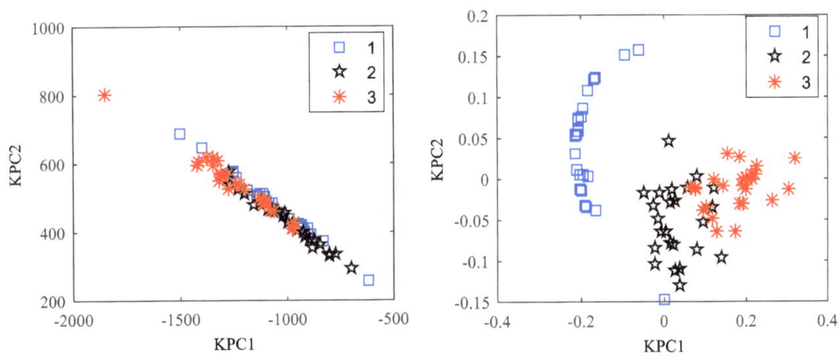

(**a**) (**b**)

Figure 5. Iris data single-kernel principal component analysis results: (**a**) polynomial kernel function ($d_1 = 1$); (**b**) RBF kernel function ($\sigma = 3.1$).

Table 1. KPCA results for Iris data.

Algorithm \ Serial Number		1	2	3	4
DKKPCA	Kernel principal component eigenvalue	2.518	0.2996	0.1014	0.0284
	Contribution %	85.431	10.165	3.441	0.963
	Cumulative contribution rate %	85.431	95.595	99.036	100.000
KPCA_Poly	Kernel principal component eigenvalue	49.566	4.0559	0.3441	0.0217
	Contribution %	91.8099	7.5126	0.6373	0.0402
	Cumulative contribution rate %	91.8099	99.3225	99.9558	100.00
KPCA_RBF	Kernel principal component eigenvalue	3.0601	0.5177	0.0946	0.0127
	Contribution %	83.0404	14.0486	2.5666	0.3444
	Cumulative contribution rate %	83.0404	97.089	99.6556	100.000

In order to compare the effect of KPCA before and after the parameter optimization of the dual-kernel function, the KPCA results before the optimization of the dual kernel are given, as shown in Figure 6. The results for parameters $\gamma = 0.5$, $d_1 = 1$, and $\sigma = 3.389$ are shown in Figure 6a, and those for $\gamma = 0.06$, $\sigma = 3.333$, and $d_1 = 1$ are shown in Figure 6b. It can be seen from these that due to the influence of the three parameters, the Iris data cannot be well separated linearly after nonlinear mapping onto the feature space, especially for classes 2 and 3. The optimized results for the dual-kernel parameters shown in Figure 4a are obviously better than those of the KPCA before optimization in Figure 6.

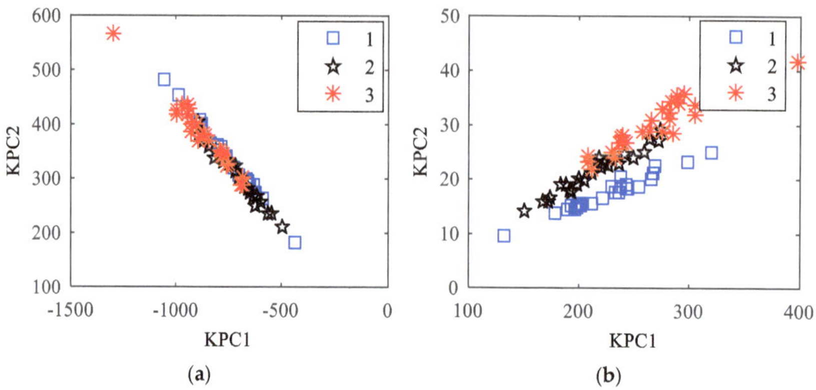

Figure 6. KPCA results for Iris data before dual-kernel function optimization: (**a**) $\gamma = 0.5$, $d_1 = 1$, and $\sigma = 3.389$; (**b**) $\gamma = 0.06$, $d_1 = 1$, and $\sigma = 3.333$.

Therefore, the DKKPCA optimized by the intelligent fusion algorithm is not only suitable for solving the nonlinear feature extraction problem, but can also provide better feature quality than the linear dimension reduction method and can greatly enhance the ability of nonlinear data processing.

4. The Simulation Failure Experiment for the Planetary Gearbox

4.1. Experimental Scheme of the Planetary Gearbox

As shown in Figure 7, the planetary gearbox fault diagnosis testbed consists of a control cabinet, a motor, a two-stage transmission box, a magnetic powder brake, etc. The structural diagram for the transmission system of the experimental platform is shown in Figure 8. The first transmission is a helical gear transmission, and the second stage is a 2K-H planetary gear transmission, which includes an inner gear ring, a solar wheel, and three planetary wheels. In this study, the fault modes for the planetary gear transmission system of a wind power system were simulated, and the speed range was 75–3000 r/min. Table 2 illustrates the equipment's technical parameters.

The motor provides power, the helical gearbox plays a deceleration role, and the planetary gearbox is a key research area. The magnetic powder brake acts as the load, and they are connected by an elastic coupling in the middle. We controlled the motor speed to adjust the shaft speed, and simulated different loads by setting the parameters of the magnetic powder brake. According to the testing plan, six measuring points, shown in Figure 9, were arranged. Four measuring points (1 to 4) were distributed on the helical gearbox, and the others (5 to 6) were on the planetary gearbox. The signal lines for the six sensors were connected to the DASP signal acquisition instrument and computer, and the entire collection system was fully built. The vibration signal was measured using six unidirectional piezoelectric acceleration sensors (CA-YD-186G) in this experiment.

Table 2. Technical parameters of main components.

Component	Name	Parameter
Helical gear box	Gear	Large gear: modules = 2, number of teeth = 77 Pinion: modules = 2, number of teeth = 55
	Bear	Deep groove ball rolling bearing 6206

Table 2. *Cont.*

Component	Name	Parameter
Planetary gearbox	Gear	Inner gear ring: modules = 2, number of teeth = 72
		Planetary gear: modules = 2, number of teeth = 27, quantity = 3
		Sun gear: modules = 2, number of teeth = 18
	Bear	Deep groove ball rolling bearing: for the planet wheel 6202, for the planet shelf = 6206, for the sun wheel 6205
Brake	The loading form is magnetic, and the loading torque is 0–100 N·m	
Motor	Converter motor	2.2 kW, the rotational speed is 1500 RPM, and the rated speed is 1410 RPM

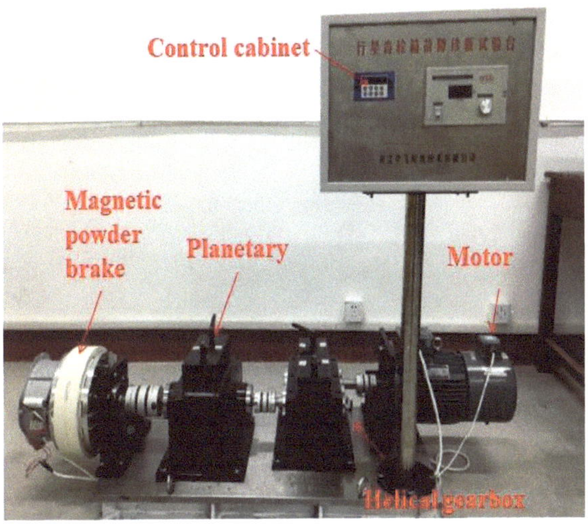

Figure 7. The planetary gear test bed.

(a) (b)

Figure 8. Structural diagrams of the transmission system of the experimental platform: (**a**) helical gear transmission; (**b**) planetary gear transmission.

Figure 9. Arrangement of measuring points.

4.2. Analysis of Vibration Signal

The vibration signals for each measuring point on the box body were measured. According to the experimental scheme, the vibration signals for the normal state and three kinds of planetary gear tooth surface wear states were mainly measured and analyzed, and the characteristic parameters were extracted. They include 21 time-domain features, such as the mean value, mean square value, maximum value, minimum value, variance, root-mean-square value, root amplitude, absolute average amplitude, skewness, kurtosis, peak, and six-order moments, and six frequency-domain features, namely, the frequency domain variance, correlation factor, power spectrum barycenter index, mean square spectrum, harmonic factor, and origin moment of spectrum. Sixty groups of training samples and sixty groups of test samples were extracted. After their standardization, the KPCA dual-kernel parameter optimization and feature extraction were carried out.

5. Feature Extraction of the Planetary Gearbox Based on the KPCA Dual-Kernel Parameter Optimized by the SFLA-PSO

5.1. Optimization of the Dual-Kernel Parameters of KPCA Based on the SFLA-PSO

In order to improve the mapping performance of KPCA, the SFLA-PSO fusion algorithm was used to optimize the parameters of the dual-kernel function for the data sets for the normal state, one-tooth wear, two-tooth wear, and three-tooth wear of the planetary gear (referred to as models A, B, C, and D, respectively). The parameters of the SFLA-PSO fusion algorithm are shown in Table 3, and the optimization was carried out after substituting the sample data. The optimization results for the kernel parameters of the different wear modes are shown in Table 4.

Table 3. The parameters of the SFLA-PSO fusion algorithm.

Algorithm	Parameters
SFLA-PSO	$N = 200$, $n_{PSO} = 20$, $n_P = 10$, $c_1 = c_2 = 2$, $F = 20$, $m = 5$, $n = 4$, $L_{max} = 10$, $T_{max} = 100$, $d = 20$, $S_{max} = 20$

Table 4. The optimization results of dual-kernel parameters of planetary gear wear state.

Type	Model	KPCA Kernel Parameter			J_{Fnsher}
		γ^*	$d_1{}^*$	σ^*	
3-type	ABC	0.005	0.8	23.37	0.3497
	BCD	0.038	0.896	5.4438	0.8970
4-type	ABCD	0.055	1.03	13.1	0.9949

The evolutionary optimization process is shown in Figure 10a,b, from which it can be seen that the fitness function J_{Fisher} can reach the minimum value and obtain the optimal parameter in 100 iteration steps. It can be seen that different flexible weight coefficients γ^* and kernel parameters $d_1{}^*$ and σ^* should be adopted when DKKPCA mapping is applied to identify the different wear models.

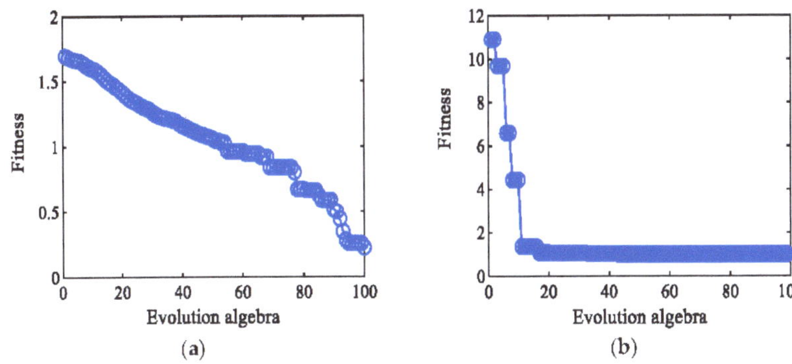

Figure 10. Evolution curves of the kernel parameter optimization process: (**a**) ABC model; (**b**) ABCD model.

5.2. KPCA Feature Extraction of the Planetary Gear Wear Based on the Dual-Kernel Optimization

Based on the test results of the normal state and the planetary gear wear in the experiment, the kernel parameters optimized by the SFLA-PSO were substituted into the DKKPCA to analyze the characteristics of the different wear models of the planetary gear. The analysis flow is shown in Figure 11.

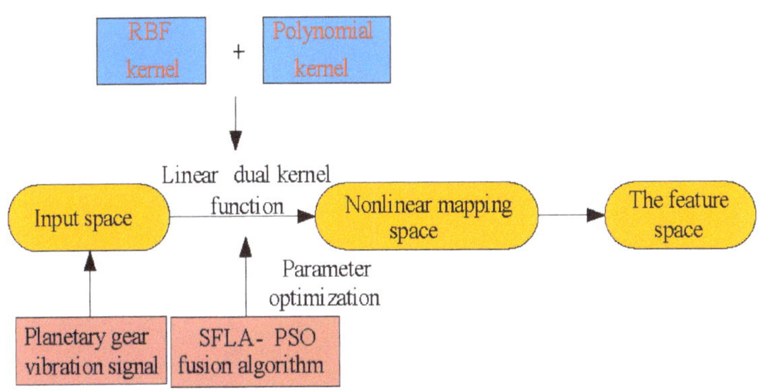

Figure 11. A feature extraction flow diagram for dual-kernel optimization.

The KPCA analysis results are shown in Figure 12; Figure 12a$_1$–c$_1$ show the scatter diagrams for the KPCA principal component before the optimization of the RBF kernel parameters. The kernel parameter σ was empirically set as follows: ABC: σ = 10.25; BCD: σ = 5.4438; ABCD: σ = 23.1. Figure 13a$_2$–c$_2$ show those for the DKKPCA principal component before the dual-kernel parameter optimization, for which the three parameters were set according to experience: ABC: γ = 0.07, σ = 23.37, and d_1 = 1.2; BCD: γ = 0.009, d_1 = 1, and σ = 5.4438; ABCD: γ = 0.2, d_1 = 2, and σ = 10. Figure 14a$_3$–c$_3$ show those for the DKKPCA principal component after the dual-kernel parameter optimization, and the three parameters are the optimization results γ^*, $d_1{}^*$, and σ^*.

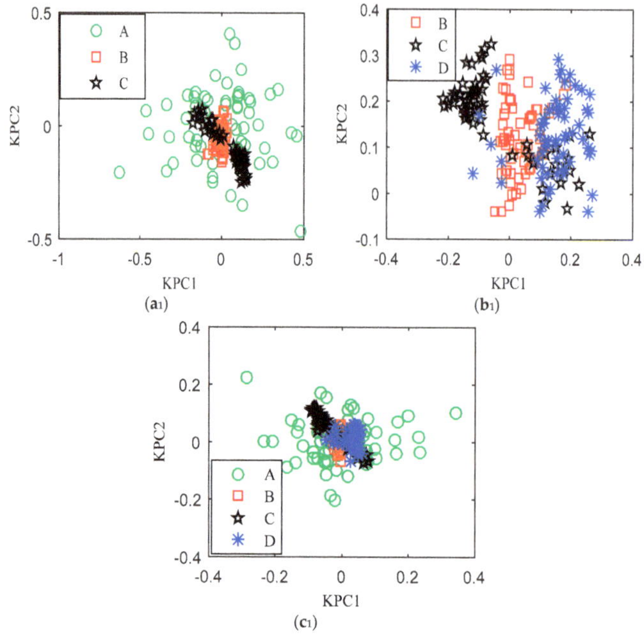

Figure 12. KPCA scatter diagram of planetary faults before single-kernel parameter optimization: A: normal state, B: one-tooth wear, C: two-tooth wear, D: three-tooth wear. (**a$_1$**) ABC: σ = 10.25; (**b$_1$**) BCD: σ = 5.4438; (**c$_1$**) ABCD: σ = 23.1.

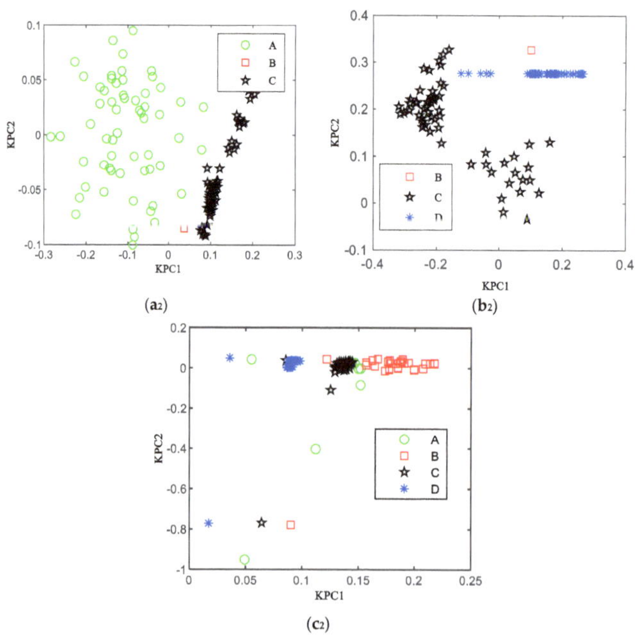

Figure 13. DKKPCA scatter diagram of planetary faults before dual-kernel parameter optimization: (**a$_2$**) γ = 0.07, d = 1.2, and σ = 23.37; (**b$_2$**) γ = 0.009, d_1 = 1, and σ = 5.443; (**c$_2$**) γ = 0.2, d = 2, and σ = 10.

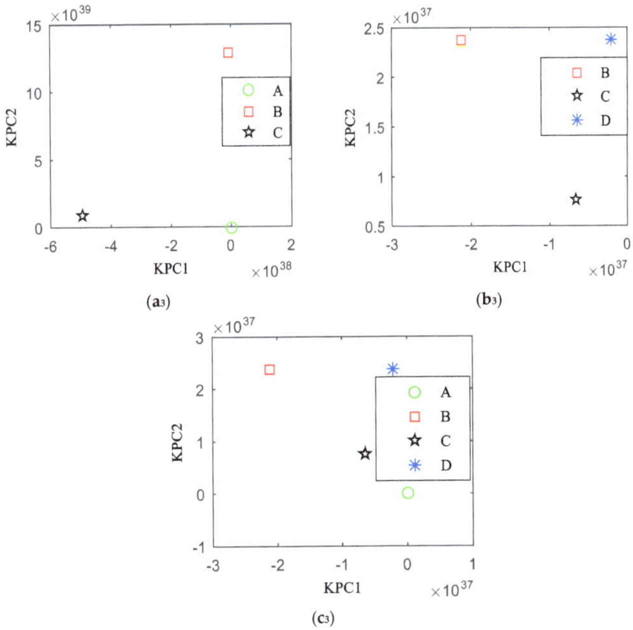

Figure 14. DKKPCA scatter diagram of planetary gear faults after dual-kernel parameter optimization: (**a3**) $\gamma = 0.005$, $d_1 = 0.8$, and $\sigma = 23.37$; (**b3**) $\gamma = 0.038$, d1 = 0.896, and $\sigma = 5.4438$; (**c3**) $\gamma = 0.055$, $d_1 = 1.03$, and $\sigma = 13.1$.

5.3. The Analysis of the Obtained Results

As can be seen from Figure 12a$_1$–c$_1$, the data points for the normal model are relatively scattered and interspersed with the three models of teeth wear. Although B, C, and D have their own clustering centers, some of the samples with the three models of wear are interwoven, and there is no obvious boundary line to accurately distinguish the wear models. Thus, it can be seen from Figure 13a$_2$–c$_2$ that the DKKPCA greatly improved the wear recognition of several planetary gears via the dual-kernel function, but some features of ABCD still intersected and were not completely distinguished, especially for the normal state (model A) and the slight wear (model B) of one tooth. Therefore, the KPCA scatter diagrams' comparison of the planetary faults using the single-kernel KPCA and DKKPCA indicate that the DKKPCA has a better nonlinear mapping performance than the single-kernel KPCA. The DKKPCA is suitable for intricate processes and nonlinear dimensionality reduction.

It can be seen from Figure 14a$_3$–c$_3$ that regardless of whether three or four models are employed, they can be highly clustered to a point after DKKPCA mapping. There is no gap in any category of B, C, and D for the three different wear models, the class spacing is large, and the categories are clear and divisible. This is because the interference from the small sample features of the contribution rate is eliminated after high-dimensional nonlinear mapping. It can be seen that when the dual-kernel function after a flexible weight linear combination is adopted for the DKKPCA, the kernel parameters are optimized to avoid the blindness of parameter selection, and the accuracy of the fault feature extraction and identification are significantly improved.

The KPCA method proposed in this paper was applied to a gearbox (JZQ-250 type) [30], for which the simulation failures included normal working conditions, a bearing outer-ring crack in the intermediate shaft, a broken tooth in the bearing cage fracture gear, and a broken tooth in the gear combined with a fracture of the bearing outer ring. The analysis results of the gearbox fault diagnosis demonstrate its effectiveness. So, the method has been verified for other conditions of other gearboxes.

6. Conclusions

KPCA is a nonlinear dimensionality reduction technique that maps data onto a high-dimensional feature space using kernel functions, enabling the detection of faults in complex and nonlinear systems. Its advantages lie in its ability to handle nonlinearity, making it suitable for intricate processes. However, KPCA's performance heavily relies on the appropriate choice of kernel function and its associated parameters, which can be challenging to determine in practice.

Aiming at the disadvantages of the KPCA single-kernel function and the advantages of the multi-kernel method, this paper proposes a KPCA with a flexible weight linear combination of the dual-kernel function (DKKPCA). In order to improve KPCA's performance and solve the optimization problem of dual-kernel parameters, an optimization model of dual-kernel parameters was constructed by referring to Fisher's criterion and defining the optimization variables and objectives. Then, taking full advantage of the SFLA-PSO fusion algorithm, which is simple, and a global search, the kernel parameters were optimized. A comprehensive comparison test was carried out using the Iris data to verify the application effect of the DKKPCA feature extraction method, which could be separated well after nonlinear mapping onto the feature space, especially for classes 2 and 3. The simulation results indicate that the DKKPCA optimized by the intelligent fusion algorithm is not only suitable for solving the nonlinear feature extraction problem, but can also provide better feature quality than the linear dimension reduction method, and greatly enhances the ability of nonlinear data processing.

A planetary gear simulation fault diagnosis experiment was conducted, and the vibration signals of the normal state and three kinds of planetary gear tooth surface wear states were mainly measured and analyzed. Then, the proposed DKKPCA method was applied to the feature extraction of multi-fault-mode coupled vibration signals of a planetary gear. The KPCA scatter diagrams for the planetary faults using the single-kernel KPCA and dual-kernel KPCA (DKKPCA) were compared and analyzed, and the results show that the DKKPCA has a better nonlinear mapping performance than the single-kernel KPCA. Similarly, a comparative analysis of the DKKPCA before and after the dual-kernel parameter optimization was completed shows that the scattering can be highly clustered to a point, and the wear damage state of a planetary gear can be accurately identified using the DKKPCA after parameter optimization; so, it has better nonlinear mapping performance than before parameter optimization. Therefore, the DKKPCA method has been adapted to feature extraction and state recognition for the nonlinear behavior of other mechanical equipment.

Author Contributions: Writing—original draft preparation and writing—review and editing, Y.H.; visualization and software, L.Y.; writing—review and editing, Y.L. All authors have read and agreed to the published version of the manuscript.

Funding: This research was funded by the Young Science Foundation of Shanxi province, China, grant number 201901D211201; the National Natural Science Foundation of China, grant numbers 52375470 and 52005455; and the Central Guidance on Local Science and Technology Development Fund of Shanxi Province, grant number YDZJSX2022C005.

Data Availability Statement: The data used to support the findings of this study are included within the article.

Conflicts of Interest: The authors declare no conflicts of interest.

References

1. Pan, H.; Zheng, J.; Yang, Y.; Cheng, J. Nonlinear sparse mode decomposition and its application in planetary gearbox fault diagnosis. *Mech. Mach. Theory* **2021**, *155*, 104082. [CrossRef]
2. Wang, C.; Li, H.; Ou, J.; Hu, R.; Hu, S.; Liu, A. Identification of planetary gearbox weak compound fault based on parallel dual-parameter optimized resonance sparse decomposition and improved momeda. *Measurement* **2020**, *165*, 108079. [CrossRef]
3. Schölkof, B.S.; Muller, K.R. Nonlinear component analysis as a kernel eigenvalue problem. *Neural Comput.* **1998**, *10*, 1299–1319. [CrossRef]
4. Xiao, Y.; Feng, L. A novel neural-network approach of analog fault diagnosis based on kernel discriminant analysis and particle swarm optimization. *Appl. Soft Comput.* **2012**, *12*, 904–920. [CrossRef]
5. Fatma, L.; Lotfi, M.; Wiem, A.; Hedi, D.; Khaled, Y. Investigating Machine Learning and Control Theory Approaches for Process Fault Detection: A Comparative Study of KPCA and the Observer-Based Method. *Sensors* **2023**, *23*, 6899.
6. Wang, Y.; Dong, R.; Wang, X.; Zhang, X. Research on Rolling Bearing Fault Diagnosis Based on Volterra Kernel Identification and KPCA. *Shock. Vib.* **2023**, *2023*, 1–9. [CrossRef]
7. Shao, R.; Hu, W.; Wang, Y.; Qi, X. The fault feature extraction and classification of gear using principal component analysis and kernel principal component analysis based on the wavelet packet transform. *Measurement* **2014**, *54*, 118–132. [CrossRef]
8. Lei, Y.; Chen, W.; Li, N.; Lin, J. A relevance vector machine prediction method based on adaptive multi-kernel dual and its application to remaining useful life prediction of machinery. *J. Mech. Eng.* **2016**, *52*, 87–93. [CrossRef]
9. Fu, H.; Ren, R.; Yan, Z.; Ma, Y. Fault diagnosis method of power transformers using multi-kernel RVM and QPSO. *High Volt. Appar.* **2017**, *53*, 131–135.
10. Deng, X.; Lei, W. Modified kernel principal component analysis using dual-weighted local outlier factor and its application ton on linear process monitoring. *ISA Trans.* **2018**, *72*, 218–228. [CrossRef]
11. Pan, C.; Li, H.; Chen, B.; Zhou, M. Fault Diagnosis method with class mean kernel principal component analysis based on combined kernel function. *Comput. Simul.* **2019**, *36*, 414–419.
12. Wang, H.; Cai, Y.; Sun, F.; Zhao, Z. Adaptive sequence learning and application of multi-scale kernel method. *Pattern Recognit. Artif. Intell.* **2011**, *24*, 72–81.
13. Lv, L.; Wang, W.; Zhang, Z.; Liu, X. A novel intrusion detection system based on an optimal hybrid kernel extreme learning machine. *Knowl.-Based Syst.* **2020**, *195*, 105648. [CrossRef]
14. Nithya, A.; Appathurai, A.; Venkatadri, N.; Ramji, D.R.; Anna, P.C. Kidney disease detection and segmentation using artificial neural network and multi-kernel k-means clustering for ultrasound images. *Measurement* **2020**, *149*, 106952. [CrossRef]
15. Ouyang, A.; Liu, Y.; Pei, S.; Peng, X.; He, M.; Wang, Q. A hybrid improved kernel LDA and PNN algorithm for efficient face recognition. *Neurocomputing* **2020**, *393*, 214–222. [CrossRef]
16. Afzal, A.L.; Asharaf, S. Deep multiple multilayer kernel learning in core vector machines. *Expert Syst. Appl.* **2018**, *96*, 149–156.
17. Li, X.; Gao, X.; Li, K.; Hou, Y. Prediction for dynamic fluid Level of oil well based on GPR with AFSA optimized combined kernel function. *J. Northeast. Univ. (Nat. Sci.)* **2017**, *38*, 11–15.
18. Xie, F.; Chen, H.; Xie, S.; Jiang, W.; Liu, B.; Li, X. Bearing state recognition based on kernel principal component analysis of particle swarm optimization. *Meas. Control Technol.* **2018**, *37*, 28–35.
19. Bernal, D.; Lázaro, J.M.; Prieto, M.A.; Llanes, S.O.; da Silva, N.A.J. Optimizing kernel methods to reduce dimensionality in fault diagnosis of industrial systems. *Comput. Ind. Eng.* **2015**, *87*, 140–149. [CrossRef]
20. Kennedy, J.; Eberhart, R. Particle Swarm Optimization. In Proceedings of the IEEE International Conference on Neural Networks; IEEE Computer Society: Washington, DC, USA, 1995; pp. 1942–1948.
21. Eusuff, M.M.; Lansey, K.E. Optimization of water distribution network design using shuffled frog leaping algorithm. *J. Water Resour. Plan. Manag.* **2003**, *129*, 210–225. [CrossRef]
22. Jaafari, A.; Zenner, E.K.; Panahi, M.; Shahabi, H. Hybrid artificial intelligence models based on a neuro-fuzzy system and metaheuristic optimization algorithms for spatial prediction of wild fire probability. *Agric. For. Meteorol.* **2019**, *266–267*, 198–207. [CrossRef]
23. Wang, Y.; Sun, W.; Liu, L.; Wang, B.; Bao, S.; Jiang, R. Fault Diagnosis of Wind Turbine Planetary Gear Based on a Digital Twin. *Appl. Sci.* **2023**, *13*, 4776. [CrossRef]
24. Kennedy, J.; Eberhart, R.C. *Swarm Intelligence*; Morgan Kaufmann Division of Academic Press: San Francisco, CA, USA, 2001; pp. 11–15.
25. Rajamohana, S.P.; Umamaheswari, K. Hybrid approach of improved binary particle swarm optimization and shuffled frog leaping for feature selection. *Comput. Electr. Eng.* **2018**, *67*, 497–508. [CrossRef]
26. Wang, H.; Sun, F.; Cai, Y.; Chen, N.; Ding, L. Multi-kernel learning method. *Acta Autom. Sin.* **2010**, *36*, 1037–1050. [CrossRef]
27. Li, J.; Qiao, J.; Yin, H.; Liu, D. *Kernel Adaptive Learning and Application in Pattern Recognition*; Publishing House of Electronics Industry: Beijing, China, 2013; pp. 23–25.
28. Chen, W.; Panahi, M.; Tsangaratos, P.; Shahabi, H.; Ilia, I.; Panahi, S.; Li, S.; Jaafari, A.; Bin Ahmad, B. Applying population-based evolutionary algorithms and a neuro-fuzzy system for modeling landslide susceptibility. *Catena* **2019**, *172*, 212–231. [CrossRef]

29. Zhao, Y. *Pattern Recognition;* Shanghai Jiao Tong University: Shanghai, China, 2013; pp. 12–16.
30. He, Y.; Wang, Z. Regularized kernel function parameter of kpca using WPSO-FDA for feature extraction and fault recognition of gearbox. *J. Vibroengineering* **2018**, *20*, 225–239. [CrossRef]

machines

Article

Efficiency of Bevel and Hypoid Gears—Test Rig Development and Experimental Investigations

Lorenz Constien *, Martin Weber, Josef Pellkofer and Karsten Stahl

Gear Research Center (FZG), TUM School of Engineering and Design, Technical University of Munich, Boltzmannstraße 15, D-85748 Garching near Munich, Germany; martin.wm.weber@tum.de (M.W.); josef.pellkofer@tum.de (J.P.); karsten.stahl@tum.de (K.S.)
* Correspondence: lorenz.constien@tum.de

Abstract: The efficiency of bevel and hypoid gears is, alongside load capacity, one of their most important design criteria. To consider the efficiency of bevel and hypoid gears during the development and design process, validated calculation methods based on experimental investigations are necessary. However, the isolated experimental investigation of the load-dependent power losses of bevel and hypoid gears has not been adequately investigated, as most of the experimental investigations consider the complete gearbox. This paper presents a test rig that allows for the experimental investigation of the efficiency of bevel and hypoid gears with a measurement uncertainty of the efficiency of $\Delta\eta \leq \pm 0.08\%$ according to the Guide to the Expression of Uncertainty in Measurement (GUM). Using the developed test rig, experimental investigations on the efficiency behavior of bevel and hypoid gears regarding the influence of the axial offset, driving direction, and microgeometry are carried out for different operating points varying in circumferential speed and load. This paper discusses the methodology and the first experimental results of a study on the efficiency of bevel and hypoid gears in detail.

Keywords: bevel gears; hypoid gears; efficiency; test rig; measuring concept; losses

1. Motivation and Introduction

In 2015, the United Nations addressed the global challenges facing the planet through the formulation of the 17 Sustainable Development Goals. Considering the increasing challenges of climate change and its consequences in the present time, it is unquestionably essential to take concerted action, especially in alignment with the UN Sustainable Development Goals. Therefore, the reduction of greenhouse gases (GHGs) and, in particular, CO_2, as a primary driver of climate change, must be taken into account by gear engineers within their research and development activities.

In this context, one priority topic is the increase in the efficiency of powertrains. For the optimization of powertrains, it is necessary to concentrate on each component separately. Bevel and hypoid gears are widely used in applications where the rotary motion needs to be transmitted between non-parallel axes. Besides axle drives within the automotive section, bevel and hypoid gears are used in marine applications, helicopter drives, and industrial applications. Due to the high amount of sliding movement and subsequent heat dissipation within the tooth contact of bevel and hypoid gears, there is high potential to increase the efficiency by considering it in the design process of bevel and hypoid gears in addition to the load carrying capacity. There are various calculation approaches regarding the estimation of the efficiency behavior of bevel and hypoid gears but currently there is no standardized calculation approach to define the load-dependent and independent gear power losses. For the validation of calculation approaches in general, experimental investigations are used. The available measurement results on the efficiency of bevel and hypoid gears are mostly based on whole gearboxes, which impedes the consideration of the gear power losses due to parallel considerations of other mechanical components like bearings and sealings. Therefore, the available measurement results do not fulfill the requirement for the accuracy of the measurement resolution to reflect small differences in the efficiency of bevel and hypoid gears.

Within this paper, a new test rig for the experimental investigation of the efficiency of bevel and hypoid gears is presented. The test rig allows for the measurement of the gear power losses of bevel and hypoid gears with a suitable measurement resolution to validate state-of-the-art calculation approaches of the gear power losses of bevel and hypoid gears. Experimental investigations carried out by one of the authors with the presented test rig [1] are discussed regarding the influence of the axial offset, the driving direction, and the microgeometry on the efficiency of bevel and hypoid gears.

2. State of the Art

The total power losses P_V of a gearbox are composed of the partial losses generated in the individual components of the gearbox. Typical components contributing to the total power losses are gear tooth mesh losses P_{VZ}, bearing losses P_{VL}, sealing losses P_{VD}, and other losses P_{VX}. Additionally, the losses can be subdivided into load-dependent and load-independent losses [2]. The total power losses P_V of a gearbox can be summarized as follows:

$$P_V = P_{VZP} + P_{VZ0} + P_{VLP} + P_{VL0} + P_{VD} + P_{VX} \tag{1}$$

P_{VZP}	W	load-dependent gear tooth mesh losses	P_{VL0}	W	load-independent bearing losses
P_{VZ0}	W	load-independent gear power losses	P_{VD}	W	sealing losses
P_{VLP}	W	load-dependent bearing losses	P_{VX}	W	other losses

In the following, the power losses contributing to the total power losses and the state-of-the-art calculation methods for the gear tooth mesh losses of bevel and hypoid gears will be discussed. Then, the state-of-the-art test rig concepts and measurement methods for gear power losses in general and specifically for bevel and hypoid gears will be presented.

The load-dependent gear tooth mesh losses occur during the meshing of two tooth flanks under load due to rolling and sliding friction. According to Niemann [2], the losses due to rolling friction can be neglected, and therefore, the load-dependent gear tooth mesh losses can be calculated based on Coulomb's law. Wirth [3] describes a methodology for determining gear power loss based on the results from a tooth contact analysis using BECAL [4]. By using an LTCA, Wirth [3] considers the changing load and velocity conditions along the path of contact. For the local coefficient of friction, Wirth [3] uses an approach by Klein [5], which allows for a local examination of the individual contact points on the gears. The total load-dependent gear power losses can be determined by summing up all locally occurring losses. A comparable approach to improving the efficiency of bevel and hypoid gears using an LTCA and a semi-empirical formulation of the friction coefficient is given by Grabovic et al. [6,7]. A combined calculation approach using an LTCA and a subsequent EHD simulation for predicting the gear power losses in bevel and hypoid gears is given by Kolivand [8], Simon [9], Mohammadpour et al. [10–12], and Ding et al. [13]. In addition to the high accuracy calculation methods using an LTCA, the efficiency of bevel and hypoid gears can be calculated with simplified approaches mainly using macrogeometry data of the gear set and operation conditions. Simplified calculation methods regarding the load-dependent gear tooth mesh losses of bevel and hypoid gears are given by Wech [14] and also within the ISO/TS 1300-20:2021 [15] and ISO/TR 14179-1:2001 [16].

Load-independent gear power losses can be grouped into hydraulic and windage losses [17]. The main influencing parameters are speed, circumferential speed, temperature of the lubricant, and the oil level [17]. Empirical calculation approaches for the hydraulic losses of gears are given by Mauz and Walter [18,19]. Simulative investigations based on computational fluid dynamic (CFD) programs are provided by [20–22]. Concli validates theoretical calculation approaches with experimental investigations of spur gears [23]. In [24], Seetharman et al. present a calculation model based on fluid mechanics for calculating idle losses, validated in [25] based on results from extensive experimental investigations that varied speed, oil level, and fundamental gear geometry parameters. Experimental investigations on the load-independent gear power losses on bevel gears were performed by Jeon [26] using a practical axle gearbox as well as by Quiban et al. [27].

Load-dependent and load-independent bearing losses can be calculated using empirical calculation approaches provided by the manufacturer, like *SKF* [28,29] or *INA* [30]. These calculation approaches are widely used within the industry and are similar to the calculation method given in ISO/TR 14179-2:2001 [31]. Calculation methods with higher accuracy and consideration of the whole shaft-bearing system and housing stiffness are described in [32,33]. Yilmaz [34] describes experimental investigations of the bearing losses.

Simplified calculation approaches for the sealing losses are given in ISO/TR 14179-1:2001 [16]. For non-contact type sealings, power losses can be assumed to be negligible [16]. Other losses P_{VX} result from disc clutches, synchronizers, and other phenomena like churning losses of the differential cage [17,35]. According to [35], these losses are typically carried out by power loss measurements using the complete gearbox.

Various methods are currently used in industry and research to experimentally determine gearbox losses; this can be done either by measuring the power loss (direct determination of the degree of loss) or by measuring the power difference (indirect determination of the degree of loss). Methods of direct power loss measurement of gearboxes typically involve the measurements at a back-to-back test rig using a pendulum gearbox suspension or the calorimetry method [1]. These main methods of measurement are summarized in Figure 1.

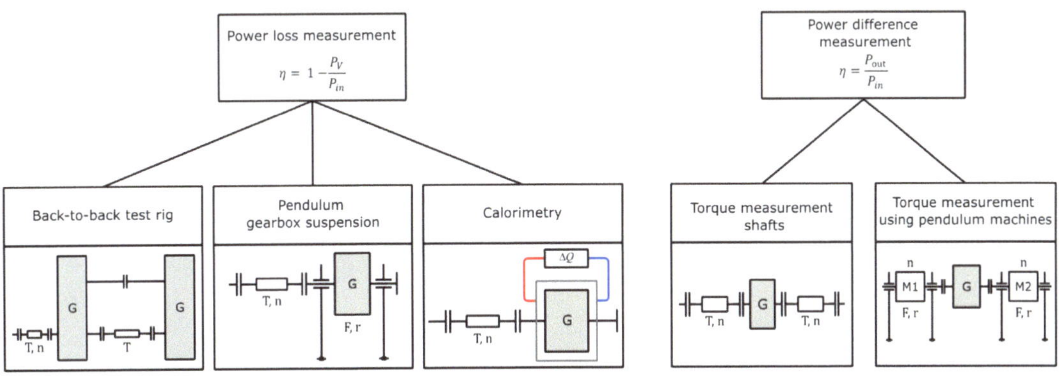

Figure 1. Measuring principles for efficiency determination according to Goebbelet et al. [36].

The mechanical back-to-back principle is used in various investigations regarding the efficiency of spur gears [37,38] as well as for high-speed applications [39]. Wech [14] used the mechanical back-to-back principle for his extensive experimental investigations of the efficiency of bevel gears. The test rig setup is based on four identical bevel-helical gearbox combinations. The spur gears are used to operate all bevel gearboxes under similar operating conditions with regard to loaded flank and driving direction. Detailed knowledge of the losses occurring in addition to the bevel gear power losses, such as bearing and spur gear power losses, is required to determine the gearing loss performance of the investigated bevel and hypoid gearings.

According to Goebbelet et al. [36], measuring the power loss of gearboxes using a pendulum gearbox suspension is only suitable for selected gearbox systems. This measurement method is described in [40] for experimental investigations at high speeds.

The calorimetry method is based on the fact that power losses due to friction are completely and irreversible transformed into heat [41]. The applicability of this method for multi-megawatt gearboxes of wind turbines was proven by Pagitsch et al. [42], although the time required to achieve the necessary thermal steady-state condition is relatively high.

When using a power difference measurement, the input and output power of the tested gearbox are measured. According to Goebbelet et al. [36], two different methods of torque measurement can be distinguished in power difference measurements. One standard method is the power difference measurement using a pendulum machine. The second standard method for power difference measurement is torque measurement using torque measurement shafts. An advantage compared to the use of pendulum machines is the variable design of the measuring area, which allows the use of additional components with losses, such as support bearings and transmission gears, without affecting the measuring accuracy. According to Homann [41], power difference measurement using torque measuring shafts is currently the most common method of power difference measurement and, at the same time, the most common method for experimental efficiency determination. An electrical tensed test rig for bevel gears using two electrical machines is described by Leighton [43], which is capable of investigating the efficiency of the gearing system. *Strama-MPS Maschinenbau GmbH & Co. KG*, Straubing, Germany [44] offers a bevel gear test rig based on the principle of a CNC machine, which enables the measurement of efficiency and the determination of load carrying capacity and wear.

3. Objective and Approach

To define the efficiency of bevel and hypoid gears during the development process, methods for calculating the efficiency of bevel and hypoid gears are needed. Experimental investigations are generally used to validate these calculation methods. However, the available measurement results on the efficiency of bevel and hypoid gears are predominantly based on the entire gearboxes, which complicates the isolation of gear power losses, as the measurements also account for other mechanical components. This highlights the need for a test rig concept and a test rig for precisely measuring the gear efficiency of bevel and hypoid gears. This test rig allows subsequent researchers to develop and validate a standardized calculation approach to the efficiency of bevel and hypoid gears, which is currently not present in the state-of-the-art.

The objective of this paper is the development of a test rig for the investigation of the efficiency of bevel and hypoid gears. By using this test rig, experimental investigations of the efficiency behavior of bevel and hypoid gearboxes with regard to the axial offset, the driving direction, and the microgeometry are carried out for varying circumferential speeds and loads.

4. Test Rig Concept

The bevel gear efficiency test rig designed in the underlying research project [1] uses two electrical machines based on the electrical tensed concept. To capture the torque loss T_V of the complete gearbox, the input and output torque, $T_{in,out}$, and the input and output speed of the gearbox, are precisely measured using optical angular measurement systems. The total gearbox torque loss can be calculated using Equation (2).

$$T_V = T_{in} - \frac{T_{out}}{i}, \tag{2}$$

T_V	Nm	total gearbox torque loss	T_{out}	Nm	output torque
T_{in}	Nm	input torque	i	-	gear ratio

To determine the load-independent gearbox torque loss, depending on the driving direction, the measuring shaft at the pinion or wheel shaft is removed, and in this way, the drag torque caused by the gearbox is measured. Within the drag torque, the load-independent gear torque loss, the load-independent bearing torque loss, and the sealing torque loss for specific operational conditions are included. The load-dependent bearing losses for specific operational conditions and no-load conditions can be evaluated experimentally for each bearing using the FZG bearing power loss test rig [34] and are checked for consistency with the calculation approach by *SKF* [29]. By doing so, the load-dependent gear torque loss T_{VZP} can be calculated using Equation (3).

$$T_{VZP} = T_{\mathrm{V}} - T_{\mathrm{V0}} - T_{\mathrm{VLP,total}}, \tag{3}$$

T_{VZP}	Nm	load-dependent gear torque loss	T_{V0}	Nm	drag torque of gearbox (load-independent)
T_{V}	Nm	total gearbox torque loss	$T_{\mathrm{VLP,total}}$	Nm	load-dependent bearing torque loss of all bearings of the gearbox

The schematic structure of the bevel gear efficiency test rig is depicted in Figure 2.

	unit	data
hypoid offset	mm	0/31.75
max. pinion torque	Nm	700
speed	min^{-1}	0 to 3 000
lubrication	-	dip lubrication / injection lubrication
lubrication temperature	°C	40 °C to 120 °C
driving Conditions		▪ pinion drives (drive/coast flank) ▪ wheel drives (drive/coast flank)

schematic structure

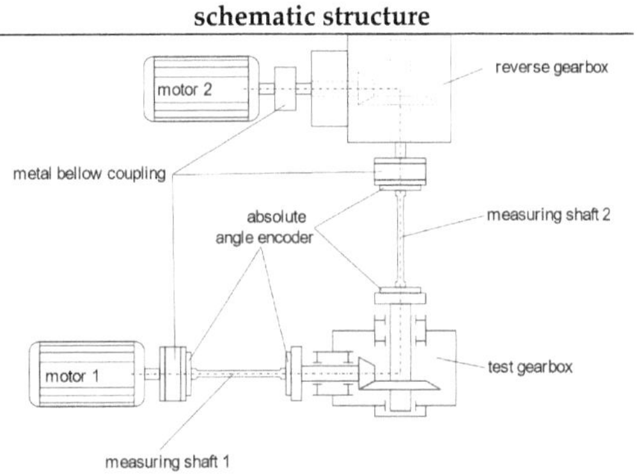

Figure 2. Schematic structure of bevel gear efficiency test rig [1].

4.1. Mechanical Structure

The bevel gear efficiency test rig consists of a test gearbox, a reverse gearbox, and two identical electric motors. The load on the test gearbox is applied by one of the two electric motors. Depending on the driving direction to be examined in the test gearbox, one of the two motors acts as a brake. The individual test rig components are connected via shafts and metal bellows couplings. The test gearbox is connected to the other test rig components through torsional flexible measuring shafts. The reverse gearbox is used for space reasons. Additionally, due to the identical gear ratio of the test gearbox and the reverse gearbox, the entire characteristic curve of the two identical electric motors can be utilized. The bearing arrangement of the pinion and the wheel of the test gear set is designed as a fixed-floating bearing arrangement. Four-point contact bearings are used as axial bearings, and cylindrical roller bearings are used for radial restraint. Applying a fixed-floating bearing arrangement, compared to a preloaded and adjusted bearing arrangement with tapered roller bearings, allows for a clear assignment and, therefore, a reproducible determination of the loads relevant for bearing losses on each respective bearing. The pinion and wheel are in an axial direction and are continuously adjustable; therefore, a precise and reproducible configuration of the contact pattern and backlash is given. The test gear set is dip-lubricated, and the gearbox bearings are supplied with a defined oil flow from an external oil unit to ensure lubrication and heat transfer. In addition, by having a defined oil flow from an external oil unit, the comparability of the lubrication conditions of the bearings between the bevel gear efficiency test rig and the FZG bearing power loss test rig is given. The oil level in the gearbox housing is kept constant using a siphon. The oil unit allows for the adjustment of a defined oil temperature in the oil sump through cooling and heating. The shafts are sealed using radial shaft seals.

4.2. Measuring Equipment

To measure the torques applied to the input and output shafts, the elastic deformation of the shafts under load is measured. For this purpose, an ultra-high accuracy absolute angle encoder of the REXA type by *Renishaw plc*, Wotton-under-Edge, UK is attached to each end of the input and output shaft. The absolute position is determined at each encoder using two readheads of the RESOLUTE type by *Renishaw plc*. Using two readheads for each encoder allows for the compensation of errors arising from eccentricities of the shaft. According to *Renishaw plc.* [45], this method enables the determination of the angular position with a maximum measurement uncertainty of $\pm$ 1 under the condition that the encoder and readheads are correctly aligned on the shaft and with each other.

The torque measurement system on the bevel gear efficiency test rig is calibrated in accordance with DIN 51309:2022 [46], based on the lever–mass system. Therefore, a particular calibration routine was developed, including three ascending and descending load phases up to the maximum test load interrupted by relaxation phases after each load step to determine short-term creep. The measuring shafts are calibrated in the same installation position in the test rig at four rotational positions, each offset by 90°.

Data processing takes place on an evaluation computer, with communication and data processing facilitated by a CompactRIO controller from *National Instruments Corp*, Austin, USA with BiSS Interface modules from the manufacturer *S.E.A. Datentechnik GmbH*, Troisdorf, Germany. The temperature measurement in the oil sump is carried out using a Pt100 temperature sensor of accuracy class A according to DIN EN IEC 60751:2023 [47]. The temperature sensor can be adjusted depending on the oil level to ensure sufficient immersion depth.

4.3. Measurement Uncertainty

As part of the research project [1], an uncertainty analysis following the Guide to the Expression of Uncertainty in Measurement (GUM) [48] was conducted to investigate the measurement uncertainty associated with the parameters captured on the bevel gear efficiency test rig. Therefore, the software *GUM-Workbench (version 2.4)* [49] and the procedure according to Sommer [50] were used.

To investigate the measurement uncertainties occurring on the bevel gear efficiency test rig, the uncertainty of the calibration of the measuring shafts is determined in the first step, and the uncertainty of the efficiency determination is calculated on this basis. As part of the calibration, the calibration factor k, which describes the ratio of the actual calibration torque T_{Cal} to the theoretically determined torque $T_{\mathrm{Meas,th}}$, is determined. The torque applied to the measuring shaft is calculated by multiplying the theoretically measured torque $T_{\mathrm{mess,th}}$ by the calibration factor k. To determine the efficiency, the measured output torque $T_{\mathrm{Mess,out}}$ and the input torque $T_{\mathrm{Mess,in}}$ are set in relation to each other. Equation (4) thus represents the system equation for the uncertainty analysis carried out according to GUM [48] for the case of a gear ratio of $i = 1$.

$$\eta = \frac{\mathrm{T_{Meas,out}}}{\mathrm{T_{Meas,in}}} = \frac{\frac{\Delta\varphi_{\mathrm{Meas,out}}}{\Delta\varphi_{\mathrm{Meas,Cal,out}}}\cdot(m_{W,out}\cdot L_{L,out} + m_L\cdot R)\cdot cos(\alpha_{out})\cdot cos(\beta_{out})}{\frac{\Delta\varphi_{\mathrm{Meas,in}}}{\Delta\varphi_{\mathrm{Meas,Cal,in}}}\cdot(m_{W,in}\cdot L_{L,in} + m_L\cdot R)\cdot cos(\alpha_{in})\cdot cos(\beta_{in})}, \quad (4)$$

η	-	Efficiency	T_{Meas}	Nm	Measured torque
$\Delta\varphi_{Meas}$	rad	Measured torsion angle	$\Delta\varphi_{Meas,Cal}$	rad	Measured torsion angle at calibration torque
m_W	kg	Mass of the calibration weights	m_L	kg	Mass of the lever
L_L	m	Length of the lever arm	R	m	Distance of the lever's center of gravity from the center of the axis
α	°	Angle between the reference axis of the lever and the horizontal	β	°	Angle between the reference axis of the lever and the normal to the measuring shaft

Since the measurement uncertainty is reduced by a factor of $1/i$ for gear ratios greater than $i = 1$, the uncertainty analysis shown here represents the worst case. Table 1 summarizes the expanded measurement uncertainties for the efficiency measured on the bevel gear efficiency test rig with a coverage probability of 95% for three load cases (20%, 60%, 100%) and exemplary efficiency values.

Table 1. Measurement uncertainty of the bevel gear efficiency test rig.

Parameter	Value	Expanded Measurement Uncertainty	Coverage Factor	Coverage Probability
η_{20}	0.59797	0.000840	2	95%
η_{60}	0.58855	0.000826	2	95%
η_{100}	0.58902	0.000828	2	95%

The expected measurement uncertainty of the efficiency is within a range of $\Delta\eta \leq \pm 0.08\%$ throughout the entire operating range of the test rig. The developed measurement technology thus allows the resolution of even slight differences in efficiency. It is, therefore, suitable for application in the context of the experimental investigations of the present study.

5. Methodology

5.1. Objects of Investigations

Within the experimental investigations, three variants of bevel and hypoid test gear sets are examined in terms of their efficiency behavior. The macrogeometries of both hypoid test gear sets are identical; they differ only in the design of the microgeometry. The basic geometry data of both macrogeometries are given in Table 2. The material of these test gear sets is 18CrNiMo7-6.

The hypoid test gear set variant G31.75 was designed with two different microgeometries. The test gear set variants G31.75kl and G31.75gr were designed by Klein [5,51] to investigate the influence of the Ease-Off design on the local flank pressure distribution. Therefore, variant G31.75gr was designed with conventional Ease-Off, meaning normal crowning values resulting in a large contact pattern and thus in lower contact stresses for the same input torque in comparison to variant G31.75kl. The G31.75kl test gear variant is designed with high crowning values, resulting in a small contact pattern. The values for lengthwise (LB) and profile crowning (HB), as well as the ease-off design and the load-free contact pattern for the drive side, calculated with the program *Klingelnberg Integrated Manufacturing of Spiral Bevel Gears (KIMoS 5)* [52], is given for the test gear variants in Figure 3. Exemplary contact patterns under load and load distribution for a pinion load of T_1 = 300 Nm calculated with *KIMoS 5* [52] are also given in Figure 3.

Table 2. Basic geometry data of the gear set variants G0 and G31.75.

Gear Type	Symbol	Unit	G0 Pinion	G0 Wheel	G31.75 Pinion	G31.75 Wheel
number of teeth	$z_{1,2}$	-	9	34	9	34
hypoid offset	a	mm	0.00		31.75	
mean normal module	m_{mn}	mm	3.57		4.03	
normal pressure angle (drive side)	α_{nD}	°	20.00		15.90	
normal pressure angle (coast side)	α_{nC}	°	20.00		24.10	
mean spiral angle	$\beta_{m1,2}$	°	33.00	33.00	44.99	21.01
face width	$b_{1,2}$	mm	26.00	26.00	31.13	26.00
profile shift coefficient	$x_{hm1,2}$	-	0.45	−0.45	0.45	−0.45

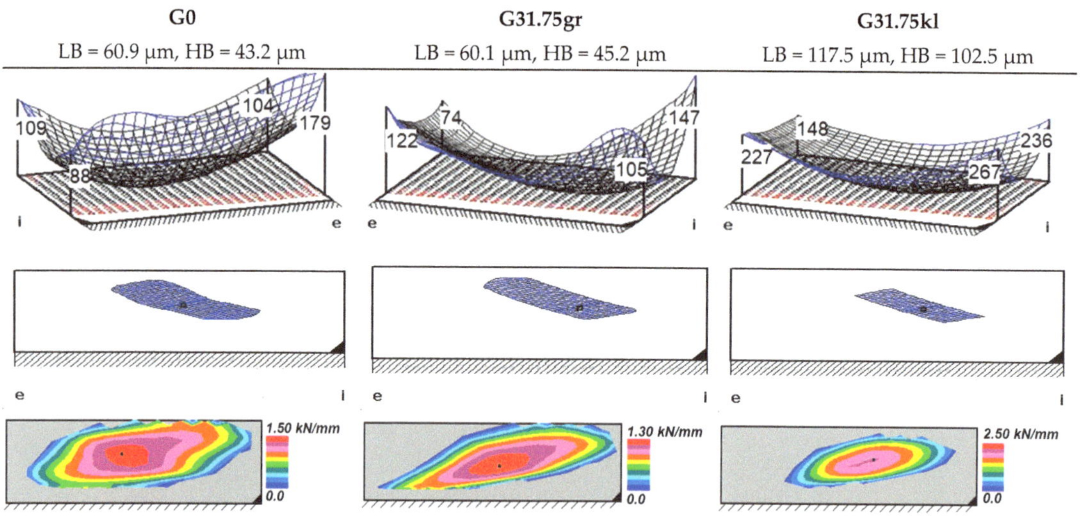

Figure 3. Ease-Off design and load-free contact pattern of test gear variants (drive side).

The quality of all used test gear sets was investigated using a 3D coordinate measurement center *Klingelnberg P40* by *Klingelnberg GmbH*, Hückeswagen, Germany. The total cumulative pitch deviation F_P was measured and classified according to DIN 3965-1:1986 [53]. The flank roughness was measured using the 3D coordinate measurement center at three teeth and on each tooth at three separate measuring paths vertical to the pitch angle. The results of these measurements are given in Table 3. Additionally, the flank topography of the test gear sets was measured, resulting in only minor, hence negligible, deviations from the nominal microgeometry.

Table 3. Quality and flank roughness of test gear variants G0, G31.75gr and G31.75kl.

Gear Type		Unit	G0 Pinion	G0 Wheel	G31.75gr Pinion	G31.75gr Wheel	G31.75kl Pinion	G31.75kl Wheel
quality of total cumulative pitch deviation F_P acc. To DIN 3965-1:1986 (drive side)		-	2	4	1	3	1	4
flank roughness (drive side)	Ra	μm	0.53	0.47	0.66	0.35	0.56	0.34
flank roughness (drive side)	Rz	μm	3.97	3.57	4.68	2.90	3.71	2.61

5.2. Considered Lubricant

For the experimental investigation, the reference oil FVA3 with an additive proportion of 4% Anglamol 99 (FVA3 A for short) was used. It is a mineral-based oil with a kinematic viscosity of $\nu_{40°C} = 95\ mm^2/s$ and $\nu_{100°C} = 10.7\ mm^2/s$. The density is $\rho_{15°C} = 885\ kg/m^3$ [54].

5.3. Experimental Procedure

For the experimental investigations of the gear power losses, the test rig, described in Section 4, was used. The test gear set was lubricated by dip lubrication; the oil level was kept constant on the level of the pinion axle and controlled to an oil temperature of $\vartheta_{Oil} = 90\ °C$. For each experimental investigation, test gear sets with run-in flank surfaces, using the same running-in procedure, were used. For each test run, the load was kept constant, and the circumferential speed was variated in ascending or descending order. After all circumferential speed levels had been investigated, the next load was set, and the procedure was repeated. After completing the test runs under load, the load-independent losses were investigated using no load. To ensure a steady-state operating condition, each test point was held constant for 15 min, though only the last 5 min were taken into account for the evaluation. For the examination of repeat accuracy, at least one repeat test was performed for each test point. The drive direction for all experimental investigations was *pinion drives wheel* on the drive side, except for the investigation of the driving direction where the operation condition *wheel drives pinion* on the drive side was applied.

For the different operating points, the circumferential speed v_t was varied according to Table 4, and the pinion torque T_1 was varied according to Table 5. The varied pinion torque T_1 results in varied flank pressures $\sigma_{H1,2}$, which are calculated using an LTCA [4,55] (see Table 5). The LTCA program [4,55] uses a manufacturing simulation to model the actual tooth, performs a tooth contact analysis using, in this case, the Boundary Elements Method (BEM), and is widely used in both industry and research for the analysis of bevel and hypoid gears.

Table 4. Variation of circumferential speed of operation points.

Circumferential Speed v_t in m/s	Speed $n_{1,2}$ in min^{-1} G0 Pinion	G0 Wheel	G31.75 Pinion	G31.75 Wheel
0.5	249	66	188	50
2	997	264	753	199
4	1994	528	1507	399
6	2991	792	2260	598

Table 5. Variation of pinion torque of operating points.

Pinion Torque T_1 in Nm	Flank Pressure $\sigma_{H1,2}$ in MPa G0	G31.75gr	G31.75kl
100	1445	1339	1577
300	1983	1855	2186
500	2430	2127	2542

6. Results and Discussion

In the following, the measurement results for the torque loss T_V of the experimental investigations on the bevel gear efficiency test rig are presented in detail, whereby the measured values describe the arithmetic mean of the measured data for each load case. As a result, for each torque stage, represented by the pinion torque T_1, the measured torque loss T_V over the circumferential speed v_t is shown. Initial and repeat tests are entered to document the repeat accuracy. The lines between the test points represent the mean values.

6.1. Influence of Axial Offset on the Efficiency of Bevel and Hypoid Gears

Figure 4 shows the measured torque losses T_V of the overall test gearbox for the test gear variant G0 at the oil sump temperature ϑ_{oil} = 90 °C using the oil FVA3 A as a solid line. The entire tooth width of the ring gear is immersed in the oil sump. The results show fundamental trends in the loss behavior of the gearbox. The gearbox loss torque T_V at the lowest load increases slightly with increasing circumferential speed v_t. At the highest load, there is a slight increase in the gearbox loss torque with increasing circumferential speed v_t. The lowest loss torques T_V occur in the no-load operating state. These increase as the load increases. The no-load loss torque T_{V0} exhibits a behavior close to linear over the entire speed range and increases with increasing circumferential speed v_t. At the lowest measured load of T_1 = 100 Nm, the no-load losses represent the decisive share of the total gearbox losses. At the higher loads investigated, the load-related losses T_{VL} make the most significant contribution to the total gearbox losses.

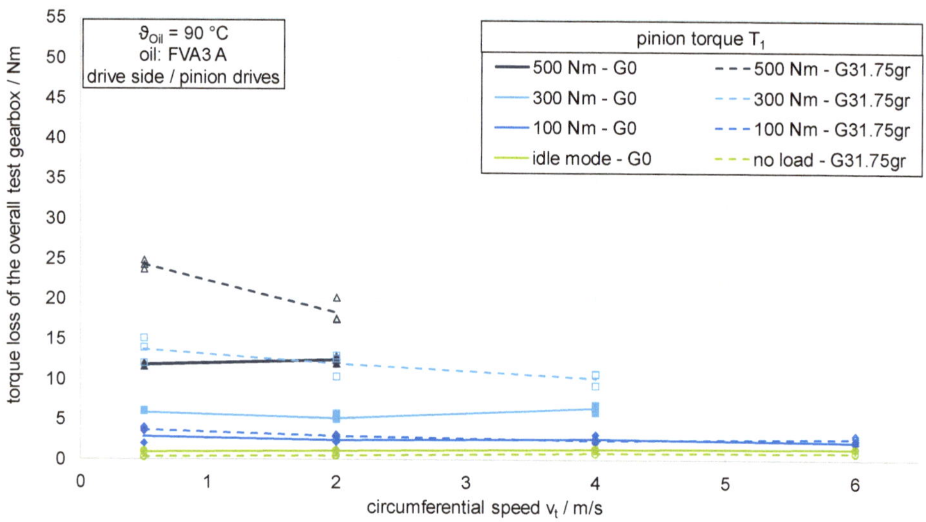

Figure 4. Influence of axial offset on the efficiency of bevel and hypoid gears.

The results for the test gear variant G31.75gr, which shows a hypoid offset of a = 31.75 mm and a similar microgeometry to the G0 variant, are shown as a dashed line. The ring gear is also immersed in the oil sump over its entire tooth width. Compared to the test gear variant G0, the losses at the same load are significantly larger in the higher load stages. In contrast to test gear variant G0, these losses decrease as the circumferential speed v_t increases. At the lowest investigated load. $T_1 = 100$ Nm, the measured losses show a similar level and a decreasing behavior with increasing circumferential speed v_t. The no-load losses of the test gear variant G31.75gr are slightly lower than those of variant G0 and show a slightly increasing curve.

According to the current state of knowledge, the difference in the losses of the two gear variants can be explained by the additional amount of sliding in the longitudinal direction of the teeth in gearing with hypoid offset compared to non-offset bevel gears. This additional amount of sliding means, among other things, that there is no pure rolling on the pitch cone, as is the case with non-offset bevel gears and cylindrical gears, but that sliding in the longitudinal direction of the teeth also occurs in this area. The effect of higher power losses for hypoid gears due to the amount of sliding increases for high torques as the contact pattern enlarges, and therefore, the amount of sliding within the gear mesh increases. This can be seen in Figure 4, where the difference in torque loss of the overall gearbox between the non-offset bevel gear and the hypoid variant increases for higher torques.

6.2. Influence of Microgeometry on the Efficiency of Bevel and Hypoid Gears

Figure 5 shows the measured torque losses T_V of the overall test gearbox for the test gear variants G31.75gr and G31.75kl. The two test gear variants differ in their microgeometry design, whereby the G31.75kl variant is characterized by larger crowning and, therefore, a smaller contact pattern than the G31.75gr variant. The oil sump temperature is 90 °C, and the entire tooth width of the ring gear is immersed in the oil sump. As expected, the measured no-load losses of the two variants match very well. Therefore, the microgeometry of the gearing has no measurable influence on the no-load losses of the gearbox with otherwise identical gear parameters. The curves of the measured loss torque of the test gear variant G31.75kl show smaller amounts compared to the G31.75gr variant, which clearly indicates that this deviation is load-dependent and increases with increasing load. The loss torque curves as a function of the circumferential speed v_t show similar trends and tendencies for both microgeometry variants and decrease with increasing speed.

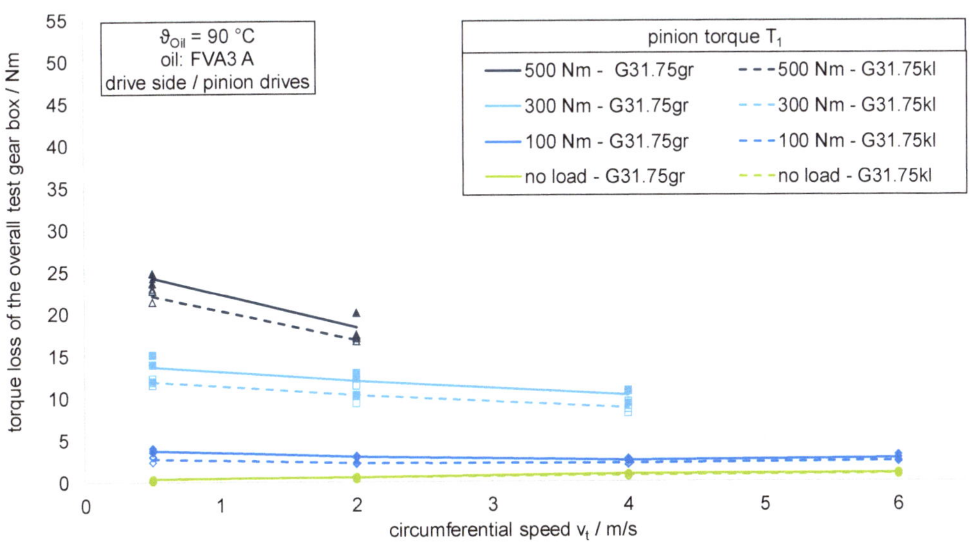

Figure 5. Influence of microgeometry on the efficiency of bevel and hypoid gears.

Due to the different microgeometry designs of the test gear variants G31.75gr and G31.75kl, different crowning and, therefore, different contact patterns result. Since the contact pattern sizes of the two gear variants differ not only in the unloaded state, as shown in Figure 3. In the Ease-Off design and load-free contact pattern of test gear variants (drive side), but also in the loaded state, there is a difference in the flank pressure. On the other hand, areas of the flank are in contact with different sliding velocities. As the efficiency behavior depends on the contact stress and the occurring sliding velocity, the two gear variants show different loss behavior.

6.3. Influence of Driving Direction on the Efficiency of Bevel and Hypoid Gears

Figure 6 compares the measured torque losses T_V of the G31.75gr variant in the operating mode "pinion drives" with the losses in operating mode "wheel drives". The oil sump temperature ϑ_{oil} is 90 °C, and the entire tooth width of the ring gear is immersed in the oil sump. Compared to an operation with a driving pinion, significantly higher losses occur with a driving wheel at all loads and speeds investigated. The curves as a function of the circumferential speed v_t show similar tendencies for both operating conditions and decrease with increasing speed. The no-load losses with a driving wheel are slightly lower than with a driving pinion.

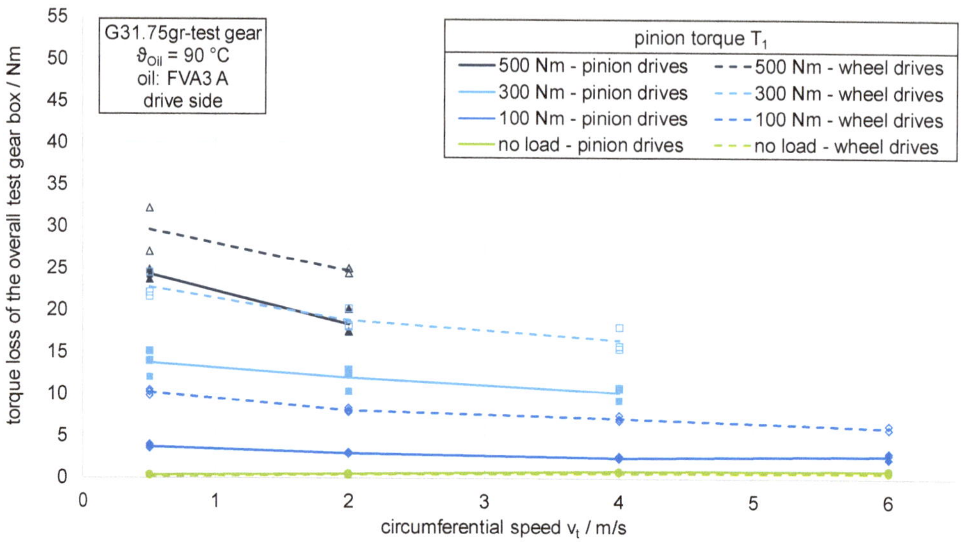

Figure 6. Influence of driving direction on the efficiency of bevel and hypoid gears.

According to Wech [14], the drive direction-dependent loss behavior can be explained by the change in the sliding movement of the tooth flanks. Due to the positive profile shift of the test gears, the addendum of the pinion and wheel are different. When the gear set is operated with a driving wheel, larger areas of the tooth flank are subjected to negative specific sliding than when the pinion is driving. According to Jurkschat [38], there are higher gear friction coefficients in the case of negative specific sliding, which leads to greater load-dependent gear power losses. Michaelis [56] identifies the poorer lubrication conditions at the start of meshing and, therefore, in the areas of highest sliding velocities when the gear is driving, in addition to the larger flank areas subjected to negative specific sliding, as the cause of higher losses with a driving wheel. The changing rotation direction for the different operation conditions may also influence the load-dependent bearing losses of the whole gearbox.

7. Summary and Outlook

The consideration of the efficiency in mechanical transmissions moves into focus due to increasing challenges caused by climate change and its consequences in the present time. Therefore, calculation methods are necessary to consider efficiency in the early stages of the design process of bevel and hypoid gears. For the validation of such calculation methods, highly accurate experimental results are required. Thus, this paper's objective is to present a test rig concept for the experimental investigation of the efficiency of bevel and hypoid gears and, subsequently, the experimental investigation of several influencing factors on the efficiency of bevel and hypoid gear by using the newly developed test rig.

The newly developed bevel gear efficiency test rig allows for the highly accurate measurement of the efficiency of bevel and hypoid gears at different operational conditions with a measuring uncertainty of $\Delta\eta \leq \pm 0.08\%$ according to GUM [48]. With a complementary experimental investigation of the load-dependent bearing losses and the measurement of the idle losses of the gearbox, an accurate value of the load-dependent gear tooth mesh losses can be obtained.

In experimental investigations using the newly developed bevel gear efficiency test rig, the influences of the axial offset, the microgeometry, and the driving direction on the efficiency were evaluated for different operational conditions. In all experimental tests, the measured efficiency of the whole gearbox is within a reasonable section for bevel and hypoid gears between 93.5% and 97.5%. The influence of speed and load on the efficiency of the test gear set could be clearly detected for each of the investigated influencing parameters. The results show that, according to the state of knowledge, the higher amount of sliding due to the axial offset leads to a decrease in the efficiency of the gearbox. Additionally, a larger contact pattern size leads to higher torque losses, shown by the experimental results for a range of input torques and speeds. A further experimental study investigated the influence of the driving direction on the efficiency level of hypoid gears. The results show that, according to the state-of-the-art, the driving condition "wheel drives pinion" leads to higher losses than "pinion drives wheel".

The next step is to evaluate the bearing losses for the respective operational conditions using the test rig concept by Yilmaz [34]. The calculation of the load-dependent gear tooth mesh losses for the bevel and hypoid gears and the use of these results for the validation of calculation methods is subject to further studies.

Author Contributions: Conceptualization, L.C. and J.P.; methodology, L.C. and J.P.; software, J.P.; validation, J.P. and M.W.; investigation, J.P., L.C. and M.W.; writing—original draft preparation, L.C., J.P. and M.W.; writing—review and editing, J.P. and K.S.; visualization, L.C., M.W. and J.P.; supervision, J.P. and K.S.; project administration, J.P.; funding acquisition, K.S. All authors have read and agreed to the published version of the manuscript.

Funding: The research project (IGF No. 19880 N; FVA 821 I) was conducted with the kind support of the FVA (Forschungsvereinigung Antriebstechnik e.V.) research association. The project was sponsored by the German Federal Ministry of Economics and Technology (BMWi) through the AiF (Arbeitsgemeinschaft industrieller Forschungsvereinigungen) in the course of a program for the support of collective industrial research (IGF) as a result of a decision by the German Bundestag.

Data Availability Statement: Data are contained within the article.

Conflicts of Interest: The authors declare no conflicts of interest.

References

1. Pellkofer, J.; Hein, M.; Stahl, K. *IGF-Nr. 19880 N—FVA 821 I: Kegelrad- und Hypoideffizienz: Berechnungsverfahren zur Vorhersage der Lastabhängigen Verzahnungsverluste von Kegelrad- und Hypoidgetrieben Basierend auf Hochgenauen Verlustmomentmessungen*; 2022 Heft 1503; Forschungsvereinigung Antriebstechnik e.V. (FVA): Frankfurt/Main, Germany, 2022.
2. Niemann, G.; Winter, H. *Maschinenelemente: Band 2: Getriebe Allgemein, Zahnradgetriebe—Grundlagen, Stirnradgetriebe. Zweite, Völlig Neubearbeitete Auflage*; Springer: Berlin/Heidelberg, Germany, 2003.
3. Wirth, C.; Thomas, J. Local efficiency grade of bevel and hypoid gears: Determination of gear efficiency by means of loaded TCA. In Proceedings of the International Conference on Gears, Munich, Germany, 5–7 October 2015; VDI Verlag: Düsseldorf, Germany, 2015; pp. 105–114.
4. Mieth, F. *FVA 223 XX CAD2BECAL: Numerische Beanspruchungsberechnung an CAD-Modellen in BECAL*; 2022 Heft 1519; Forschungsvereinigung Antriebstechnik e.V. (FVA): Frankfurt/Main, Germany, 2022.
5. Klein, M. Zur Fresstragfähigkeit von Kegelrad- und Hypoidgetrieben. Ph.D. Thesis, Lehrstuhl für Maschinenelemente, Technische Universität München, München, Germany, 27 February 2012.
6. Grabovic, E.; Artoni, A.; Gabiccini, M.; Ciulli, E. Exploration of trade-offs between NVH and efficiency in bevel gear design. *Forsch Ingenieurwes* **2023**, *87*, 933–947. [CrossRef]
7. Grabovic, E.; Artoni, A.; Gabiccini, M.; Guiggiani, M.; Mattei, L.; Di Puccio, F.; Ciulli, E. Friction-Induced Efficiency Losses and Wear Evolution in Hypoid Gears. *Machines* **2022**, *10*, 748. [CrossRef]
8. Kolivand, M.; Li, S.; Kahraman, A. Prediction of mechanical gear mesh efficiency of hypoid gear pairs. *Mech. Mach. Theory* **2010**, *45*, 1568–1582. [CrossRef]
9. Simon, V. Improvements in the mixed elastohydrodynamic lubrication and in the efficiency of hypoid gears. *Proc. Inst. Mech. Eng. Part J J. Eng. Tribol.* **2020**, *234*, 795–810. [CrossRef]
10. Mohammadpour, M.; Rahnejat, H.; Theodossiades, S. Elastohydrodynamics of Hypoid Gears in Axle Whine Conditions. In *Elastohydrodynamics of Hypoid Gears in Axle Whine Conditions*; SAE International: Warrendale, PA, USA, 2012.
11. Mohammadpour, M.; Theodossiades, S.; Rahnejat, H.; Saunders, T. Non-Newtonian mixed elastohydrodynamics of differential hypoid gears at high loads. *Meccanica* **2014**, *49*, 1115–1138. [CrossRef]

12. Mohammadpour, M.; Theodossiades, S.; Rahnejat, H. Elastohydrodynamic lubrication of hypoid gear pairs at high loads. *Proc. Inst. Mech. Eng. Part J J. Eng. Tribol.* **2012**, *226*, 183–198. [CrossRef]
13. Ding, H.; Li, H.; Chen, S.; Shi, Y.; Wang, Y.; Rong, K.; Lu, R. Energy loss and mechanical efficiency forecasting model for aero-engine bevel gear power transmission. *Int. J. Mech. Sci.* **2022**, *231*, 107569. [CrossRef]
14. Wech, L. Untersuchungen zum Wirkungsgrad von Kegelrad- und Hypoidgetrieben. Ph.D. Thesis, Institut für Konstruktion, Technische Universität München, München, Germany, 7 July 1987.
15. *ISO/TS 10300-20:2021*; Calculation of Load Capacity of Bevel Gears: Part 20: Calculation of Scuffing Load Capacity—Flash Temperature Method. ISO International Organization for Standardization: London, UK, 2021.
16. *ISO/TR 14179-1:2001*; Gears—Thermal Capacity: Part 1: Rating Gear Drives with Thermal Equilibrium at 95 °C Sump Temperature. ISO International Organization for Standardization: London, UK, 2001.
17. Geiger, J. Wirkungsgrad und Wärmehaushalt von Zahnradgetrieben bei Instationären Betriebszuständen. Ph.D. Thesis, Lehrstuhl für Maschinenelemente, Technische Universität München, München, Germany, 2014.
18. Mauz, J. Hydraulische Verluste bei Tauch- und Einspritzschmierung von Zahnradgetrieben. Ph.D. Thesis, Universität Stuttgart, Stuttgart, Germany, 1985.
19. Walter, P. Anwendungsgrenzen für die Tauchschmierung von Zahnradgetrieben. Ph.D. Thesis, Universität Stuttgart, Stuttgart, Germany, 1982.
20. Liu, H.; Link, F.; Lohner, T.; Stahl, K. Computational fluid dynamics simulation of geared transmissions with injection lubrication. *Proc. Inst. Mech. Eng. Part C J. Mech. Eng. Sci.* **2019**, *233*, 7412–7422. [CrossRef]
21. Liu, H.; Jurkschat, T.; Lohner, T.; Stahl, K. Detailed Investigations on the Oil Flow in Dip-Lubricated Gearboxes by the Finite Volume CFD Method. *Lubricants* **2018**, *6*, 47. [CrossRef]
22. Gorla, C.; Concli, F.; Stahl, K.; Höhn, B.R.; Klaus, M.; Schultheiß, H.; Stemplinger, J.P. CFD Simulations of Splash Losses of a Gearbox. *Adv. Tribol.* **2012**, *2012*, 616923. [CrossRef]
23. Concli, F.; Gorla, C.; Stahl, K.; Höhn, B.R.; Michaelis, K.; Schultheiß, H.; Stemplinger, J.P. Load Independent Power Losses of Ordinary Gears: Numerical and Experimental Analysis. In Proceedings of the Politecnico di Torino (Hg.) 2013—5th World Tribology Congress, Torino, Italy, 8–13 September 2013.
24. Seetharaman, S.; Kahraman, A. Load-Independent Spin Power Losses of a Spur Gear Pair: Model Formulation. *J. Tribol.* **2009**, *131*, 022201. [CrossRef]
25. Seetharaman, S.; Kahraman, A.; Moorhead, M.D.; Petry-Johnson, T.T. Oil Churning Power Losses of a Gear Pair: Experiments and Model Validation. *J. Tribol.* **2009**, *131*, 022202. [CrossRef]
26. Jeon, S.I. Improving Efficiency in Drive Lines: An Experimental Study on Churning Losses in Hypoid Axle. Ph.D. Thesis, Imperial College London, London, UK, 2010.
27. Quiban, R.; Changenet, C.; Marchesse, Y.; Ville, F.; Belmonte, J. Churning losses of spiral bevel gears at high rotational speed. *Proc. Inst. Mech. Eng. Part J J. Eng. Tribol.* **2020**, *234*, 172–182. [CrossRef]
28. SKF Gruppe. *SKF Wälzlagerkatalog*; SKF: Gothenburg, Sweden, 1994.
29. SKF Gruppe. *SKF Wälzlagerkatalog*; SKF: Gothenburg, Sweden, 2014.
30. Schaeffler Technologies AG & Co., KG. *Wälzlager—Technische Grundlagen und Produktdaten zur Gestaltung von Wälzlagerungen. Herzogenaurach*; Schaeffler Technologies AG & Co.: Berlin, Germany, 2018.
31. *ISO/TR 14179-2:2001*; Gears—Thermal Capacity: Part 2: Thermal Load-Carrying Capacity. ISO International Organization for Standardization: London, UK, 2001.
32. Wang, D. Berechnung der Wälzlagerreibung Aufgrund Weiterentwickelter Rheologischer Fluidmodelle. Ph.D. Thesis, Universität Hannover, Hannover, Germany, 2015.
33. Schleich, T. Temperatur- und Verlustleistungsverhalten von Wälzlagern in Getrieben. Ph.D. Thesis, Lehrstuhl für Maschinenelemente, Technische Universität München, München, Germany, 2013.
34. Yilmaz, M.; Lohner, T.; Michaelis, K.; Stahl, K. Bearing Power Losses with Water-Containing Gear Fluids. *Lubricants* **2020**, *8*, 5. [CrossRef]
35. Hinterstoißer, M. Zur Optimierung des Wirkungsgrades von Stirnradgetrieben. Ph.D. Thesis, Lehrstuhl für Maschinenelemente, Technische Universität München, München, Germany, 2014.
36. Goebbelet, J.; Mierswa, D.; Weck, M. *Wirkungsgradmessung an Getrieben und Getriebeelementen*; Forschungsvereinigung Automobiltechnik: Berlin, Germany, 1982.
37. Siglmuller, F.; Weinberger, U.; Gotz, J.; Sedlmair, M.; Lohner, T.; Stahl, K. Scaling of planetary gear stages according to gear loss similarity. In Proceedings of the International Conference on Gears, Garching, Monaco, 8–20 September 2019; VDI Verlag: Düsseldorf, Germany, 2019; pp. 115–126.
38. Jurkschat, T.E. Erweiterte Bestimmung Lastabhängiger Verluste von Stirnradgetrieben. Ph.D. Thesis, Technische Universität München, München, Germany, 2020.
39. Handschuh, R.; Kilmain, C. Experimental Study of the Influence of Speed and Load on Thermal Behavior of High-Speed Helical Gear Trains. In Proceedings of the American Helicopter Society, 61st Annual Forum and Technology Display, NASA/TM—2005-213632, Grapevine, TX, USA, 1–3 June 2005.
40. Schweigert, D.; Gerlach, M.E.; Hoffmann, A.; Morhard, B.; Tripps, A.; Lohner, T.; Otto, M.; Ponick, B.; Stahl, K. On the Impact of Maximum Speed on the Power Density of Electromechanical Powertrains. *Vehicles* **2020**, *2*, 365–397. [CrossRef]

41. Homann, J.; Eckstein, L. Kalorimetrisches Verfahren zur Wirkungsgradbestimmung von Getrieben. *ATZ Automob. Z* **2014**, *116*, 68–74. [CrossRef]
42. Pagitsch, M.; Jacobs, G.; Schelenz, R.; Bosse, D.; Liewen, C.; Reisch, S.; Deicke, M. Feasibility of large-scale calorimetric efficiency measurement for wind turbine generator drivetrains. *J. Phys. Conf. Ser.* **2016**, *753*, 72011. [CrossRef]
43. Leighton, M.; Surányi, M. Specialised Gear Rig for the Assessment of Loaded Transmission Error, Line of Action and Summarized Mesh Point. In *Specialised Gear Rig for the Assessment of Loaded Transmission Error, Line of Action and Summarized Mesh Point*; Leighton, M., Surányi, M., Eds.; SAE International: Warrendale, PA, USA, 2023.
44. Strama-MPS Maschinenbau GmbH & Co. KG. Bevel Gear Test Rig TS-30. Available online: https://www.strama-mps.de/en/solutions/test-rigs/bevel-gear-test-rig-1 (accessed on 10 September 2024).
45. Renishaw plc. *RESOLUTE™ Absolute Optical Encoder System: Data Sheet*; L-9518-0013-01-A; Renishaw: Gloucester, UK, 2023.
46. *DIN 51309:2022-08*; Werkstoffprüfmaschinen: Kalibrierung von Drehmomentmessgeräten für Statische Drehmomente. DIN Deutsches Institut für Normung e.V.: Berlin, Germany, 2022.
47. *DIN EN IEC 60751:2023-06*; Industrielle Platin Widerstandsthermometer und Platin Temperatursensoren. DIN Deutsches Institut für Normung e.V.: Berlin, Germany, 2023.
48. *JCGM 100:2008*; Evaluation of Measurement Data—Guide to the Expression of Uncertainty in Measurement. JCGM Joint Committee for Guides in Metrology: Paris, France, 2008.
49. Metrodata GmbH. *GUM Workbench—Benutzerhandbuch für Version 1.3, 2.3 und 2.7. Weil am Rhein*; Metrodata GmbH: Braunschweig, Germany, 2010.
50. Sommer, K.-D.; Siebert, B.R.L. Praxisgerechtes Bestimmen der Messunsicherheit nach GUM (Practical Determination of the Measurement Uncertainty under GUM). *Tm Tech. Mess.* **2004**, *71*, 52–66. [CrossRef]
51. Pellkofer, J.; Boiadjiev, I.; Kadach, D.; Klein, M.; Stahl, K. New calculation method for the scuffing load-carrying capacity of bevel and hypoid gears. *Proc. Inst. Mech. Eng. Part C J. Mech. Eng. Sci.* **2019**, *233*, 7328–7337. [CrossRef]
52. Klingelnberg GmbH. KIMoS: Geometry, Process, and Tool Design for Bevel Gears. Available online: https://klingelnberg.com/en/business-divisions/bevel-gear-technology/software/kimos (accessed on 23 August 2024).
53. *DIN 3965:2023-04*; Toleranzen für Kegelradverzahnungen: Teil 1: Grundlagen. DIN Deutsches Institut für Normung e.V.: Berlin, Germany, 1986.
54. Forschungsvereinigung Antriebstechnik, e.V. *FVA Heft 660 Referenzöle Datensammlung*; Forschungsvereinigung Antriebstechnik e.V. (FVA): Frankfurt/Main, Germany, 2007.
55. Wagner, W.; Schumann, S.; Schlecht, B. Enhanced loaded tooth contact analysis of hypoid gears within a multi-body-system simulation. *Forsch Ingenieurwes* **2022**, *86*, 461–470. [CrossRef]
56. Michaelis, K. Die Integraltemperatur zur Beurteilung der Fresstragfähigkeit von Stirnradgetrieben. Ph.D. Thesis, Lehrstuhl für Maschinenelemente, Technische Universität München, München, Germany, 1987.

machines

MDPI

Article

Effect of Roundness Error of the Grooves on the Inner Ring Runout of Angular Contact Ball Bearings

Di Cui [1], Yongjian Yu [1,2,3,4,*], Yujun Xue [1,2,3,4], Pengge Guo [5], Hongbiao Han [1,3] and Haichao Cai [1]

1 School of Mechanical and Electrical Engineering, Henan University of Science and Technology, Luoyang 471003, China; cuidi4066@stu.haust.edu.cn (D.C.); yjxue@haust.edu.cn (Y.X.); hanhongbiao@haust.edu.cn (H.H.); chc1226@haust.edu.cn (H.C.)
2 Longmen Laboratory, Luoyang 471000, China
3 Henan Key Laboratory for Machinery Design and Transmission System, Henan University of Science and Technology, Luoyang 471003, China
4 Henan Collaborative Innovation Center for High-End Bearings, Luoyang 471003, China
5 Luoyang Bearing Group Company Ltd., Luoyang 471000, China; lycjpk@163.com
* Correspondence: yuyongjian@haust.edu.cn

Abstract: In this paper, a prediction model of the inner ring runout of angular contact ball bearings is established according to the geometric and kinematic relationships of the bearing, considering factors such as the roundness error of the inner and outer grooves, the dimensional error of the balls, and the change of the contact angle between the balls and the grooves. The correctness of the model is verified through experiments. The effects of the order and amplitude of the roundness error of the inner groove and the order and amplitude of the roundness error of the outer groove on the inner ring runout are analyzed. The coupling effect of the roundness error of the inner and outer grooves on the inner ring runout is further analyzed. The results show that the inner ring runout changes periodically with a change to the roundness error order of the grooves, which increases with an increase in the roundness error amplitude. Under the coupling of the roundness error of the inner and outer grooves, the magnification of the inner ring runout increases as a whole. When there are specific relationships between the roundness error orders of the grooves and the number of balls, the magnification of the axial or radial runout changes significantly.

Keywords: angular contact ball bearings; roundness error; rotational accuracy; axial runout; radial runout

1. Introduction

Angular contact ball bearings are widely used in high-precision machine tools, aerospace and other precision machinery fields, due to their good load-carrying ability and high rotational accuracy [1–3]. The rotational accuracy of the bearing affects the working accuracy and reliability of the whole machine [4–6]. However, roundness error will be produced inevitably during the machining process of bearing parts, and the interaction of the roundness error of bearing parts will have an important effect on the rotational accuracy of the bearing. The inner ring runout is an important evaluation index reflecting the rotational accuracy of the bearing, including axial runout and radial runout [7]. Therefore, it is of great significance to study the effect of the roundness error of the bearing parts on the inner ring runout in order to improve the level of design and production of high-precision angular contact ball bearings.

Noguchi, S. et al. [8–10] established a mathematical model of the non-repetitive runout of ball bearings and the factors affecting the non-repeatable runout of bearings were analyzed. The research indicated that the geometrical error of the grooves, the dimensional error of the balls and the number of balls have important effects on the non-repeatable runout. Okamoto, J. [11] developed an apparatus that enabled the setting of the outer ring of the ball bearing in a prescribed profile and they analyzed the loci of the center of the shaft during rotation. The results showed that the loci vary with the relationship between the profile of the raceway and the number of balls. Tada, S. [12] analyzed the effects of the corrugation of the grooves and balls on the non-repeatable runout of the bearings. The research showed that the harmonic order of the grooves is nonlinearly related to the non-repeatable runout of the bearing. Yang, Z. [13,14] et al. studied the effects of the geometrical errors of the bearing parts on the non-repeatable runout of the bearing and the research indicated that the specific relationship between the corrugation of the grooves and the number of the balls significantly affects the non-repeatable runout. Chen G. et al. [15] established a simulation model for the rotational accuracy of cylindrical roller bearings and analyzed the effects of the bearing raceway roundness error and roller diameter error on the rotational accuracy of the bearing. Their results showed that the clearance and runout periodically vary with the increase in the harmonic number of raceways. Ma, F. et al. [16] developed a numerical simulation model for calculating the axis orbit of spherical roller bearings and researched the effects of single off-sized roller and multiple off-sized rollers on the axis orbit of the bearing, The research indicates that the diameter error of the rollers largely affects the radial runout of the inner ring. Yu, Y. et al. [17–19] proposed the prediction methods of the radial runout of the inner and outer rings of cylindrical roller bearings, considering the dimension and form errors of the bearing parts and the change of contact positions. The research indicated that the roundness error of the inner raceway and outer raceway and the motion error of rollers have great effects on the rotational accuracy of the bearing. Zha, J. et al. [20] studied the relationship between the elliptical form error and rotation accuracy for hydrostatic journal bearings. The results revealed that the effect of the shaft elliptical form error on rotation accuracy is six times larger than the bearing bush. Hu, G. et al. [21] investigated the formative mechanism of dynamic rotational error. Their research indicated that the SRB system produced an obvious asynchronous error motion when the difference between the order and the number of rolling elements is one. Zhang, X. et al. [22] analyzed the rotational accuracy and its influencing factors of angular contact ball bearings at high speed. The analysis results showed that the raceway curvature coefficient and the number of balls are positively correlated with the bearing rotation accuracy. Zhang, P. et al. [23] studied the effect of the roundness error of the raceway on the error motion of bearings by considering nonlinear roller force. The research showed that the wave number of the roundness error of the raceway has a major effect on the error motion.

Scholars have achieved a large number of research results; however, the research on the inner ring runout of angular contact ball bearings still needs to be improved. The existing models only considered the roundness error of a single groove and did not consider the factor of the change of the contact angle, which could only analyze the effect of the roundness error of a single groove on the bearing runout. However, the coupling effect of the roundness error of the inner and outer grooves, and the change of the contact angle due to the roundness error of the grooves and the dimensional error of the balls, among other factors, have significant effects on the runout of the bearing. In this paper, a prediction model of the inner ring runout of the angular contact ball bearings is established according to the geometric and kinematic relationships of the bearing, considering factors such as the roundness error of the inner and outer grooves, the dimensional error of the balls and the change of the contact angle between the balls and the grooves. The model realizes the prediction of the axial and radial runout of the inner ring with the known structural parameters of the bearing, the roundness error of the grooves, and the dimensional error of the balls. The correctness of the model is verified through experiments. The effects of the order and amplitude of the roundness error of the inner groove and the order and amplitude of the roundness error of the outer groove on the inner ring runout are analyzed. The coupling effect of the roundness error of the inner and outer grooves on the inner ring runout is further analyzed. These provide theoretical support for the design of high-precision angular contact ball bearings.

2. The Prediction Model of the Inner Ring Runout of the Bearing

This prediction model considers the roundness error of the inner and outer grooves, the dimensional error of the balls, and the change in the contact angle between the balls and the grooves. It also realizes the prediction of the axial and radial runout of the inner ring when the structural parameters of the bearing, the roundness error of the grooves, and the dimensional error of the balls are known.

The processes of this model are as follows: (1) the inner ring is rotated according to the set step, and the balls are rotated to the new position; (2) according to the roundness error of the grooves, the equations of the contour curves of the grooves are obtained; (3) the contact angles between the balls and the grooves are changed, due to the existence of the roundness error of the grooves and the dimensional error of the balls. The coordinates of the expected contact points on the grooves are calculated at this point; (4) all balls are moved in the direction of the contact angle until they are in contact with the outer groove, then the coordinates of the center of the balls are calculated; and (5) the inner ring is given different positions to find the optimal stable position and obtain the optimal center coordinate of the inner ring. The displacement of the center of the inner ring is calculated according to the coordinates of the optimal center of the inner ring. The inner ring is rotated one step further and the above processes are repeated. After the inner ring has been rotated for several weeks, the difference between the maximum and minimum displacement of the inner ring in the axial or radial direction is the axial or radial runout of the inner ring of the bearing.

2.1. The Geometric and Kinematic Relationships of the Bearing

When the inner ring of the bearing is rotated counterclockwise at an angle α_i, the balls rotate clockwise around their axes and are also rotated around the axis of the bearing [24]. The rotation angle, orbital angle, and position angle of the balls are as follows:

$$\omega_b = \frac{d_m^2 - D_w^2 \cos^2 \alpha}{2 d_m D_w} \alpha_i \tag{1}$$

$$\omega_m = 0.5\alpha_i (1 - D_w \cos \alpha / d_m) \tag{2}$$

$$\beta_j = 2\pi(j-1)/Z + \omega_m \tag{3}$$

where ω_b, ω_m are the angle of rotation and revolution of the balls, d_m is the diameter of the bearing pitch circle, D_w is the diameter of the balls, α is the contact angle, Z is the number of balls, β_j is the position angle of the jth ball, and where j is the serial number of the ball (j = 1, 2, . . ., Z).

The overall coordinate system of the bearing, XYZ, is fixed at the center of the outer ring. The local coordinate system $Y_{bj}O_{bj}Z_{bj}$ is fixed to the center of rotation of the outer ring, the Y_{bj} axis passes through the center of rotation of the jth ball, as shown in Figure 1.

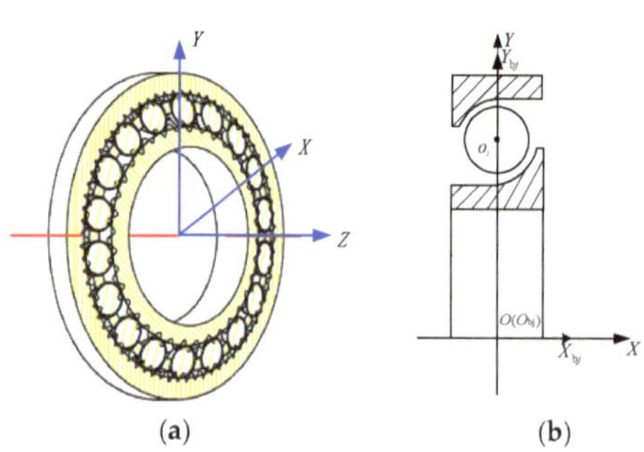

Figure 1. Bearing coordinate system: (**a**) overall coordinate system; (**b**) localized coordinate system.

2.2. Equations of the Contour Curves of the Grooves

Fourier series is composed of a series of trigonometric functions and it is often used to characterize complex surface contours [25,26]. Here, the Fourier series is used to characterize the contour curves of the grooves, as shown in Equation (4).

$$\Delta S(\theta) = \sum_{m=2}^{\infty} C_m \cos(m\theta + \psi_m) \tag{4}$$

where ΔS is roundness error of the grooves, m is the roundness error order, which characterizes the contour shape of the bearing parts, C_m is the amplitude of the mth-order roundness error, which characterizes the peak value of the deviation of the contour of the bearing parts from the ideal circle, ψ_m is the initial phase angle of the mth-order roundness error, and θ is the position angle.

The roundness error $\Delta S_{\mathrm{i}}(\theta_{\mathrm{i}})$ of the inner groove and $\Delta S_{\mathrm{e}}(\theta_{\mathrm{e}})$ of the outer groove are expressed as:

$$\begin{cases} \Delta S_{\mathrm{i}}(\theta_{\mathrm{i}}) = \sum_{m_{\mathrm{i}}=2}^{\infty} C_{\mathrm{i}m_{\mathrm{i}}} \cos(m_{\mathrm{i}}(\theta_{\mathrm{i}} - \alpha_{\mathrm{i}}) + \psi_{\mathrm{i}m_{\mathrm{i}}}) \\ \Delta S_{\mathrm{e}}(\theta_{\mathrm{e}}) = \sum_{m_{\mathrm{e}}=2}^{\infty} C_{\mathrm{e}m_{\mathrm{e}}} \cos(m_{\mathrm{e}}\theta_{\mathrm{e}} + \psi_{\mathrm{e}m_{\mathrm{e}}}) \end{cases} \tag{5}$$

where θ_{i} is the position angle of any point on the inner groove, θ_{e} is the position angle of any point on the outer groove, m_i and m_e are the roundness error orders of the inner and outer grooves, respectively, $C_{\mathrm{i}m_{\mathrm{i}}}$ and $C_{\mathrm{e}m_{\mathrm{e}}}$ are the amplitudes of the m_{i}th-order and m_{e}th-order roundness error, respectively, $\psi_{\mathrm{i}m_{\mathrm{i}}}$ and $\psi_{\mathrm{e}m_{\mathrm{e}}}$ are the phase angles of the m_{i}th-order and m_{e}th-order roundness error, respectively.

After considering the roundness error, the radius of curvature of the inner groove contour $r_{\mathrm{i}}(\theta_{\mathrm{i}})$, the radius of curvature of the outer groove contour $r_{\mathrm{e}}(\theta_{\mathrm{e}})$, as well as the diameters of the bottom of the inner groove contour $d_{\mathrm{i}}(\theta_{\mathrm{i}})$, and the bottom of the outer groove contour $d_{\mathrm{e}}(\theta_{\mathrm{e}})$ are expressed as:

$$r_{\mathrm{i}}(\theta_{\mathrm{i}}) = r_{\mathrm{i}} + \Delta r_{\mathrm{i}} - \Delta S_{\mathrm{i}}(\theta_{\mathrm{i}}) \tag{6}$$

$$r_{\mathrm{e}}(\theta_{\mathrm{e}}) = r_{\mathrm{e}} + \Delta r_{\mathrm{e}} + \Delta S_{\mathrm{e}}(\theta_{\mathrm{e}}) \tag{7}$$

$$d_{\mathrm{i}}(\theta_{\mathrm{i}}) = d_{\mathrm{i}} + \Delta d_{\mathrm{i}} + \Delta S_{\mathrm{i}}(\theta_{\mathrm{i}}) + \Delta S_{\mathrm{i}}(\theta_{\mathrm{i}} + \pi) \tag{8}$$

$$d_{\mathrm{e}}(\theta_{\mathrm{e}}) = d_{\mathrm{e}} + \Delta d_{\mathrm{e}} + \Delta S_{\mathrm{e}}(\theta_{\mathrm{e}}) + \Delta S_{\mathrm{e}}(\theta_{\mathrm{e}} + \pi) \tag{9}$$

where r_{i} and r_{e} are the radii of the curvature of the inner and outer grooves, respectively, Δr_{i} and Δr_{e} are the dimensional errors of the radii of the curvature of the inner and outer grooves, respectively, and d_{i}, d_{e}, Δd_{i} and Δd_{e} are the diameters of the bottom of the inner and outer grooves and their dimensional errors, respectively.

2.3. Coordinates of Contact Points on Grooves

The contact points on the grooves and the contact angles are changed due to the existence of the roundness error of the grooves and the dimensional error of the balls. The jth ball will be in contact with point C_j on the inner groove and point D_j on the outer groove if the jth ball can be in contact with both the inner and outer grooves. In the bearing, the positions of the expected contact points on the grooves and the jth ball in the axial plane are as shown in Figure 2. The position of the jth ball in the radial plane is as shown in Figure 3. The three-dimensional coordinates of C_j and D_j are expressed by the following derivation.

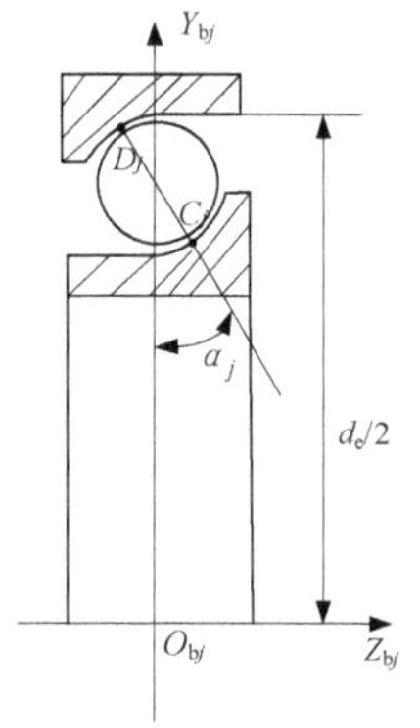

Figure 2. Position of the *j*th ball in the axial plane.

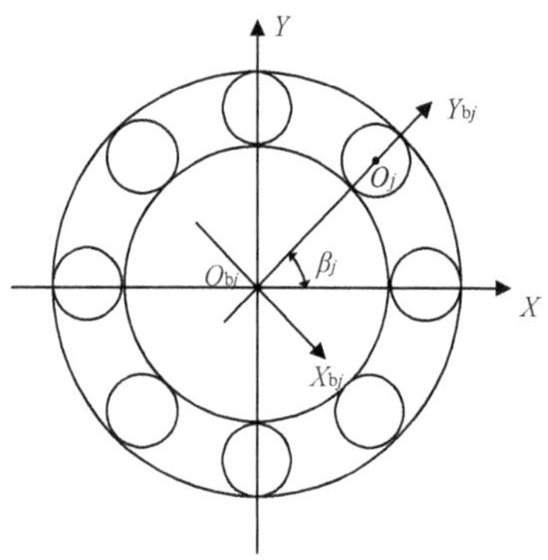

Figure 3. Position of the *j*th ball in radial plane.

The coordinates of the center of curvature o_i of the inner groove in the $Y_{\mathrm{b}j}O_{\mathrm{b}j}Z_{\mathrm{b}j}$ are:

$$\begin{cases} Y_{o_\mathrm{i}} = d_\mathrm{i}/2 + r_\mathrm{i} \\ Z_{o_\mathrm{i}} = 0 \end{cases} \tag{10}$$

Similarly, the coordinates of the center of curvature o_e of the outer groove in the $Y_{\mathrm{b}j}O_{\mathrm{b}j}Z_{\mathrm{b}j}$ are:

$$\begin{cases} Y_{o_\mathrm{e}} = d_\mathrm{e}/2 - r_\mathrm{e} \\ Z_{o_\mathrm{e}} = 0 \end{cases} \tag{11}$$

The contact angles are changed due to the existence of the roundness error of the grooves and the dimensional error of the balls. The contact angle of the *j*th ball is calculated by Equation (12).

$$\alpha_j = \arccos\left[1 - \frac{d_\mathrm{e}(\beta_j) - d_\mathrm{i}(\beta_j) - 2(D_\mathrm{w} + \Delta D_{\mathrm{r}j})}{2(r_\mathrm{i}(\beta_j) + r_\mathrm{e}(\beta_j) - (D_\mathrm{w} + \Delta D_{\mathrm{r}j}))}\right] \tag{12}$$

where $\Delta D_{\mathrm{r}j}$ is the dimensional error of the *j*th ball.

The distance from point D_j on the outer groove to the center of curvature o_e of the outer groove is equal to the sum of the outer groove radius of curvature and the roundness error at that point. The coordinates of point D_j in the $Y_{\mathrm{b}j}O_{\mathrm{b}j}Z_{\mathrm{b}j}$ plane are as follows:

$$\begin{cases} Y_{D_j} = Y_{o_\mathrm{e}} + r_\mathrm{e}(\beta_j)\cos\alpha_j \\ Z_{D_j} = Z_{o_\mathrm{e}} - r_\mathrm{e}(\beta_j)\sin\alpha_j \end{cases} \tag{13}$$

On the overall coordinate system XYZ, the three-dimensional coordinates of point D_j on the outer groove can be obtained:

$$\begin{cases} X_{D_j} = (Y_{o_e} + r_e(\beta_j)\cos\alpha_j)\cos\beta_j \\ Y_{D_j} = (Y_{o_e} + r_e(\beta_j)\cos\alpha_j)\sin\beta_j \\ Z_{D_j} = Z_{o_e} - r_e(\beta_j)\sin\alpha_j \end{cases} \quad (14)$$

Similarly, the C_j point on the inner groove can be expressed as:

$$\begin{cases} X_{C_j} = (Y_{o_i} - r_i(\beta_j)\cos\alpha_j)\cos\beta_j \\ Y_{C_j} = (Y_{o_i} - r_i(\beta_j)\cos\alpha_j)\sin\beta_j \\ Z_{C_j} = Z_{o_i} + r_i(\beta_j)\sin\alpha_j \end{cases} \quad (15)$$

2.4. Coordinates of the Center of the Balls

When the inner ring is rotated at a certain angle, all the balls are rotated to a new position and then moved to the outer groove until the balls contact the outer groove (as shown in Figure 4). Through the geometrical relationship of the bearing, the center coordinates of the jth ball in contact with the outer groove are obtained.

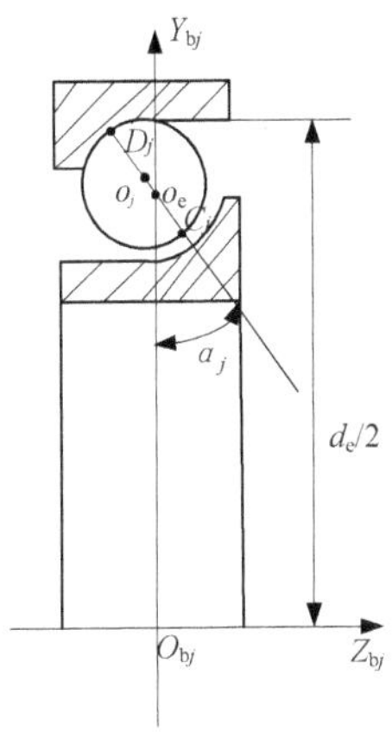

Figure 4. Coordinates of the center of the ball in contact with the outer groove.

When the inner ring is rotated by an angle α_i, the balls are rotated to a certain position. The jth ball is in contact with the outer groove point D_j. The center o_j of the ball and the curvature center o_e of the outer groove are both on the contact normal $D_j o_e$. At this time, the distance between the two center points is calculated using Equation (16).

$$\overline{o_e o_j} = \overline{D_j o_e} - \overline{D_j o_j} \quad (16)$$

where $\overline{D_j o_j}$ is the radius of the jth ball contour at this time, calculated using Equation (17). $\overline{D_j o_e}$ is the radius of curvature at point D_j on the outer groove contour at this time.

$$\overline{D_j o_j} = D_w/2 + \Delta D_{rj}/2 \quad (17)$$

$$\overline{D_j o_e} = r_e(\beta_j) \quad (18)$$

When the jth ball is in contact with the outer groove, the distance from the center of the ball to the curvature center of the outer groove is known. The coordinates of the center of the jth ball in the local coordinate system $Y_{bj}O_{bj}Z_{bj}$ are derived, according to the geometrical relationship (Equation (19)).

$$\begin{cases} Z_{o_j} = Z_{o_e} - \overline{o_j o_e} \sin \alpha_j \\ Y_{o_j} = Y_{o_e} + \overline{o_j o_e} \cos \alpha_j \end{cases} \quad (19)$$

The coordinates of the center of the ball in the global coordinate system XYZ are deduced:

$$\begin{cases} X_{o_j} = (Y_{o_e} + \overline{o_j o_e} \cos \alpha_j) \cos \beta_j \\ Y_{o_j} = (Y_{o_e} + \overline{o_j o_e} \cos \alpha_j) \sin \beta_j \\ Z_{o_j} = Z_{o_e} - \overline{o_j o_e} \sin \alpha_j \end{cases} \quad (20)$$

2.5. Coordinates of the Rotation Center of the Inner Ring

In order to find the coordinates of the optimal rotation center of the inner ring, the inner ring needs to be translated in the direction of the ball, according to the set step size during the rotation process. Some balls will not be in contact with the inner groove due to the existence of the roundness error of the grooves and the dimensional error of the balls. At this time, when at any given position of the inner ring, there are three possible contact states between the inner groove and each ball: contact, separation, or interference. The contact state between the ball and the inner groove is determined by calculating the shortest distance between the surface of the ball and contact point C_j of the inner groove.

The inner ring is displaced in the X, Y, and Z directions and the coordinates of contact point C_j on the inner groove are changed to:

$$\begin{cases} X'_{C_j} = (Y_{o_i} - r_i(\beta_j) \cos \alpha_j) \cos \beta_j + \Delta x \\ Y'_{C_j} = (Y_{o_i} - r_i(\beta_j) \cos \alpha_j) \sin \beta_j + \Delta y \\ Z'_{C_j} = Z_{o_i} + r_i(\beta_j) \sin \beta_j + \Delta z \end{cases} \quad (21)$$

The coordinates of center O_{bj} of the jth ball are known and the distance between the center of the ball and the contact point C_j on the inner groove can be obtained. The difference between this distance and the radius of the ball contour is the distance between the surface of the ball and the contact point C_j on the inner groove.

$$L = \sqrt{(X_{o_j} - X'_{C_j})^2 + (Y_{o_j} - Y'_{C_j})^2 + (Z_{o_j} - Z'_{C_j})^2} - (D_w + \Delta D_{rj})/2 \quad (22)$$

ε is set to the allowable interference error. When $|L| < \varepsilon$, the inner groove is in contact with the ball. When $|L| > \varepsilon$, the inner groove is separated from the ball. When $|L| < -\varepsilon$, the inner groove interferes with the ball.

The contact state of the inner groove can be obtained under several inner ring positions and the stable position of the inner ring is determined from the criteria of the stable contact state of the inner ring. The stable position of the inner ring should meet the following criteria:

1. None of the balls interfered with the inner groove. When one ball interferes with the inner ring groove, it indicates that the position of the inner ring is not the stable position of the inner ring;
2. The number of balls that contacted the inner groove is not less than three. Owing to the point contact between the ball and the inner ring groove, the number of balls contacted in the inner groove must be more than or equal to three when the inner groove is in a stable state. Otherwise, the force on the inner ring will not be able to keep the inner ring stable.
3. The balls that made contact with the inner groove were distributed in at least three different quadrants or two symmetrical quadrants. When the contact balls are distributed in three different quadrants, each ball that made contact with the inner groove is equivalent to a fulcrum; three fulcrums located in three quadrants cause the inner

ring to be in a stable state. When the contact ball is located in two quadrants, these quadrants must be symmetrical, i.e., either quadrants one and three or quadrants two and four.

Through the above criteria, the coordinates of the rotation center of the inner ring that satisfy the force balance constraints and geometric constraints are obtained. When there are multiple inner ring positions that satisfy the criteria, the average value will be used as the center coordinates of the inner ring.

3. Experimental Validation of the Prediction Model of the Inner Ring Runout of the Bearing

In order to verify the correctness of the above prediction model, five sets of ZYS-B7008C/P4 bearings were selected to carry out the measurement of the bearing parts and the measurement of the inner ring runout at the National Bearing Quality Inspection Center. The contour and roundness of the grooves, as well as the axial and radial runout of the inner ring, were measured.

3.1. Measurement of the Contours and Roundness of the Grooves

The contours of the grooves were measured with the Talyrond Model 51 roundness gauge, as shown in Figure 5. During the measurement, the ring was fixed to the table and the measuring head was mounted on the rotating spindle and was in contact with the surface of the groove. The measuring head was driven by the rotating spindle for a week, and 1024 points on each groove were measured. The raw data of the measured contours were stored.

Figure 5. Roundness gauge to measure the grooves.

The raw data were subjected to spectral analysis to obtain the harmonic orders, amplitudes, and phase angles on the inner and outer grooves of the bearing. The harmonic components with the higher amplitudes were isolated for further investigation. Tables 1 and 2 show the data for the middle cross-section of the measured B7008C-1 bearing, for example. The harmonic orders and their corresponding amplitudes and phase angles were substituted into Equation (4) to construct the equations of the contour curves of the grooves. The greater the selected harmonic components, the closer the fitted contour curves are to the actual contour.

Table 1. Harmonic components on the inner groove.

Orders	0	1	2	3	4	5	6	7	8	9	10
Amplitudes/μm	0.378	0.555	0.144	0.034	0.023	0.010	0.014	0.014	0.009	0.003	0.001
Phase/rad	0	−0.463	0.491	2.008	−0.582	0.189	−1.376	3.070	3.019	−0.871	−0.817

Table 2. Harmonic components on the outer groove.

Orders	0	1	2	3	4	5	6	7	8	9	10
Amplitudes/µm	0.708	1.688	0.251	0.037	0.030	0.009	0.011	0.004	0.004	0.010	0.002
Phase/rad	3.142	0.689	0.364	2.008	−2.075	−2.087	0.351	−2.480	−1.646	−1.727	−1.979

A comparison of the fitted contour curves of the grooves with the original contour curves of the grooves is given in Figure 6, where it can be seen that the fitted contour curves of the grooves reflect the true contour curves well.

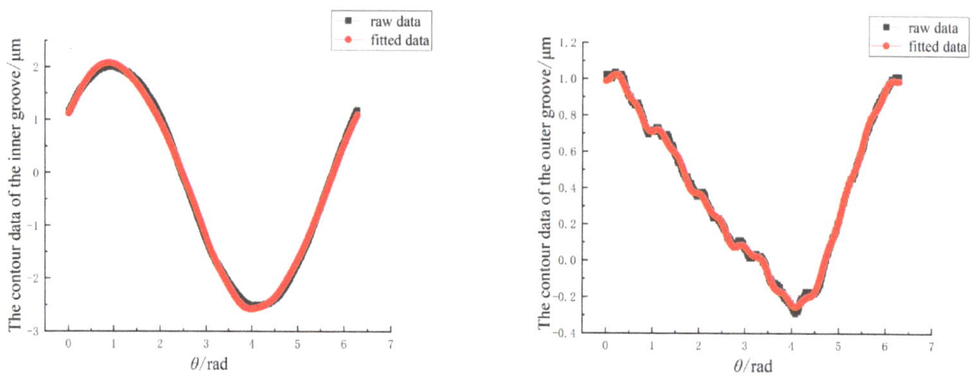

(**a**) (**b**)

Figure 6. Comparison of fitted contour data and raw contour data: (**a**) the inner groove; (**b**) the outer groove.

The roundness values of the measured cross-sections were determined by the difference between the maximum and minimum radius of the measured contours of the grooves, derived by the least squares method. The measured data are shown in Table 3. The roundness value of the outer groove is in the range of 0.28 µm~0.57 µm, and the variation of the roundness value at three positions of the outer groove is in the range of 0.02 µm~0.18 µm. The roundness value of the inner groove is in the range of 0.15 µm~0.57 µm, and the variation of the roundness value at three positions of the inner groove is in the range of 0.03 µm~0.07 µm.

Table 3. Roundness value of B7008C/P4 bearings.

Bearing Number	B7008C-1			B7008C-2			B7008C-3			B7008C-4			B7008C-5		
Radial cross-sections	1	2	3	1	2	3	1	2	3	1	2	3	1	2	3
Roundness value of the outer groove/µm	0.4	0.35	0.53	0.37	0.33	0.44	0.35	0.28	0.39	0.54	0.55	0.53	0.56	0.42	0.57
Roundness value of the inner groove/µm	0.53	0.57	0.56	0.36	0.3	0.3	0.22	0.15	0.16	0.21	0.15	0.2	0.18	0.18	0.15

3.2. Measurement of the Inner Ring Runout and Validation of the Prediction Model

Measurement of the inner ring runout of the bearing was carried out on the assembled bearings using the B024 bearing inner ring runout gauge, as shown in Figure 7. The axial and radial runouts of the inner ring of the measured bearings were measured. The measurement principle is shown in Figure 8. During measurement, the outer ring of the bearing is fixed and the small axial load *a* is applied to the inner ring. The two measuring heads of the instrument are in contact with the inner circular surface and the end face of the inner ring. The inner ring is rotated for several weeks and the axial and radial runouts of the inner ring are determined according to the maximum and minimum values on the micrometers of the gauge.

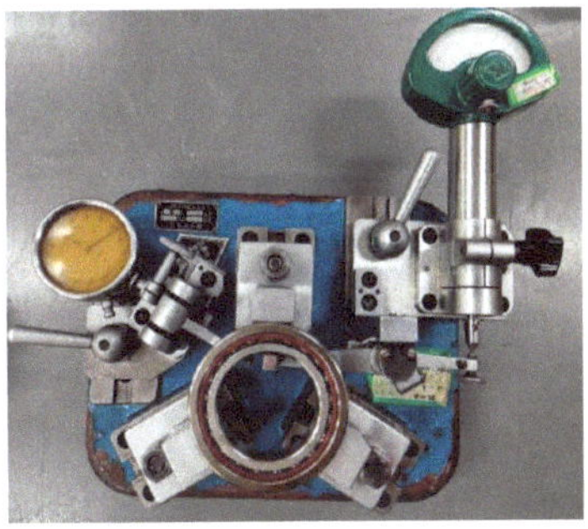

Figure 7. B024 bearing inner ring runout gauge.

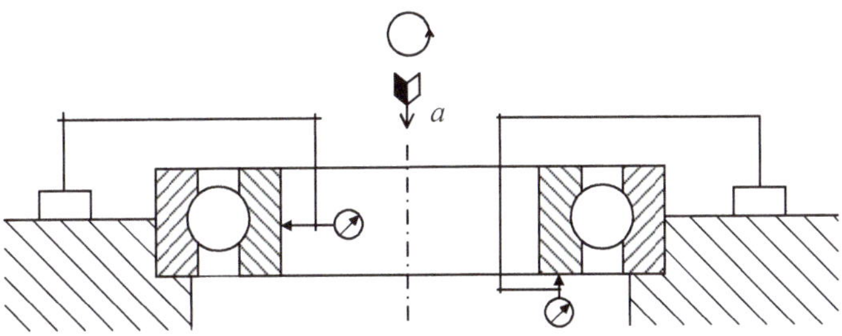

Figure 8. Measurement principle of the inner ring runout.

The structural parameters of the bearings as well as the dimensional differences of the balls were also measured, as shown in Table 4.

Table 4. Main parameters of B7008C bearing.

Parameters	Numerical Value
The bottom diameter of the inner groove d_i/mm	46.964
The bottom diameter of the outer groove d_e/mm	61.019
The radius of curvature of the inner groove r_i/mm	3.99
The radius of curvature of the outer groove r_e/mm	3.78
Original contact angle α	15
Diameter of the balls D_w/mm	7.001 (±0.0002)
Number of balls Z	18

The structural parameters and the fitted equations of the contour curves of the grooves were substituted into the prediction model and the predicted values of the inner ring runout of the measured bearings were obtained. Taking the measured B7008C-1 bearing as an example, in the model, the inner ring is rotated 1° per step and rotated for one week. The minimum displacement of the center of the inner ring in the axial direction is obtained as 0.20205 mm and the maximum displacement is 0.20405 mm, meaning that the predicted value of the axial runout is 2.00 μm. In the radial direction, the minimum displacement of the center of the inner ring is 0.00072 mm and the maximum displacement is 0.00184 mm; therefore, the predicted value of the radial runout is 1.12 μm.

The measurement results and the model prediction results are shown in Table 5. From the tabular data, it can be seen that the prediction results were close to the measurement results, which verified the correctness of the prediction model established in this paper.

Table 5. Measured and predicted results of B7008C bearing.

The Inner Ring Runout/μm	B7008C-1		B7008C-2		B7008C-3		B7008C-4		B7008C-5	
	Measured Value	Predicted Value	Measured Value	Predicted Value	Measured Value	Predicted Value	Measured Value	Predicted Value	Measured Value	Predicted Value
K_{ia}	1	1.12	1	0.96	1	0.88	1	1.20	1	1.20
S_{ia}	2	2.00	2	1.95	2	1.75	3	3.30	2	1.92

4. Results and Analysis

4.1. Effect of the Roundness Error Order of the Inner Groove on the Inner Ring Runout

Figure 9 shows the effect of the roundness error order of the inner groove on the inner ring runout. In the figure, the number of balls Z is 18 and the roundness error amplitudes of the inner groove are 0.15 μm, 0.2 μm, and 0.25 μm. From the figure, it can be seen that the axial runout and the radial runout of the inner ring show periodic changes with changes to the roundness error order of the inner groove; the period is Z. For the axial runout, when the roundness error order is nZ (n is a positive integer), the axial runout increases significantly and reaches the maximum value. The ratio of the runout value to the roundness error amplitude value (referred to as the magnification) is between 7.33 and 7.6. When the roundness error order is $nZ/3$ (n is a positive integer and is not a multiple of 3), the axial runout reaches a peak value, and the magnification is between 2.32 and 2.47. When the roundness error order is $nZ/2$ (n is an odd positive integer), the axial runout reaches a larger peak value and the magnification is between 4.2 and 4.33. In the remaining cases, the magnification of the axial runout is between 0.28 and 1.14.

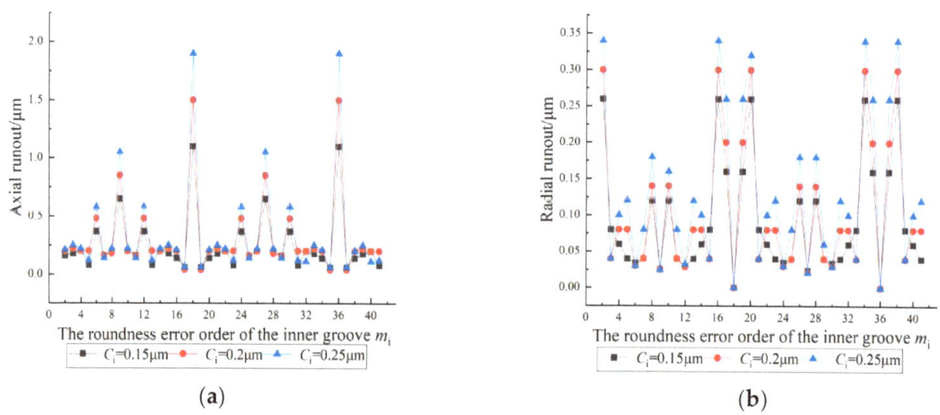

Figure 9. Effect of the roundness error order of the inner groove on the inner ring runout: (**a**) axial runout; (**b**) radial runout.

For the radial runout, when the roundness error order is ($nZ/2 \pm 1$) (n is a positive integer), the radial runout reaches a peak value: the magnification is between 0.6 and 0.88 when n is an odd positive integer, and between 1 and 1.07 when n is an even positive integer. When the roundness error order is ($nZ \pm 2$) (n is a positive integer), the radial runout reaches the maximum value and the magnification is between 1.36 and 1.73. When the roundness error order is $nZ/3$ (n is a positive integer other than a multiple of 3) or $nZ/2$ (n is an odd positive integer), the radial runout decreases to a low value. When the roundness error order is nZ (n is a positive integer), the radial runout is 0 μm and achieves the minimum value. In the rest of the order cases, the magnification of the radial runout is between 0 and 0.53.

After considering the effect of the roundness error order of the inner groove on the inner ring runout, it was found that when the roundness error order of the inner ring has the specific relationship with the number of balls, the magnification of the axial runout or radial runout of the inner ring of the bearing changed significantly. Therefore, the roundness error order of the inner groove should be distributed in reasonable intervals by utilizing the harmonic control theory. The inner ring runout can be effectively reduced by avoiding these specific roundness error orders.

4.2. Effect of the Roundness Error Amplitude of the Inner Groove on the Inner Ring Runout

Figure 10 shows the effect of the roundness error amplitude of the inner groove on the inner ring runout. In the figure, the number of balls Z is 18, the roundness error orders of the inner groove are 4, 5, 6, 7, 8, 9, 10, 11, 12, 18, 24, 27, 30, and 36, and the roundness error amplitude of the inner groove varies in the range of 0.5 to 1.5 μm. As shown in the figure, when the roundness error orders are 4, 5, 7, 8, 10, and 11, the inner ring runout increases with an increase in the roundness error amplitudes of the inner groove. The magnifications of the axial runout are distributed between 0.3 and 0.5 and the magnifications of the radial runout are distributed between 0.21 and 0.5. When the roundness error orders are $nZ/3$ (n is a positive integer other than a multiple of 3), $nZ/2$ (n is an odd positive integer), and nZ (n is a positive integer), the axial runout increases significantly with an increase in the roundness error amplitudes of the inner groove. The magnifications range from 1.94 to 2.1, 3.83 to 4.0, and 7.53 to 7.6, respectively. The radial runout does not change with an increase in the roundness error amplitude of the inner groove and the magnifications of the radial runout fluctuate only slightly between 0 and 0.05 when the roundness error order is $nZ/3$ or $nZ/2$. There is no radial runout when the roundness error order is nZ.

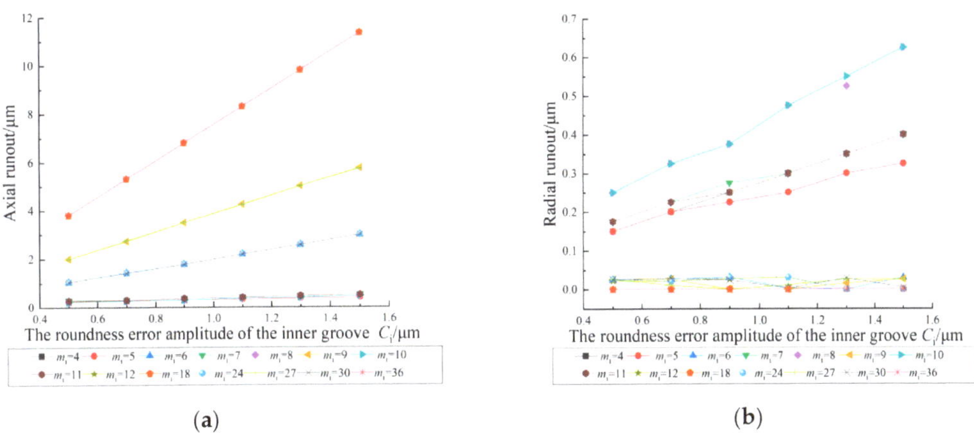

Figure 10. Effect of the roundness error amplitude of the inner groove on the inner ring runout: (**a**) axial runout; (**b**) radial runout.

These results show that the effect of the roundness error amplitude of the inner groove on the inner ring runout is significant. This is because the increase in the roundness error amplitude increases the distance between the peaks and valleys of the groove, the center of the inner ring displacement is increased, and the runout of the inner ring is obvious. Therefore, the inner groove runout can be reduced by appropriately reducing the roundness error amplitude of the inner groove. However, when the roundness error orders are $nZ/3$ (n is a positive integer other than a multiple of 3), $nZ/2$ (n is an odd positive integer), and

nZ (n is a positive integer), the radial runout is relatively stable and not easily affected by the roundness error amplitude of the inner groove.

4.3. Effect of the Roundness Error Order of the Outer Groove on the Inner Ring Runout

Figure 11 shows the effect of the roundness error order of the outer groove on the inner ring runout. In the figure, the number of balls Z is 18 and the roundness error amplitudes of the outer groove are 0.15 μm, 0.2 μm, and 0.25 μm. From the figure, it can be seen that the roundness error order of the outer groove and the roundness error order of the inner groove have the same effect on the inner ring runout. When the roundness error orders are $nZ/3$ (n is a positive integer other than a multiple of 3), $nZ/2$ (n is an odd positive integer), and nZ (n is a positive integer) in these specific orders, the magnifications of the axial runout increase significantly. When the roundness error orders are $(nZ/2 \pm 1)$, $(nZ \pm 2)$ (n is a positive integer) in these specific orders, the magnifications of the radial runout increase significantly. Therefore, the roundness error order of the outer groove should also be controlled in reasonable intervals to reduce the inner ring runout of the bearing.

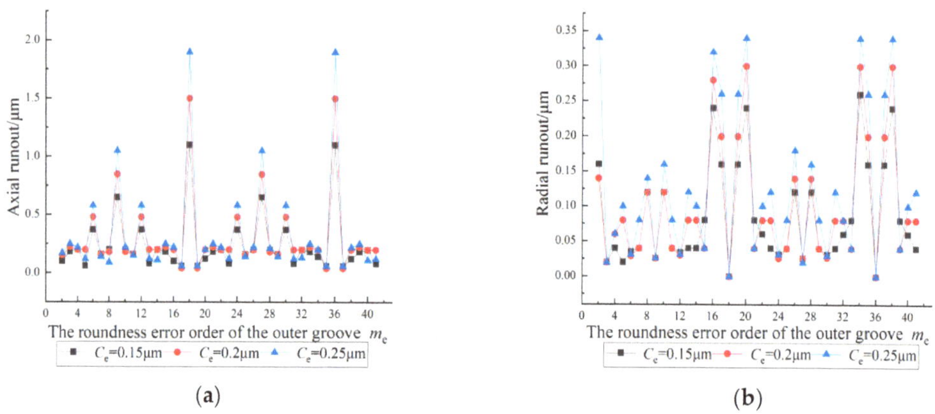

Figure 11. Effect of the roundness error order of the outer groove on the inner ring runout: (**a**) axial runout; (**b**) radial runout.

4.4. Effect of the Roundness Error Amplitude of the Outer Groove on the Inner Ring Runout

Figure 12 shows the effect of the roundness error amplitude of the outer groove on the inner ring runout. In this case, the number of balls Z is 18 and the roundness error orders of the outer groove are 4, 5, 6, 7, 8, 9, 10, 11, 12, 18, 24, 27, 30, and 36, and the roundness error amplitude of the outer groove varies in the range of 0.5 to 1.5 μm. From the figure, it can be seen that the roundness error amplitude of the outer groove and the roundness error amplitude of the inner groove have the same effect on the inner ring runout. The inner ring runout increases with an increase in the roundness error amplitude of the outer groove. When the roundness error orders are $nZ/3$ (n is a positive integer other than a multiple of 3), $nZ/2$ (n is an odd positive integer), and nZ (n is a positive integer), the magnifications of the axial runout increase significantly with an increase in the roundness error amplitude, but the magnifications of the radial runout do not change with an increase in roundness error amplitude. When the roundness error orders are $nZ/3$ and $nZ/2$, the radial runout is obviously reduced and only a slight radial runout is produced. There is no radial runout when the roundness error order is nZ.

These results show that the effect of the roundness error amplitude of the outer groove on the inner ring runout is significant. Therefore, the inner ring runout can also be reduced by appropriately reducing the roundness error amplitude of the outer groove.

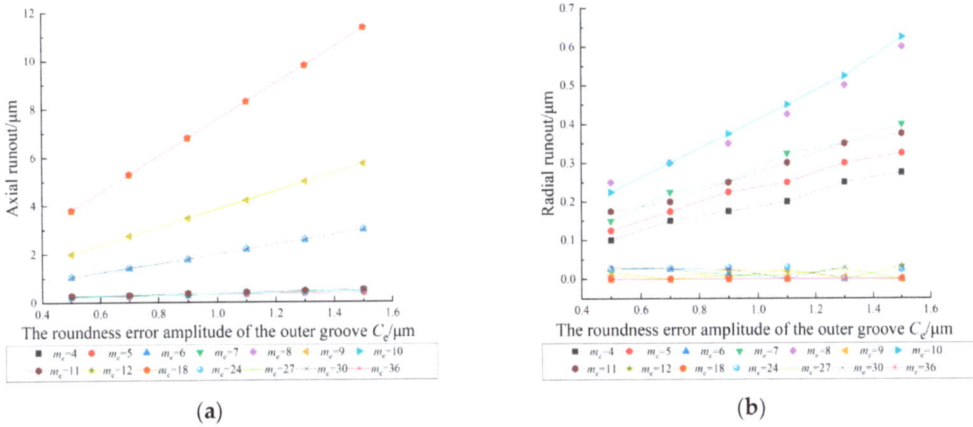

Figure 12. Effect of the roundness error amplitude of the outer groove on the inner ring runout: (**a**) axial runout; (**b**) radial runout.

4.5. Coupling Effect of the Roundness Error of the Inner and Outer Grooves on the Inner Ring Runout

Figures 13–24 show the coupling effect of the roundness error of the inner and outer grooves on the inner ring runout. In this case, the number of balls Z is 18 and the roundness error amplitude of both the inner and outer grooves is 0.2 μm. Figures 13–18 show that the roundness error orders of the outer groove are 2, 3, 6, 7, 9, and 10, and the range of the roundness error order of the inner groove is 2~41. Figures 19–24 show that the roundness error orders of the inner groove are 2, 3, 6, 7, 9, and 10, and the range of the roundness error order of the outer groove is 2~41. It can be seen that under the coupling effect of the roundness error of the inner and outer grooves, the inner ring runout shows a periodic change with the period of Z. In addition, compared with the effect of the roundness error of a single groove, the magnifications of the inner ring runout increase as a whole when the roundness error of the inner and outer grooves are coupled.

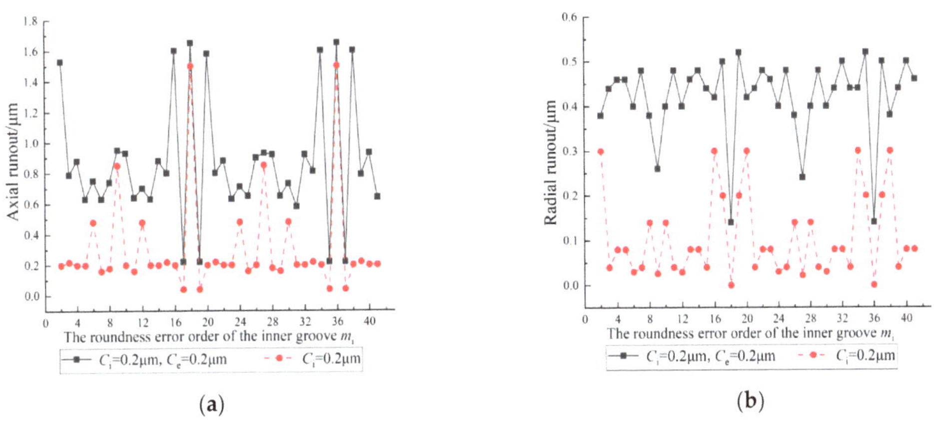

Figure 13. Coupling effect of the roundness error of the grooves on the inner ring runout (outer groove order is 2): (**a**) axial runout; (**b**) radial runout.

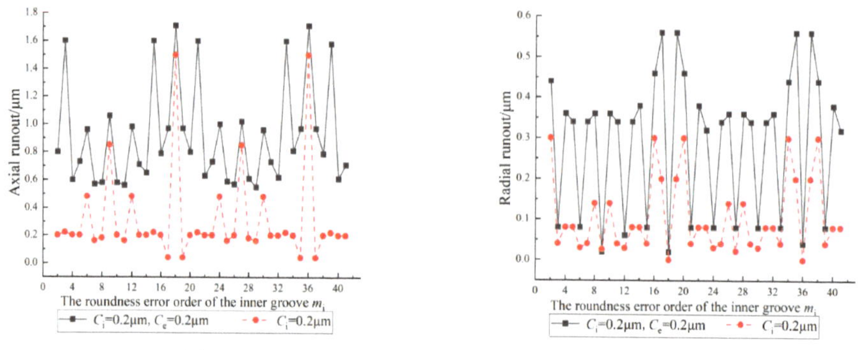

(**a**) (**b**)

Figure 14. Coupling effect of the roundness error of the grooves on the inner ring runout (outer groove order is 3): (**a**) axial runout; (**b**) radial runout.

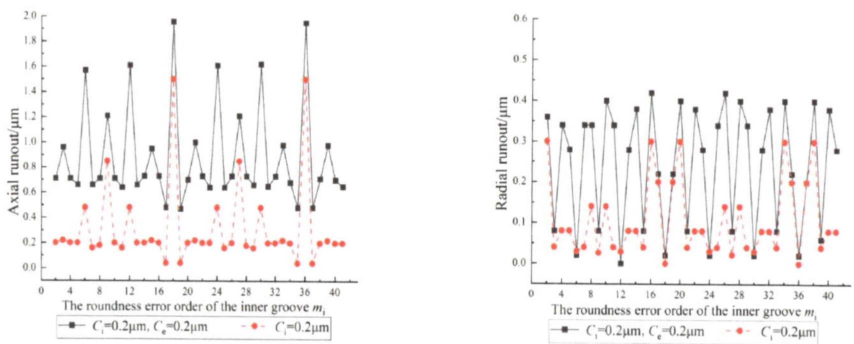

(**a**) (**b**)

Figure 15. Coupling effect of the roundness error of the grooves on the inner ring runout (outer groove order is 6): (**a**) axial runout; (**b**) radial runout.

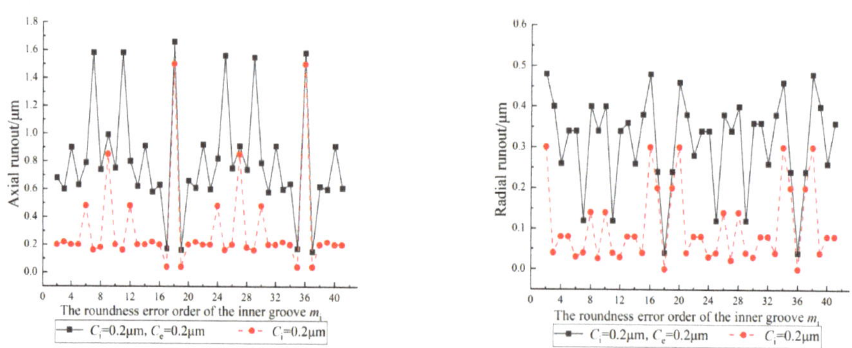

(**a**) (**b**)

Figure 16. Coupling effect of the roundness error of the grooves on the inner ring runout (outer groove order is 7): (**a**) axial runout; (**b**) radial runout.

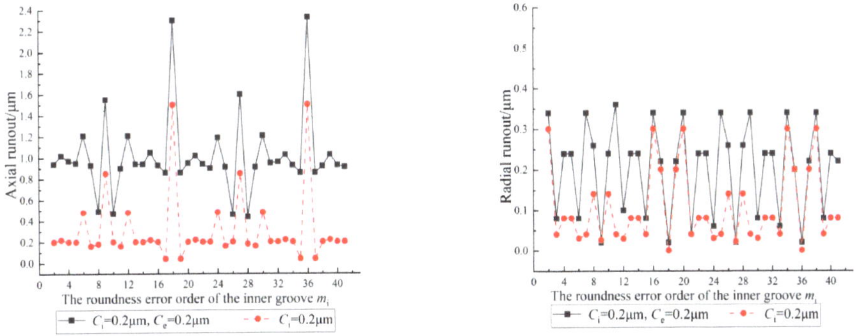

(**a**) (**b**)

Figure 17. Coupling effect of the roundness error of the grooves on the inner ring runout (outer groove order is 9): (**a**) axial runout; (**b**) radial runout.

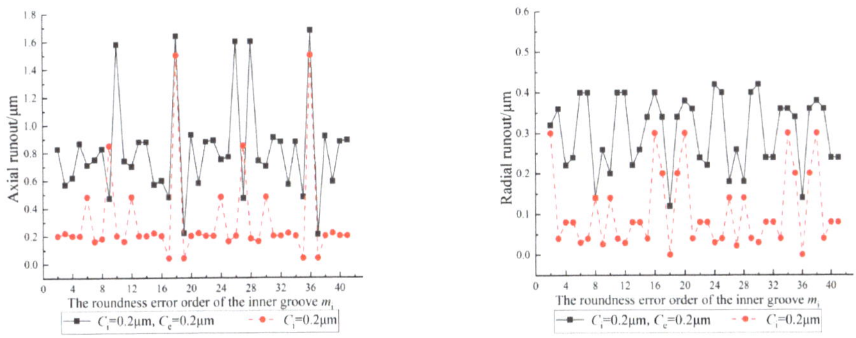

(**a**) (**b**)

Figure 18. Coupling effect of the roundness error of the grooves on the inner ring runout (outer groove order is 10): (**a**) axial runout; (**b**) radial runout.

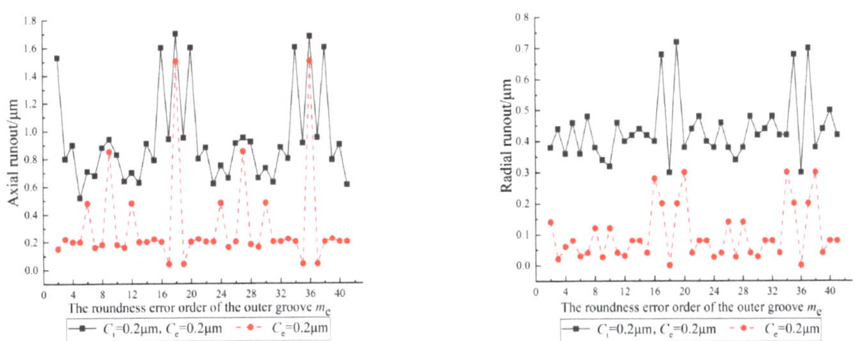

(**a**) (**b**)

Figure 19. Coupling effect of the roundness error of the grooves on the inner ring runout (inner groove order is 2): (**a**) axial runout; (**b**) radial runout.

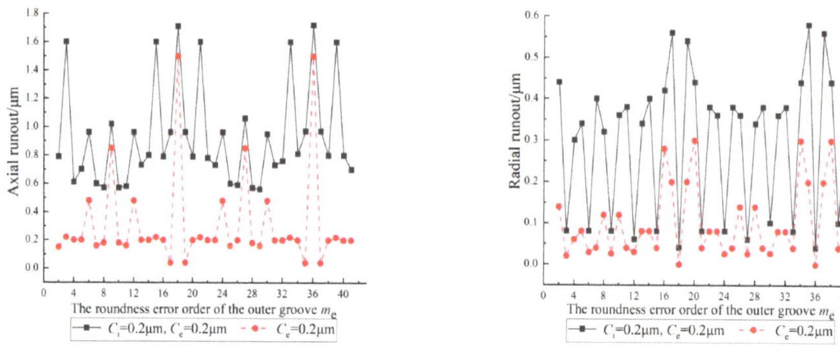

(**a**) (**b**)

Figure 20. Coupling effect of the roundness error of the grooves on the inner ring runout (inner groove order is 3): (**a**) axial runout; (**b**) radial runout.

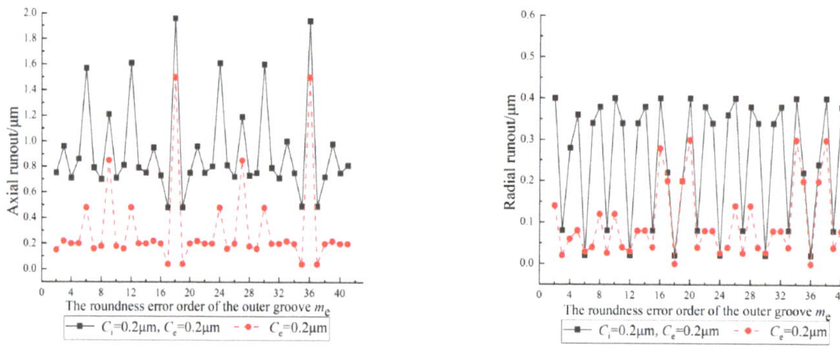

(**a**) (**b**)

Figure 21. Coupling effect of the roundness error of the grooves on the inner ring runout (inner groove order is 6): (**a**) axial runout; (**b**) radial runout.

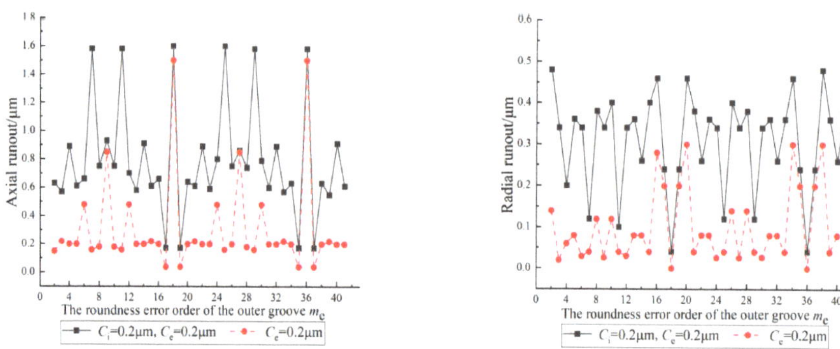

(**a**) (**b**)

Figure 22. Coupling effect of the roundness error of the grooves on the inner ring runout (inner groove order is 7): (**a**) axial runout; (**b**) radial runout.

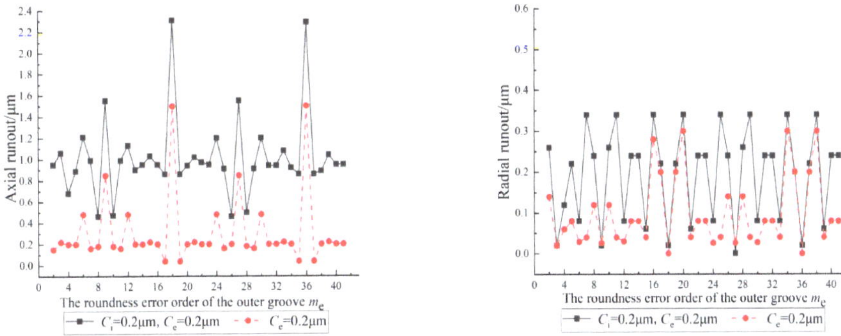

(a) (b)

Figure 23. Coupling effect of the roundness error of the grooves on the inner ring runout (inner groove order is 9): (**a**) axial runout; (**b**) radial runout.

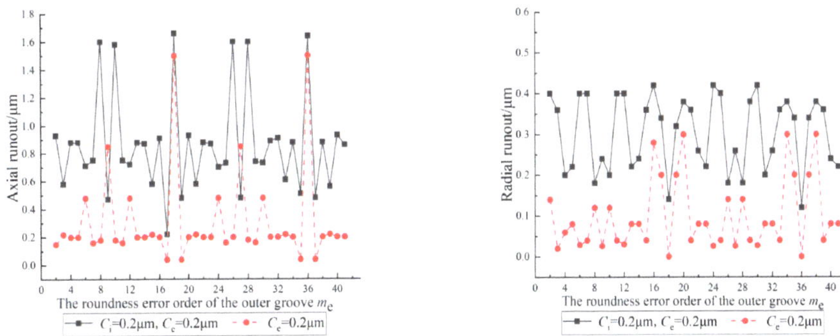

(a) (b)

Figure 24. Coupling effect of the roundness error of the grooves on the inner ring runout (inner groove order is 10): (**a**) axial runout; (**b**) radial runout.

From Figures 13–18, it can be seen that when the roundness error order of the inner groove is nZ (n is a positive integer), the axial runout reaches the maximum value and the radial runout reaches the minimum value. When the roundness error order of the inner groove is the same as that of the outer groove, the axial runout rises to a peak value, while the radial runout decreases. When the roundness error order of the inner groove is ($nZ \pm m_e$), the axial runout rises to a peak value again and the radial runout decreases to a low point. When the roundness error order of the inner groove is ($nZ \pm 1$) or ($nZ \pm 2$) (n is a positive integer), the radial runout increases significantly and reaches the maximum value.

From Figures 19–24, it can be seen that when the roundness error order of the outer groove is nZ (n is a positive integer), the axial runout reaches the maximum value and the radial runout reaches the minimum value. When the roundness error order of the outer groove is the same as that of the inner groove, the axial runout rises to a peak value while the radial runout decreases. When the roundness error order of the outer groove is ($nZ \pm m_i$), the axial runout rises to a peak value again and the radial runout decreases to a low point. When the roundness error order of the outer groove is ($nZ \pm 1$) or ($nZ \pm 2$) (n is a positive integer), the radial runout increases significantly and reaches the maximum value.

5. Conclusions

In order to find the effect law of the roundness error of the grooves on the inner ring runout of the angular contact ball bearings, a prediction model of the inner ring runout of the bearing is established. The effects of the order and amplitude of the roundness error of the inner groove and the order and amplitude of the roundness error of the outer groove on the inner ring runout are analyzed. The coupling effect of the roundness error of the inner and outer grooves on the inner ring runout is further analyzed. The results show that:

1. The inner ring runout changes periodically with a change in the roundness error order of the grooves; the period is the number of balls.
2. The inner ring runout increases with an increase in the roundness error amplitude and the roundness error order affects the magnitude of the increase.
3. Under the coupling of the roundness error of the inner and outer grooves, the magnification of the inner ring runout increases as a whole.
4. When there are specific relationships between the roundness error orders of the grooves and the number of balls, the magnification of the axial or radial runout changes significantly.

According to the tolerance ranges in which the geometric errors of the bearing parts are located, the bearing parts can be classified into different accuracy levels. Obviously, the bearings assembled with bearing parts of different accuracy levels have different accuracy levels. Therefore, a prediction model for the distribution of the rotational accuracy of angular contact ball bearings will be developed in the future to study the dependent relationship between the distribution of the geometric error of the bearing parts and the distribution of the rotational accuracy of the bearings.

Author Contributions: Conceptualization, D.C.; methodology, D.C.; validation, D.C.; resources, Y.Y.; writing—original draft, D.C.; writing—review & editing, Y.Y., Y.X. and H.H.; supervision, H.C.; project administration, P.G. All authors have read and agreed to the published version of the manuscript.

Funding: This work was supported by the Special Project of the Ministry of Industry and Information Technology (Approval No. TC220H05V), the Open Fund Program of Henan Provincial Key Laboratory of High-Performance Bearing Technology (Approval No. ZYSKF202408), and the Key Scientific Research Project of Higher Education Institutions in Henan Province (Approval No. 22A460018).

Data Availability Statement: The authors confirm that the data supporting the findings of this study are available within the article.

Conflicts of Interest: Pengge Guo is employed by Luoyang Bearing Group Company Ltd., which has no role in the design of the study; in the collection, analyses, or interpretation of data; in the writing of the manuscript, or in the decision to publish the results. The authors declare no conflicts of interest.

References

1. Li, X.; Yu, K.; Ma, H.; Cao, L.; Luo, Z.; Li, H.; Che, L. Analysis of varying contact angles and load distributions in defective angular contact ball bearing. *Eng. Fail. Anal.* **2018**, *91*, 449–464. [CrossRef]
2. Bodewig, A.H.; Pape, F.; Poll, G. Optimizing Stainless Steel Bearings: Enhancement of Stainless Steel Bearing Fatigue Life by Low-Temperature Forming. *Metals* **2024**, *14*, 512. [CrossRef]
3. Pape, F.; Coors, T.; Wang, Y.; Poll, G. *Fatigue Life Calculation of Load-Adapted Hybrid Angular Contact Ball Bearings*; Springer: Singapore, 2019; pp. 401–414.
4. Takabi, J.; Khonsari, M. On the thermally-induced seizure in bearings: A review. *Tribol. Int.* **2015**, *91*, 118–130. [CrossRef]
5. Gao, S.; Wang, L.; Zhang, Y. Modeling and dynamic characteristic analysis of high speed angular contact ball bearing with variable clearance. *Tribol. Int.* **2023**, *182*, 108330. [CrossRef]
6. Singh, S.; Howard, C.Q.; Hansen, C.H. An extensive review of vibration modelling of rolling element bearings with localised and extended defects. *J. Sound Vib.* **2015**, *357*, 300–330. [CrossRef]
7. Rejith, R.; Kesavan, D.; Chakravarthy, P.; Murty, S.N. Bearings for aerospace applications. *Tribol. Int.* **2023**, *181*, 108312. [CrossRef]
8. Noguchi, S.; Tanaka, K.; Ono, K. Theoretical analysis of a ball bearing used in HDD spindle motors for reduction of NRRO. *IEEE Trans. Magn.* **1999**, *35*, 845–850. [CrossRef]
9. Noguchi, S.; Ono, K. Reduction of NRRO in ball bearings for HDD spindle motors. *Precis. Eng.* **2004**, *28*, 409–418. [CrossRef]

10. Noguchi, S.; Obinata, S.; Saito, Y.; Akamatsu, Y.; Sakaguchi, T. The influence of the unequal orbital intervals of balls on the NRRO of rotational frequency of the cage in a ball bearing. *Toraibarojisuto/J. Jpn. Soc. Tribol.* **2005**, *50*, 90–96.
11. Okamoto, J. Study on run-out of ball bearings-relation between unroundness of race and locus of shaft in rotation. *J. Jpn. Soc. Tribol.* **2001**, *46*, 578–584.
12. Tada, S. Three-dimensional analysis of non-repeatable runout (NRRO) in ball bearing. *KOYO Eng. J.* **2002**, *161*, 31–37.
13. Yang, Z.; Hong, J.; Liu, J.; Zhu, Y. Theoretical model to analyze the non-repetitive run-out (NRRO) of ball bearings. *Adv. Sci. Lett.* **2011**, *4*, 2522–2527. [CrossRef]
14. Yang, Z.; Hong, J.; Liu, J.; Ding, Y.; Wang, M.Y. Theoretical method to reduce the non-repetitive run-out (NRRO) of angular contact ball bearings. In Proceedings of the 2011 IEEE International Symposium on Assembly and Manufacturing (ISAM), Tampere, Finland, 25–27 May 2011; pp. 1–6.
15. Chen, G.; Wang, B.; Mao, F. Effects of raceway roundness and roller diameter errors on clearance and runout of a cylindrical roller bearing. *Proc. Inst. Mech. Eng. Part J J. Eng. Tribol.* **2013**, *227*, 275–285. [CrossRef]
16. Ma, F.; Ji, P.; Li, Z.; Wu, B.; An, Q. Influences of off-sized rollers on mechanical performance of spherical roller bearings. *Proc. Inst. Mech. Eng. Part K J. Multi-Body Dyn.* **2015**, *229*, 344–356. [CrossRef]
17. Yu, Y.; Chen, G.; Li, J.; Xue, Y.; Pang, B. Prediction Method for the Radial Runout of Inner Ring in Cylindrical Roller Bearings. *Math. Probl. Eng.* **2017**, *2017*, 6584561. [CrossRef]
18. Yu, Y.; Chen, G.; Li, J.; Xue, Y. A method to predict the radial runout of outer ring in cylindrical roller bearings. *Adv. Mech. Eng.* **2017**, *9*, 168781401773325. [CrossRef]
19. Yu, Y.; Li, J.; Xue, Y. Influence of roundness errors of bearing components on rotational accuracy of cylindrical roller bearings. *Sci. Rep.* **2022**, *12*, 6794. [CrossRef]
20. Zha, J.; Chen, Y.; Zhang, P. Relationship between elliptical form error and rotation accuracy of hydrostatic journal bearing. *Ind. Lubr. Tribol.* **2017**, *69*, 905–911. [CrossRef]
21. Hu, G.; Chen, Y.; Cui, L.; Jin, G.; Wang, T.; Qi, H.; Tian, Y. Investigation on modeling and formation mechanism of dynamic rotational error for spindle-rolling bearing system. *Appl. Sci.* **2020**, *10*, 5753. [CrossRef]
22. Zhang, X.; Meng, Q.; Ma, J.; Zhang, Y.; Wen, B.; Guo, Z. Rotation Accuracy Analysis of High-speed Angular Contact Ball Bearings. *Des. Eng.* **2020**, 38–49. Available online: https://www.semanticscholar.org/paper/Rotation-Accuracy-Analysis-of-High-speed-Angular-Xu-Qingguo/b91bb1ee79a91c732b734927e9f3f4bef1605dae (accessed on 31 July 2024). [CrossRef]
23. Zhang, P. Accuracy prediction and roller number optimization of journal roller bearing based on averaging coefficient considering raceway roundness error. *J. Braz. Soc. Mech. Sci. Eng.* **2023**, *45*, 203. [CrossRef]
24. Wang, W.; Hu, L.; Zhang, S.; Zhao, Z.; Ai, S. Modeling angular contact ball bearing without raceway control hypothesis. *Mech. Mach. Theory* **2014**, *82*, 154–172. [CrossRef]
25. Cho, N.; Tu, J. Roundness modeling of machined parts for tolerance analysis. *Precis. Eng.* **2001**, *25*, 35–47. [CrossRef]
26. Viitala, R.; Widmaier, T.; Hemming, B.; Tammi, K.; Kuosmanen, P. Uncertainty analysis of phase and amplitude of harmonic components of bearing inner ring four-point roundness measurement. *Precis. Eng.* **2018**, *54*, 118–130. [CrossRef]

machines

MDPI

Article

Method of Dynamic Modeling and Robust Optimization for Chain Transmission Mechanism with Time-Varying Load Uncertainty

Taisu Liu [1,2,*], Yuan Liu [1], Peitong Liu [1] and Xiaofei Du [1,3]

1 School of Mechanical Engineering, Nanjing Institute of Technology, Nanjing 211167, China; liuyuan@njit.edu.cn (Y.L.); liupeitong@njit.edu.cn (P.L.); duxiaofei@njit.edu.cn (X.D.)
2 School of Mechanical Engineering, Nanjing University of Science and Technology, Nanjing 210094, China
3 School of Mechanical Engineering, Southeast University, Nanjing 211189, China
* Correspondence: liuts@njit.edu.cn; Tel.: +86-025-86118255

Abstract: Time-varying driving loads and uncertain structural parameters affect the transmission accuracy of chain transmission mechanisms. To enhance the transmission accuracy and placement consistency of these mechanisms, a robust optimization design method based on Karhunen–Loeve expansion and Polynomial Chaos Expansion (KL-PCE) is proposed. First, a dynamic model of the chain transmission mechanism, considering multiple contact modes, is established, and the model's accuracy is verified through experiments. Then, based on the KL-PCE method, a mapping relationship between uncertain input parameters and output responses is established. A robust optimization design model for the chain transmission process is formulated, with transmission accuracy and consistency as objectives. Finally, case studies are used to verify the effectiveness of the proposed method. Thus, the transmission accuracy of the chain transmission mechanism is improved, providing a theoretical foundation for the design of chain transmission mechanisms under time-varying load uncertainties and for improving the accuracy of other complex mechanisms.

Keywords: chain transmission mechanism; K-L expansion; PCE; placement consistency; robust optimization

1. Introduction

Chain transmission mechanisms are widely used in mechanical structures, such as machine tools, special equipment, industrial machinery, agricultural machinery, construction machinery, aerospace, and medical devices. They offer high reliability and efficient transmission, with strong adaptability to heavy loads and harsh environments. However, due to random uncertainties in structural, physical, and drive parameters, the delivery precision and positional consistency of the transmitted objects are affected during operation. To enhance the robustness of chain transmission mechanisms during the design phase, robust optimization design methods can reduce the sensitivity of the mechanism's movement to random uncertainties while ensuring motion accuracy.

In previous studies, researchers have conducted a series of investigations on chain drive mechanisms. Li et al. [1] proposed a feasible solution for the robust optimization of multi-gap chain transmission mechanisms by using a data-driven modeling framework based on deep neural networks, but without considering the influence of time-varying loads. Zhuang et al. [2] optimized the design of mechanical chain drives using wireless sensor network data algorithms, thereby improving the reliability and stability of scraper conveyors, and providing a theoretical reference for the robust optimization of chain transmission mechanisms. Chai et al. [3] introduced a new method based on multi-disciplinary design optimization, and then established system optimization objectives and subsystems to obtain the kinematic principles and optimal parameter values of chain drives, without considering the effect of parameter uncertainty. To analyze the operational stability of chain transmission scraper conveyors, Jiang et al. [4] established a combined simulation model for the chain transmission mechanism, analyzing the system's dynamic characteristics and failure causes. In order to reduce the polygonal effect in chain transmission mechanisms, Hong et al. [5] proposed a compensation method to optimize the design of the chain transmission mechanism, thereby effectively improving the transmission performance of the chain transmission system. To enhance the control and guiding capabilities of the in-wheel motor steering transmission mechanical chain transmission and improve the transmission efficiency of the in-wheel engine steering mechanical chain transmission, Da et al. [6] proposed a control method for an in-wheel motor-guided transmission mechanical chain gearbox, thereby enhancing the robustness and adaptability of the hub motor steering transmission mechanical chain transmission. To analyze the noise characteristics and structural design of chain transmission systems under complex working conditions, such as high-speed and overload, Cheng et al. [7] studied the system design and noise characteristics of a new dual-clutch automatic transmission with a double-tooth chain, achieving structural optimization design.

It can be seen that the above references have conducted research on the parameter optimization design of chain transmission mechanisms, which can provide a solid foundation for the robust optimization of chain transmission mechanisms. There has been relatively little research on the robust optimization design of chain transmission mechanisms. However, in other engineering fields, relevant research on robust optimization design has been conducted, which can provide references for the robust design of chain transmission mechanisms. Addressing the problem of improving the aerodynamic performance of drone wings and bodies, Wang et al. [8] used the Gradient-Enhanced Polynomial Chaos Expansion (GPCE) method to construct statistical moments related to the mean and variance, and established a gradient-based robust optimization design model, validating its effectiveness. To avoid the curse of dimensionality in the robust optimization design process, Zhang et al. [9] combined the polynomial chaos expansion method and proposed a technique called R-Opt, a robust aerodynamic optimization method that improved average aerodynamic performance and reduced the standard deviation of the objective. Najlawi et al. [10] developed a combined multi-objective imperialist competitive algorithm and Monte Carlo method for the multi-objective robust optimization design of sewing machine needle rods and take-up rods (NBTTL), significantly reducing the sensitivity of the mechanism's performance to design parameters. For the challenges of strong non-linearity and long experimental and simulation prediction times in the ammunition loading mechanism, Zhang et al. [11] proposed a data-driven modeling method based on artificial neural networks, and performed a robust optimization design for the ammunition loading process using the data-driven model. In robot mechanism research, current robust optimization studies mainly focus on planar closed-loop or open-chain mechanisms [12,13]. For example, Wu and Rao [14] proposed a tolerance allocation optimization technique through interval uncertainty analysis, which effectively saves manufacturing costs while ensuring positioning accuracy. Du et al. [15] proposed a planar mechanism robust optimization method based on dual-loop Monte Carlo simulation, improving the robustness of motion accuracy. To solve the general design problems of complex engineering systems, several uncertainty optimization methods have been developed [16–19]. For instance, Lee [20] and others introduced the so-called point estimation method in robust optimization to improve structural efficiency. Chatterjee [21] and others proposed a universal robust design framework integrating surrogate modeling techniques to address complex and highly non-linear problems. Lee and Rahman [22] proposed a new robust optimization method based on polynomial chaos expansion, considering the correlation of uncertain variables. Zafar et al. [23] considered the time-varying reliability of systems and proposed a dynamic robust optimization method. Jiang et al. [24] combined parallel efficient global optimization with adaptive Kriging techniques, introducing time-varying uncertainty in robust optimization. Cheng et al. [25] proposed a robust balancing optimization method, improving the robustness of structural dynamic characteristics by introducing the overlap coefficient between the interval boundary angles. These general uncertainty optimization techniques provide algorithmic insights for the robust optimization of chain transmission mechanisms.

In summary, previous studies have shown relatively little research on the robust design of chain transmission mechanisms considering time-varying load uncertainties. Therefore, in order to achieve a robust design of chain transmission mechanisms considering time-varying load uncertainties, this paper proposes a robust optimization design method for chain transmission mechanisms based on K-L expansion and PCE. The main contributions and innovations are as follows:

(1) A dynamic model of the chain transmission mechanism considering multiple clearances was established, and the accuracy of the simulation model was verified through experiments.
(2) A robust optimization design method for the chain transmission mechanism based on K-L expansion and PCE was proposed. First, the K-L expansion method was used for the dimensionality reduction of the random driving loads. Then, based on the PCE method, a fast uncertainty propagation analysis of the chain transmission mechanism was conducted, and a robust optimization design model for the chain transmission mechanism was established.
(3) The effectiveness of the proposed method was verified through engineering examples, providing theoretical reference for the robust optimization design of complex engineering chain transmission problems.

The structure of this paper is as follows. Section 2 presents the dynamic modeling and experimental validation of the chain transmission mechanism. Section 3 discusses the robust optimization design of the chain transmission mechanism considering random loads. Section 4 provides a case study analysis, and Section 5 presents the conclusions.

2. Dynamic Modeling of Chain Transmission Mechanism

2.1. Model Description of Chain Transmission Mechanism

The research object of this paper is a chain transmission mechanism in a practical engineering application, as shown in Figure 1. It consists of an open-chain transmission system with non-interconnected ends (including rollers, chain links, pins, sprockets), a push plate, a transmission cylinder, chain track, the conveyed object, and a receiving cylinder. Driven by a transmission motor, the sprocket rotates and drives the chain forward. Ultimately, the push plate pushes the conveyed object into the receiving cylinder. The chain transmission system is a multi-degree-of-freedom system composed of multiple chain links connected by pins, where the pins' two ends are rollers that contact the sprocket to achieve transmission, while the track constrains the rollers' motion direction.

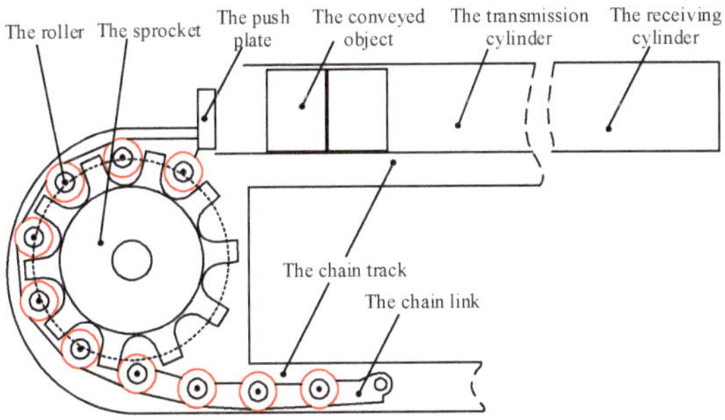

Figure 1. Schematic diagram of chain transmission mechanism (in the plane).

Based on the relative positions and connection relationships of the components of the chain transmission mechanism, the topological relationship is analyzed and shown in Figure 2. In the figure, the dashed lines represent the omitted chain links and rollers. The sprocket is installed on the box with a rotating pair, and there is a contact relationship between the sprocket and the roller. The roller is connected to the chain link through a rotating pair. The first chain link is a pushing plate, which contacts the conveyed object and pushes it forward. The track is fixed on the box and contacts the roller to constrain its movement.

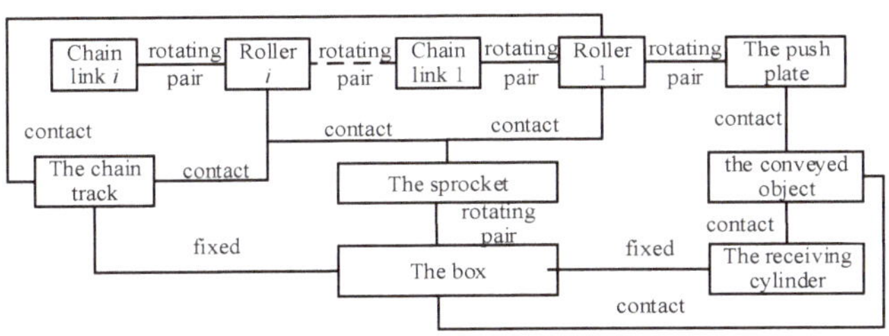

Figure 2. Topological relationship of the chain transmission mechanism.

2.2. Basic Assumptions

Based on the structural description, in order to establish a relatively accurate dynamic model of the chain transmission mechanism, the following assumptions need to be made first:

(1) The system operates under steady-state conditions, and external disturbances are assumed to be negligible;
(2) The chain transmission system is treated as a rigid body system, with no deformation of components during operation;
(3) The friction between chain links, sprockets, and rollers is assumed to be constant for simplification;
(4) The relationship between the rotational motion of the sprocket and the translational motion of the chain is idealized, assuming no slipping or stretching of the chain;
(5) The chain transmission mechanism is treated as a planar system, considering only the motion within the plane and rotation about its axis of rotation.

2.3. Motion and Contact Analysis of Open Chain System

From the topological relationship, it can be seen that there are numerous contact collisions in the model, including the contact collisions between the rollers and the sprocket teeth, between the rollers and the track, between the chain links, between the push plate and the conveyed object, etc. In order to establish a more accurate dynamic model of the chain transmission mechanism, it is necessary to analyze the above-mentioned contact collision processes.

The characteristics of the sprocket include the number of teeth n_t, the original radius of the sprocket R_s, and the pitch angle α. A schematic diagram of the chain system is shown in Figure 3. In the figure, $\boldsymbol{r}_s = [x_s, y_s]^T$ and represents the position vector of the sprocket's center in the global coordinate system; $\boldsymbol{r}_{rc} = [x_{rc}, y_{rc}]^T$ and represents the position vector of the roller i in the global coordinate system $O - XY$; and $\boldsymbol{S}_{rc}$ represents the vector from the sprocket center to the roller center. The sprocket coordinate system $O_s - X_sY_s$ is located at the center of the sprocket, with the teeth numbered from $n_i = 1$ counterclockwise, and n_i represents the n-th tooth. For convenience, a local coordinate system $O_s - \xi_{sn_i}\eta_{sn_i}, (n_i = 1, \cdots, n_t)$ for the tooth groove is established, and the rotation angle relative to the sprocket coordinate system at the origin O_s is denoted as $\theta_s = (n_i - 1)\alpha$. The coordinate transformation matrix from the $O_s - \xi_{sn_i}\eta_{sn_i}$ to the $O_s - X_sY_s$ is given by

$$\boldsymbol{A}_{sn_i} = \begin{bmatrix} \cos\theta_s & -\sin\theta_s \\ \sin\theta_s & \cos\theta_s \end{bmatrix} \tag{1}$$

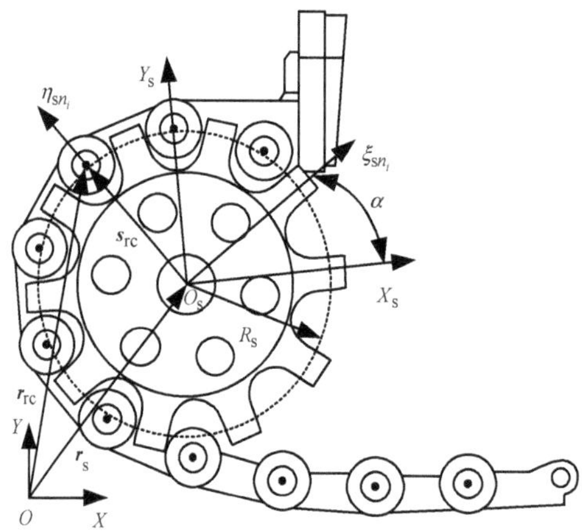

Figure 3. Schematic diagram of chain system.

For convenience in calculation, all contact relationships between the rollers and the sprocket teeth are defined in the tooth groove coordinate system. When a vector is represented in the tooth groove coordinate system, any vector $(\cdot)$ is expressed as $(\cdot)''$. For example, the vector from the center of the sprocket to the center of the roller is defined as

$$\boldsymbol{s}_{rc} = \boldsymbol{r}_{rc} - \boldsymbol{r}_s \tag{2}$$

and represented in the tooth groove coordinate system as

$$\boldsymbol{s}''_{rc} = \boldsymbol{A}^T_{sn_i}\boldsymbol{s}_{rc} \tag{3}$$

The schematic diagram of the sprocket tooth profile is shown in Figure 4. To reduce the impact during the meshing process between the roller and the sprocket tooth groove, the sprocket tooth groove is typically composed of several continuous curved surfaces [26,27]. To accurately represent the contact relationship between the roller and the sprocket tooth groove, the tooth groove is divided into three regions: the tooth groove positioning curve (bottom curve) bc, and the top transition curve ab and cd. The center of the curvature is defined as o_{ab}, o_{bc}, and o_{cd}, and the radius of the circular arcs in the respective segments ab and cd are R_1, and in the respective segment bc is R_2, respectively.

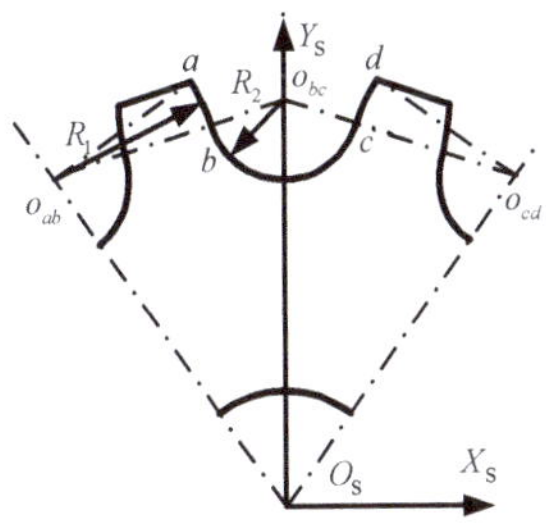

Figure 4. Geometrical diagram of sprocket tooth groove.

2.3.1. Geometric Relationship Between the Sprocket Tooth Groove and the Roller Contact

Under normal operating conditions, the contact between the roller and the sprocket occurs at different positions on the sprocket tooth groove. Therefore, to accurately calculate the kinematic and dynamic relationship between the roller and the sprocket tooth groove, it is necessary to describe the contact relationship between the roller and the different arc segments separately.

As shown in Figure 5, the schematic diagram illustrates the contact relationship between the roller and the sprocket tooth groove positioning curve. The eccentric vector $\boldsymbol{e}_{bc}$ is the vector between the center of the roller and the center of the bottom arc, expressed as

$$\boldsymbol{e}_{bc} = \boldsymbol{s}_{\mathrm{rc}} - \boldsymbol{s}_{o_{bc}} \tag{4}$$

where $\boldsymbol{s}_{o_{bc}}$ is the vector from the center of the bottom arc to the center of the sprocket.

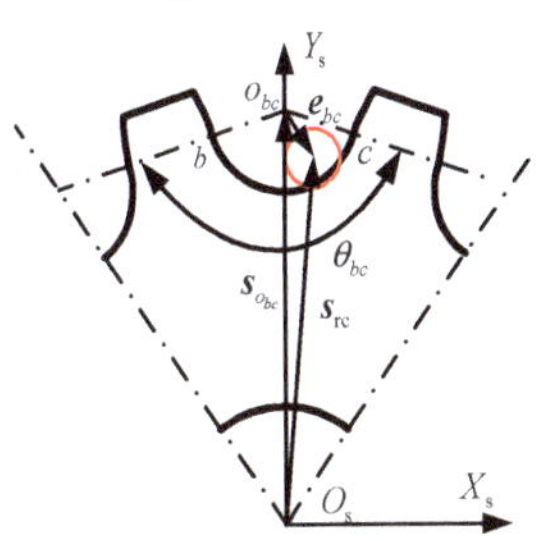

Figure 5. Contact relationship between roller and positioning curve of sprocket tooth groove.

In Figure 5, θ_{bc} represents the angle of arc segment bc relative to the center o_{bc}. Let θ_2 represent the angle between $\boldsymbol{e}_{bc}$ and $-\boldsymbol{s}_{o_{bc}}$, then

$$\theta_2 = \arccos\frac{-\boldsymbol{s}_{o_{bc}}\boldsymbol{e}_{bc}}{\left\|-\boldsymbol{s}_{o_{bc}}\right\|\left\|\boldsymbol{e}_{bc}\right\|} \tag{5}$$

The contact constraint occurring at the bottom arc is

$$-\frac{1}{2}\theta_{bc} < \theta_2 < \frac{1}{2}\theta_{bc} \tag{6}$$

If the above equation is not satisfied, check whether contact occurs at other positions; if θ_2 is positive, check if contact occurs on the arc segment cd; if θ_2 is negative, check if contact occurs on the other arc segment ab. If the above equation is satisfied, calculate the penetration δ_{bc} between the roller and the sprocket tooth groove positioning curve according to Equation (7)

$$\delta_{bc} = \|\boldsymbol{e}_{bc}\| - (R_2 - R_r) \tag{7}$$

where R_r is the radius of the roller.

If both Equation (6) and the condition $\delta_{bc} \geq 0$ are satisfied, the contact between the roller and the sprocket tooth groove occurs on the arc segment *bc*; otherwise, there is no contact between the roller and the sprocket tooth groove on the arc segment *bc*.

Figure 6 is a schematic diagram of the contact relationship between the roller and the top transition curve of the sprocket tooth groove. The top transition curve of the sprocket tooth groove consists of an arc segment *ab* and another arc segment *cd*, so both arc segments need to be analyzed separately.

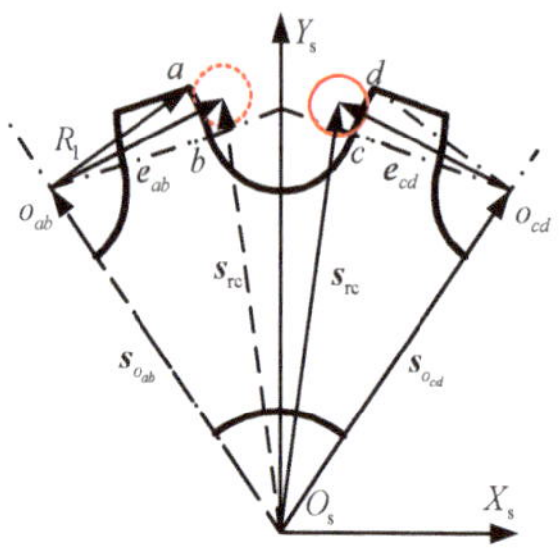

Figure 6. Contact relationship between roller and top curve of sprocket tooth groove.

In Figure 6, $\boldsymbol{e}_{ab}$ and $\boldsymbol{e}_{cd}$ represent the eccentric vectors from the center of the arc segment *ab* to the center of the roller, and from the center of the arc segment *cd* to the center of the roller, respectively. $\boldsymbol{s}_{o_{ab}}$ and $\boldsymbol{s}_{o_{cd}}$ represent the position vectors from the center of the arc segment ab and the center of the arc segment *cd* to the center of the sprocket, respectively.

Based on the different arc segments, the positional relationship between the roller and the top transition curve of the sprocket tooth groove can be described as follows:

(1) When the roller contacts the top transition curve of the sprocket tooth groove, assuming contact occurs at arc segment *ab*, we obtain

$$\boldsymbol{e}_{ab} = \boldsymbol{s}_{\mathrm{rc}} - \boldsymbol{s}_{o_{ab}} \tag{8}$$

Let T_e represent the angle between $\boldsymbol{s}_{o_{ab}a}$ and $\boldsymbol{e}_{ab}$, where $\boldsymbol{s}_{o_{ab}a}$ represents the position vector from the point o_{ab} to the point a, then

$$\theta_1 = \arccos\left(\frac{\boldsymbol{s}_{o_{ab}a}\boldsymbol{e}_{ab}}{\|\boldsymbol{s}_{o_{ab}a}\|\|\boldsymbol{e}_{ab}\|}\right) \tag{9}$$

Let R_r represent the angle between $\boldsymbol{s}_{o_{ab}b}$ and $\boldsymbol{e}_{ab}$, where $\boldsymbol{s}_{o_{ab}b}$ represents the position vector from the point R_m to the point b, then

$$\theta_1' = \arccos\left(\frac{\boldsymbol{s}_{o_{ab}b}\boldsymbol{e}_{ab}}{\|\boldsymbol{s}_{o_{ab}b}\|\|\boldsymbol{e}_{ab}\|}\right) \tag{10}$$

The contact constraint occurring at arc segment *ab* is

$$\begin{cases} 0 \leq \theta_1 < \theta_{ab} \\ 0 \leq \theta_1' < \theta_{ab} \end{cases} \tag{11}$$

where θ_{ab} represents the angle corresponding to arc segment *ab* relative to the center $T_{\max}$ of the circle.

If Equation (11) is satisfied, the penetration δ_{ab} between the roller and the top transition curve of the sprocket tooth groove for arc segment *ab* is calculated as

$$\delta_{ab} = R_1 - (\|\boldsymbol{e}_{ab}\| - R_{\mathrm{r}}) \tag{12}$$

If both Equation (11) and the condition $\delta_{ab} \geq 0$ are satisfied, the roller contacts the top transition curve of the sprocket tooth groove at arc segment *ab*. Otherwise, there is no contact between the roller and the top transition curve of the sprocket tooth groove at arc segment *ab*.

(2) When the roller contacts the top transition curve of the sprocket tooth groove at arc segment *cd*, the contact constraint for arc segment *cd* is given by

$$\begin{cases} 0 < \theta_3 \leq \theta_{cd} \\ 0 < \theta_3' \leq \theta_{cd} \end{cases} \tag{13}$$

where θ_{cd} represents the angle corresponding to arc segment *cd* relative to the center o_{cd} of the circle.

If Equation (13) is satisfied, the penetration between the roller and the top transition curve of the sprocket tooth groove for arc segment *cd* is calculated as

$$\delta_{cd} = R_1 - (\|\boldsymbol{e}_{cd}\| - R_{\mathrm{r}}) \tag{14}$$

If both Equation (13) and the condition $\delta_{cd} \geq 0$ are satisfied, the roller contacts the top transition curve of the sprocket tooth groove at arc segment *cd*; otherwise, there is no contact between the roller and the top transition curve of the sprocket tooth groove at arc segment *cd*.

2.3.2. Contact Between Chain Links

In order to use the chain links to drive an object forward, it is necessary to add planar constraints between the chain links. At the initial moment, the chain links are in the chain box. When the chain links rotate over the sprocket and reach above the sprocket, under the action of planar constraints, the planes of the two-chain links will come into contact. The critical condition for contact is when the upper surfaces of the two-chain links are parallel. As shown in Figure 7, for two adjacent chain links O_{ck} and $O_{c(k+1)}$, the angle between the two coordinate axes $O_{ck}Y_{ck}$ and $O_{c(k+1)}Z_{c(k+1)}$ is γ, and the condition for contact between the two chain links is as follows

$$\gamma \leq \frac{\pi}{2} \tag{15}$$

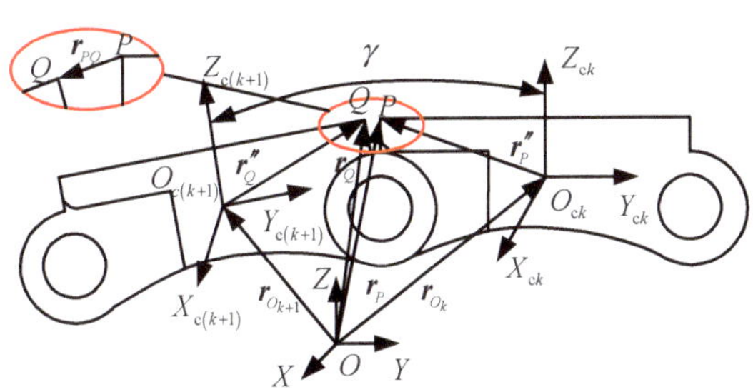

Figure 7. Contact relationship between adjacent chain links.

In Figure 7, points P and Q are the points at the top of the chain link contact plane. $\boldsymbol{r}_{PQ}$ is the position vector from point P to point Q; $\boldsymbol{r}_Q$ is the position vector from the origin of the global coordinate system to point Q; and $\boldsymbol{r}_P$ is the position vector from the origin of the global coordinate system to point P. When the two-chain links come into contact, the contact area is a planar rectangular region. Since the connection between the chain link and the roller is a rotating pair, the contact between the two-chain links will always occur at the upper end first. Due to the influence of the rotational angle, in order to ensure the rationality of the penetration calculation, the penetration depth at the upper end is taken as half of the actual penetration depth (as shown in Figure 7). Therefore, the penetration is given by

$$\boldsymbol{\delta}_{\mathrm{p}} = \frac{\boldsymbol{r}_{PQ}}{2} = \frac{\boldsymbol{r}_Q - \boldsymbol{r}_P}{2} \tag{16}$$

Based on Equation (16), the magnitude of the penetration is given by

$$\delta_{\mathrm{p}} = \sqrt{\boldsymbol{\delta}_{\mathrm{p}}^{\mathrm{T}} \boldsymbol{\delta}_{\mathrm{p}}} \tag{17}$$

Furthermore, based on the magnitude and direction of the penetration, the contact collision force between the chain links can be calculated.

2.4. Contact Force Model

The motion constraints between the roller and the sprocket, as well as between the roller and the track, are established through contact theory [28,29]. The contact force model for the arc surface is described using the Lankarani–Nikravesh model, as shown in Equation (18). The planar contact force [30] is described as in Equation (19), and the tangential friction force [31] is modeled using Coulomb's model, as shown in Equation (20). The kinematic relationship between the head and tail components, as well as the last component of the closed-loop transmission chain, is established using the cut-joint method [32].

$$F_{\mathrm{n}} = Kh^{n} + C\dot{h} \tag{18}$$

$$F_{\mathrm{np}} = K_{\mathrm{p}} \delta_{\mathrm{p}} \tag{19}$$

$$F_{\mathrm{t}} = -c_{\mathrm{f}} c_{\mathrm{d}} F_{\mathrm{n}} \frac{v_{\mathrm{t}}}{\|v_{\mathrm{t}}\|} \tag{20}$$

where, F_{n} is the normal collision force; K is the stiffness coefficient of the two contacting bodies; h is the penetration depth of the contact bodies; n is the contact index; C is the contact damping coefficient; $\dot{h}$ is the relative collision velocity at the contact point; δ_{p} is the planar contact penetration depth; K_{p} is the equivalent stiffness of the planar contact; F_{t} is the tangential friction force; c_{f} is the coefficient of sliding friction; c_{d} is the dynamic correction factor; and v_{t} is the relative tangential velocity.

2.5. Dynamic Model and Experimental Validation of Chain Transmission Mechanism

Based on the kinematic relationships, forces, and constraints of the chain transmission mechanism, and using the principle of virtual work [31], the dynamic equations of the chain transmission mechanism are established as follows: (see Appendix A for modeling details)

$$\begin{cases} \boldsymbol{M}(\boldsymbol{q},t)\ddot{\boldsymbol{q}} + \boldsymbol{C}(\boldsymbol{q},\dot{\boldsymbol{q}},t)\dot{\boldsymbol{q}} + \boldsymbol{\Phi}_{\boldsymbol{q}}^{\mathrm{T}}\lambda = \boldsymbol{Q} \\ \boldsymbol{\Phi}(\boldsymbol{q},t) = \boldsymbol{0} \end{cases} \tag{21}$$

where, t is the motion time of the transmission process, $\boldsymbol{q}$, $\dot{\boldsymbol{q}}$, and $\ddot{\boldsymbol{q}}$ represent the relative displacement, relative velocity, and relative acceleration of the rotational joints in the chain transmission mechanism, respectively. $\boldsymbol{\Phi}(\boldsymbol{q},t) = \boldsymbol{0}$ is the constraint equation of the chain transmission mechanism, $\boldsymbol{\Phi}_q$ denotes the derivative of $\boldsymbol{\Phi}(q,t)$ with respect to q, and $(\cdot)^{\mathrm{T}}$ represents the transpose of the matrix. λ is the Lagrange multiplier, $\boldsymbol{M}(q,t)$ is the mass matrix of the chain transmission mechanism, $\boldsymbol{C}(q,\dot{q},t)$ is the generalized damping matrix of the chain transmission mechanism, and Q is the external force acting on the chain transmission mechanism.

To verify the accuracy and effectiveness of the simulation model, experimental tests are conducted on the chain transmission process. The experimental setup mainly consists of a chain transmission mechanism test rig, a PLC control system, a digital data acquisition system, a laser displacement sensor, etc., as shown in Figure 8. The PLC control system serves as the drive execution system for the chain transmission mechanism and records the motor current in real time. The digital acquisition device reads and records the displacement of the conveyed object as measured by the laser displacement sensor, while also receiving synchronization signals from the PLC.

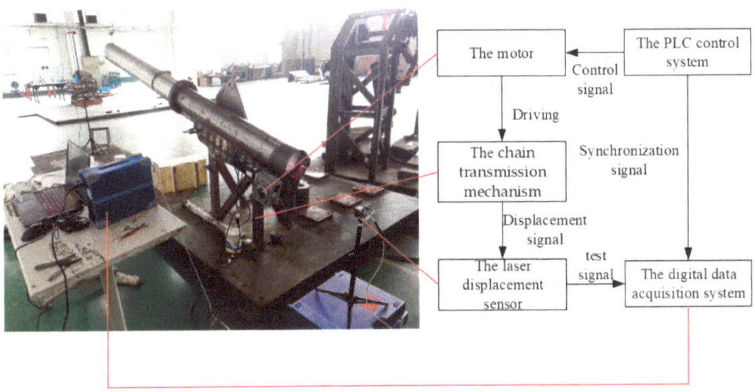

Figure 8. The test schematic diagram of the propellant transport process.

The experimental process is as follows:

(1) According to the schematic diagram in Figure 8, arrange the experimental setup and connect the testing system. The data acquisition system is connected to the PLC control system to ensure the synchronization of the data signals. The displacement sensor is mounted at the rear of the chain transmission mechanism using a bracket. The sensor captures the laser signal of the moving conveyed object, and the displacement sensor is connected to the data acquisition system.

(2) Power up the control system. Under the drive of the motor, the chain transmission mechanism moves the conveyed object forward. The PLC control system collects the current, while the data acquisition system collects the signals from the displacement sensor and the PLC's synchronization signals.

(3) To minimize the impact of external disturbances on parameter uncertainty, perform the test three times under the same conditions.

(4) Data collection and processing: Export data from the PLC control system and data acquisition software (TranAX_3.4.1.1222, the USA). Extract valid data during the motion process of the conveyed object to conduct subsequent dynamic simulations and experimental comparisons. Then, take the average of the three data sets under the same operating conditions.

Through bench tests, the current i_{q} of the chain transmission mechanism is obtained. Based on the motor electromagnetic torque equation, the motor torque T_{e} can be expressed as

$$T_{\mathrm{e}} = K_{\mathrm{T}} i_{\mathrm{q}} \tag{22}$$

where K_T is the motor torque constant. The motor torque is transmitted to the sprocket shaft through the worm gear transmission mechanism, with a transmission ratio of k. Therefore, the torque on the sprocket is given by

$$T_s = kT_e \tag{23}$$

The current driven by the motor during the transmission process is obtained through bench tests. The motor torque is then converted into the sprocket torque using Equation (23). As shown in Figure 9, for ease of analysis, the sprocket torque is approximated and represented by the red line in the figure. The converted sprocket torque is then input into the simulation dynamic model of the chain transmission mechanism, and the simulation results for the displacement of the transmitted object are compared with experimental results, as shown in Figure 10. From Figure 10, it can be seen that the simulation results are in good agreement with the experimental data. This validates the correctness and effectiveness of the model. In subsequent calculations, this simulation model will be used to replace the real physical model for analysis.

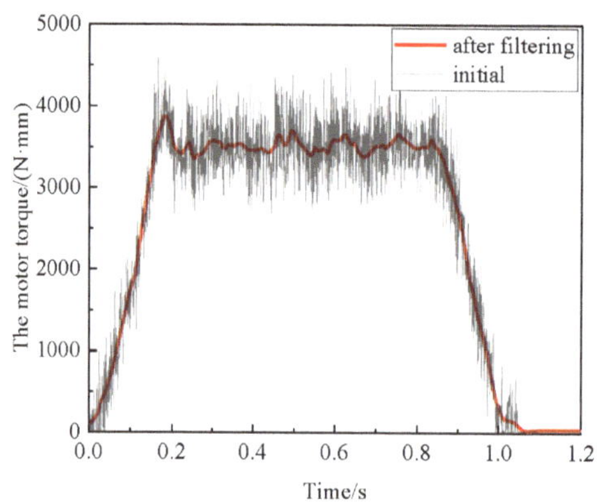

Figure 9. The torque curve for the sprocket.

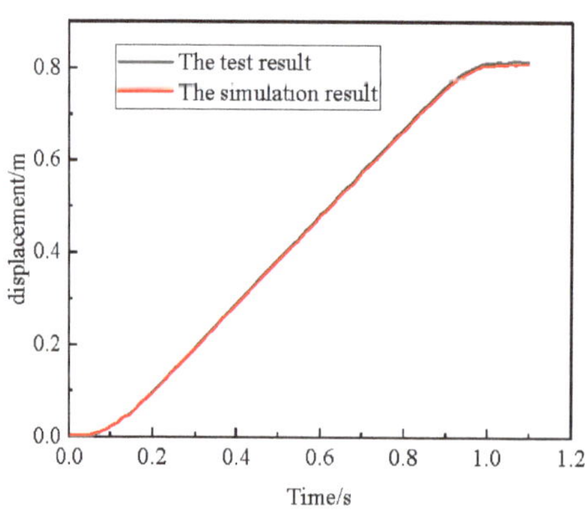

Figure 10. The comparison of simulation displacement and experiment displacement for the modular charge.

3. Robust Optimization Design Model for Chain Transmission Process Considering Random Loads

3.1. Random Parameters of the Chain Transmission Process

Based on design requirements and engineering experience, the random parameters considered in this paper for the chain transmission mechanism include the following.

(1) Structural dimension parameters: roller radius R_r, sprocket tooth groove radius R_1, sprocket transition arc radius R_2, initial position of the transmitted object relative to the end face of the receiving cylinder L_m, transmission cylinder radius R_t, and the inner diameter of the receiving cylinder R_s. To enhance the robustness of the chain transmission mechanism, the above structural dimension parameters are treated as design parameters, while considering the influence of parameter random fluctuations.

(2) Physical property parameters: the coefficient of friction between the transmitted object and the transmission track and receiving cylinder μ_s, and the mass of the transmitted object m_m. These parameters also have an influence on the positioning consistency of the transmitted object but are difficult to design. Therefore, they are not considered as design parameters, and only the effect of random fluctuations in these parameters is taken into account.

(3) Time-varying random loads: Based on the multiple torque curves after tests (Figure 11), it is observed that although the same motor speed is given, the driving load exhibits time-varying uncertainty. The fluctuations in the load have a significant effect on the positioning accuracy of the chain transmission mechanism. Therefore, the effect of time-varying random loads is considered.

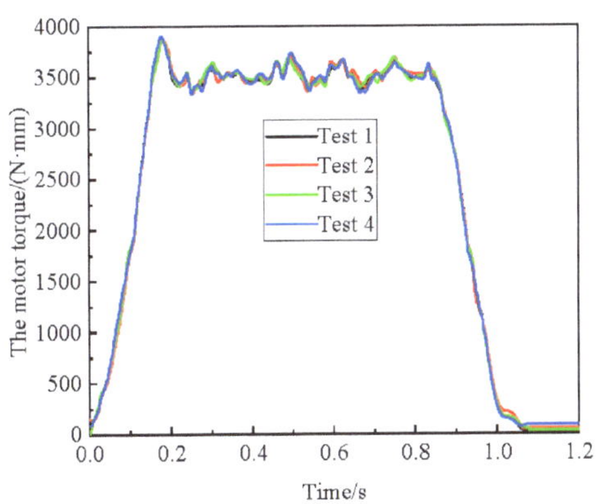

Figure 11. The torque design curve for the sprocket.

To achieve robust design of the chain transmission mechanism, the parameters are classified into controllable and uncontrollable parameters based on whether the uncertain input parameters can be controlled. Controllable parameters refer to those that can be easily altered during the design and manufacturing process, while uncontrollable parameters are those that are difficult or impossible to change during design and manufacturing. Among the controllable parameters, there are random variables $\boldsymbol{X}_v = [R_1, R_2, R_t, R_s, R_r, L_m]$ and non-random variables $\boldsymbol{X}_c = []$. Similarly, the uncontrollable parameters are also classified into random variables $\boldsymbol{P}_v = [m_m, \mu_s]$ and non-random variables $\boldsymbol{P}_c = [R_m]$.

3.2. Random Driving Load Analysis Based on Karhunen–Loeve Expansion

In multiple chain transmission processes under the same working conditions, the driving load exhibits certain fluctuations, and thus, can be treated as a dynamic random process. To ensure the accuracy of the effect of the driving load on the positioning accuracy of the conveyed object, the sampling time interval is generally set to 0.001 s. As a result, the entire random process of the driving load is discretized into thousands of random variables, which poses significant challenges for the subsequent uncertainty propagation of this dynamic uncertain parameter. To address this, this paper adopts the Karhunen–Loeve expansion (K-L expansion) method [33], which reduces the dimensionality of the time series random driving load with correlation, thereby providing strong support for the subsequent uncertainty propagation. It should be noted that the K-L expansion relies on the covariance matrix, which requires knowledge of the complete covariance structure in the problem. In the theory of random processes, the K-L expansion represents the random dynamic load as an infinite linear combination of orthogonal functions based on the spectral decomposition of the covariance function. Considering the probability space of the driving load, the covariance function of the random process $P(t)$ within the bounded interval [0, T] is $C(t_m, t_n)$, and the process can be expressed as

$$P(t) = \mu_P(t) + \sum_{i=1}^{\infty} \sqrt{\gamma_i}\xi_i\varphi_i(t) \tag{24}$$

where, $t \in [0, T]$, $\mu_P(t)$ is the mean function of the driving load random process, and ξ_i represents i-th mutually independent standard random variable. γ_i and $\varphi_i(t)$ are the i-th eigenvalue and eigenfunction of the covariance function of the driving load random process, respectively. According to Mercer's theorem [33], the covariance function has the following spectral decomposition:

$$C(t_m, t_n) = \sum_{i=1}^{\infty} \gamma_i\varphi_i(t_m)\varphi_i(t_n) \tag{25}$$

The eigenvalues γ_i and eigenfunctions $\varphi_i(t)$ can be obtained by solving the following second-kind Fredholm integral equation

$$\int_0^T C(t_m, t_n)\varphi_i(t_n)dt_m = \gamma_i\varphi_i(t_n) \tag{26}$$

and satisfies

$$\sum_{j=1}^{\infty} \gamma_j = |T| \tag{27}$$

The eigenfunctions in Equation (26) form a complete orthogonal set that satisfies the equation.

$$\int_0^T \varphi_i(t)\varphi_j(t) = \delta_{ij} \tag{28}$$

In Equation (28), δ_{ij} is the Kronecker delta function. For uncorrelated standard random variables ξ_i, they can be obtained by the following equation:

$$\xi_i = \frac{1}{\sqrt{\gamma_i}}\int_0^T [P(t) - \mu_P(t)]\varphi_i(t)dt \tag{29}$$

In the analysis of random driving loads in the chain transmission mechanism, the eigenvalues are often arranged in descending order, and the random process is approximated by a finite number of terms. For example, the truncated random process after M terms can be expressed as

$$\widetilde{P}(t) = \mu_P(t) + \sum_{i=1}^{M} \sqrt{\gamma_i}\xi_i\varphi_i(t) \tag{30}$$

Its covariance expansion function is

$$C(t_m, t_n) = \sum_{i=1}^{M} \gamma_i \varphi_i(t_m) \varphi_j(t_n) \tag{31}$$

The error $\varepsilon_\mu(t)$ in the mean function of the approximate random process $\widetilde{P}(t)$ of the load and the mean squared error $\bar{\varepsilon}_U^2(t)$ can be expressed as

$$\varepsilon_\mu(t) = \mu_P(t) - \widetilde{\mu}_P(t) \equiv 0 \tag{32}$$

$$\bar{\varepsilon}_U^2(t) = \frac{1}{T}\int_T \left(\frac{\sum\limits_{i=1}^{\infty} \sqrt{\gamma_i}\xi_i \varphi_i(t)}{\sigma(t)} - \frac{\sum\limits_{i=1}^{M} \sqrt{\gamma_i}\xi_i \varphi_i(t)}{\sigma(t)} \right) dt \le 1 - \frac{1}{|T|}\sum_{i=1}^{\infty} \lambda_j \tag{33}$$

Therefore, the approximation degree κ of the load random process can be defined as

$$\kappa = \frac{1}{|T|}\sum_{j=1}^{M} \lambda_j \tag{34}$$

3.3. Chain Transmission Mechanism Surrogate Model Based on Polynomial Chaos Expansion

Due to the high non-linearity of the chain transmission mechanism model and the long computation time for each run, multiple model evaluations are required during the robust optimization design process, which significantly increases the computational cost. To reduce the computational cost while ensuring accuracy, this paper uses the Polynomial Chaos Expansion (PCE) model to establish a surrogate model for the chain transmission mechanism. It should be noted that high-order polynomial terms in PCE will rapidly increase computational complexity, especially when there are many random variables in the problem, the number of expanded terms will increase exponentially. Moreover, PCE assumes that random variables follow certain known probability distributions (usually normal or uniform), and its effectiveness may be poor for uncertainties with heavy tailed distributions (such as Laplace distributions).

The idea of constructing a surrogate model for the chain transmission mechanism using PCE [34] is to expand the output response of the chain transmission mechanism into a form of orthogonal polynomials corresponding to e uncertainty input parameters. Let the i-th input parameter of the chain transmission mechanism be denoted as $x_i = [a_i, b_i], (i = 1, 2, \cdots, e)$, and its probability distribution function corresponding to the standard orthogonal polynomial $\Psi(\widetilde{x}_i)$. Here, $\widetilde{x}_i$ represents the definition of the input parameter $\widetilde{x}_i$ in the corresponding chaos polynomial space, and the corresponding value range is $\widetilde{x}_i = [c_i, d_i], (i = 1, 2, \cdots, e)$.

The input parameters x_i of the chain transmission mechanism and their corresponding definitions $\widetilde{x}_i$ in the chaos polynomial space satisfy the following relationship:

$$\widetilde{x}_i = \frac{d_i - c_i}{b_i - a_i} x_i + \frac{c_i b_i - a_i d_i}{b_i - a_i} \tag{35}$$

Based on the idea of sparse chaos polynomials, the output response $y = f(\boldsymbol{x})$ of the chain transmission mechanism is expanded into a series of finite order s, expressed as a sum of orthogonal polynomials of the input parameters, as follows:

$$y = f(\boldsymbol{x}) = \sum_{g=1}^{s} \beta_g \phi_{\boldsymbol{u}_g}(\widetilde{\boldsymbol{x}}) \tag{36}$$

$$\phi_{u_g}(\widetilde{x}) = \prod_{h=1}^{e} \Psi_{u_{gh}}(\widetilde{x}_h) \tag{37}$$

where, β_g is the polynomial expansion coefficient, $\phi_{u_g}(\widetilde{x})$ is the multi-variate polynomial, u_{gh} is the degree of the polynomial $\Psi_{u_{gh}}(\widetilde{x}_h)$, s is the total number of chaos polynomial expansion terms, and $\boldsymbol{u}_g$ is the set of polynomial degrees.

According to the orthogonality property of the orthogonal polynomials, the following can be obtained:

$$\left\langle \phi_{u_g}(\widetilde{x}), \phi_{u_h}(\widetilde{x}) \right\rangle = \left\langle \phi_{u_g}(\widetilde{x})^2 \right\rangle \delta_{gh} \tag{38}$$

where $\langle\cdot\rangle$ denotes the inner product with the joint probability density function as the weight function, δ_{gh} is the Kronecker delta function, and if $g = h$, then $\delta_{gh} = 1$; otherwise, $\delta_{gh} = 0$.

Based on Equation (36), the PCE model of the chain transmission mechanism can be obtained. However, due to the large number of PCE terms, the computational load has increased. To address this issue, reference [35] proposes a sparse chaos polynomial expansion strategy to reduce the number of PCE terms. The Elastic Net (EN) method is used to identify the terms in the PCE model that are important to the result, thus improving computational efficiency while maintaining accuracy. The polynomial coefficient $\widetilde{\beta}_{\mathrm{EN}}$ is then represented as

$$\widetilde{\beta}_{\mathrm{EN}} = \arg\min_{\beta}\left(\left\|Y - \boldsymbol{\phi}_{\boldsymbol{u}}^{\mathrm{T}}\widetilde{\beta}\right\|_2 + \lambda_1\left\|\widetilde{\beta}\right\|_1 + \lambda_2\left\|\widetilde{\beta}\right\|_2\right) \tag{39}$$

where $\underset{\beta}{\operatorname{argmin}}(\cdot)$ denotes the minimization functional, λ_1 and λ_2 are adjustment parameters, Y is the output, $\boldsymbol{\phi}_{\boldsymbol{u}}^{\mathrm{T}}$ is the transpose of the multi-variate polynomial set, and $\widetilde{\beta}$ represents the set of polynomial coefficients. The Least Angle Regression (LAR) algorithm [36] is used to improve the computational efficiency of the EN method. The final SPCE surrogate model is then expressed as

$$y = \boldsymbol{\phi}_{\boldsymbol{B}}^{\mathrm{T}}(\widetilde{x})\widetilde{\beta}_{\mathrm{EN}} \tag{40}$$

where $\boldsymbol{\phi}_{\boldsymbol{B}}^{\mathrm{T}}(\widetilde{x})$ is the set of multi-variate polynomials retained after EN selection.

3.4. Robust Optimization Design Model for Chain Transmission Mechanism

Considering the effect of the random variation of uncontrollable parameters $\boldsymbol{P}_{\mathrm{v}}$, and given the value of the non-random variation of the uncontrollable parameters $\boldsymbol{P}_{\mathrm{c}}$, with controllable parameters $\boldsymbol{X}$ (including both $\boldsymbol{X}_{\mathrm{v}}$ and $\boldsymbol{X}_{\mathrm{c}}$) as design parameters, setting the target position as y_0, the actual position y_{end} of the transmitted object at arrival and the difference from the target position is $Y_{\mathrm{m}} = y_{\mathrm{end}} - y_0$. The optimization objective is to minimize the mean $\mu_{Y_{\mathrm{m}}}$ of Y_{m} and the variance σ_y of y_{end}, with constraints on the arrival speed, arrival position, and the range of uncertain input parameters. The robust optimization design model for the position consistency of the chain transmission mechanism is established as follows:

$$\begin{cases} \text{find} & \boldsymbol{X}_{\mathrm{v}}, \boldsymbol{X}_{\mathrm{c}} \\ \min & (\mu_{Y_{\mathrm{m}}}, \sigma_y) \\ \text{s.t.} & \begin{cases} \dot{y}_{\mathrm{end}} \approx 0, \\ \boldsymbol{X}_{\mathrm{l}} \le \boldsymbol{X} \le \boldsymbol{X}_{\mathrm{u}}, \\ y_{end} - a\sigma_y \ge y_{\min} \end{cases} \end{cases} \tag{41}$$

where, $\dot{y}_{\mathrm{end}}$ represents the arrival speed of the conveyed object, $\boldsymbol{X}_{\mathrm{l}}$ and $\boldsymbol{X}_{\mathrm{u}}$ represent the sets of the lower and upper bounds of the design parameters, respectively, a represents the constraint index for measuring robustness, and $y_{\min}$ represents the minimum critical value for the conveyed object arrival.

From Equation (41), it can be seen that the robust optimization design objective for the conveyed object arrival consistency requires the displacement of the conveyed object to reach a specific position, while ensuring that the variance of the arrival position of the conveyed object is minimized. This is a typical multi-objective optimization problem. For the multi-objective robust optimization problem of the conveyed object arrival, there is generally no unique optimal solution, but rather a set of effective solutions for the conveyed object arrival and its variance, known as the Pareto optimal solution set. The Pareto optimal solution set refers to a set of solutions where at least one objective function is better than the other solution outside the set, while the other objective functions of the solution are not worse than those of solutions outside the set. Based on the combination of the Pareto optimal solution set, multiple objectives can be considered comprehensively to select the appropriate solutions.

Commonly used multi-objective optimization algorithms include the Non-dominated Sorting Genetic Algorithm II (NSGA-II) and the Multi-objective Particle Swarm Optimization (MOPSO) algorithm. The latter is widely used due to its advantages of fast runtime and good convergence of the solution set. This paper uses the widely applied MOPSO algorithm to solve the multi-objective robust optimization problem.

The computational steps in this paper are as follows:

(1) Obtain multiple sets of dynamic driving loads through experiments and quantify the uncertainties, obtaining random dynamic load characteristics, such as mean, variance, autocorrelation function, and other stochastic features.
(2) Solve the Fredholm integral equation using Equation (26) and arrange the obtained eigenvalues in descending order. Based on the given approximation degree, select the top M eigenvalues and their corresponding eigenfunctions.
(3) For the reduced-dimensional random dynamic loads and uncertain input parameters, perform sampling and input them into the dynamic model to calculate the output response.
(4) Use Equation (40) to establish the surrogate model of the chain transmission mechanism, relating the uncertain input parameters to the output response.
(5) Combine Equation (41) and the surrogate model of the chain transmission mechanism to establish the robust optimization design model for the chain transmission mechanism.
(6) Apply the MOPSO algorithm to solve the robust optimization design model of the chain transmission mechanism and obtain the optimal arrival displacement and its corresponding design parameters.

4. Case Study Analysis

As an example, a chain transmission mechanism used in engineering is taken, and its dynamic model is established for the study of robust optimization design. The input parameters are shown in Table 1. The statistical characteristics of the load uncertainty are obtained from multiple sets of experimental data.

Table 1. Uncertainty input parameters of the propellant transport process.

Serial Number	Parameters	Nominal Value	Error Type	Error	Design Range
1	m_m/kg	3	Normal	0.03	—
2	μ_s	0.2	Uniform	[0.19, 0.21]	—
3	R_1	12.7	Uniform	[0, 0.002]	[12.6, 12.8]
4	R_2	27.29	Uniform	[0, 0.002]	[25, 30]
5	R_t	82	Uniform	[0, 0.002]	[80, 84]
6	R_s	82	Uniform	[0, 0.002]	[80, 84]
7	R_m	78	—	—	—
8	R_r/mm	12.5	Uniform	[0, 0.002]	[12.4, 12.6]
9	L_m/mm	0	Uniform	[0, 0.1]	[0, 5]

Notes: "—" indicates no data.

Considering both computational accuracy and efficiency, the uncertainty input parameters in Table 1, along with the motor drive parameters after dimensionality reduction through K-L expansion, are sampled using the optimized Latin Hypercube Sampling technique to generate 500 combinations of uncertain input parameters within the parameter ranges. These are input into the simulation dynamic model of the chain transmission mechanism to obtain the corresponding displacement output results. Based on the surrogate model method described in Section 3.3, a surrogate model for the chain transmission mechanism is established. Subsequently, 50 sets of parameters are randomly sampled and input into both the original dynamic model and the PCE surrogate model. A comparison of the results is shown in Figure 12. The comparison of calculation accuracy between the proposed method and other methods is shown in Table 2. The commonly used multiple correlation coefficient R^2 is employed to validate the accuracy of the surrogate model. The computed value of $R^2 = 0.9558$ with the proposed method indicates that the surrogate model obtained by PCE has a higher accuracy than others and meets engineering requirements, making it suitable for use in the robust optimization design solution.

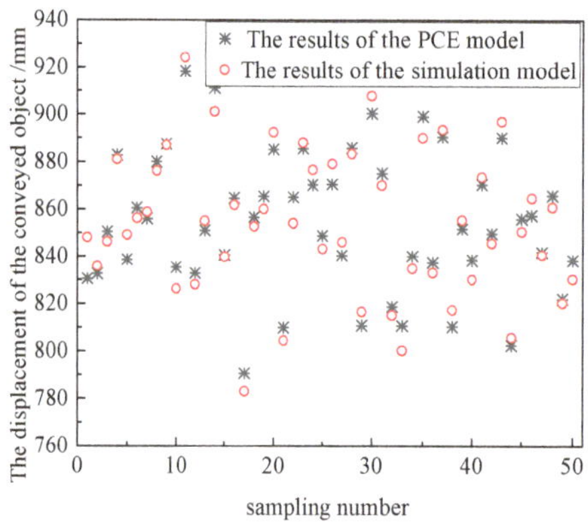

Figure 12. The test result of the propellant transport mechanism's PCE model.

Table 2. The comparison of calculation accuracy between the proposed method and other methods.

Methods	Correlation Coefficient R^2
PCE	0.1528
KL-PCE	0.9558
KL-Kriging	0.8547
KL-RBF	0.9234

Considering both computational efficiency and accuracy, the parameters for the MOPSO algorithm are set as follows: the number of particles is 100, and the number of iterations is 300. The robust target parameter values are set to $y_0 = 820$ mm, $a = 3$, $y_{\min} = 800$ mm. The final Pareto optimal solution set is obtained, and the results are shown in Figure 13.

Based on the Pareto optimal solution set, different optimized parameter combinations can be obtained, providing convenience for designers to make selections. Table 3 shows a portion of the Pareto solution set corresponding to the optimized input parameter combinations. It can be observed that the consistency of the optimized design parameters is relatively good.

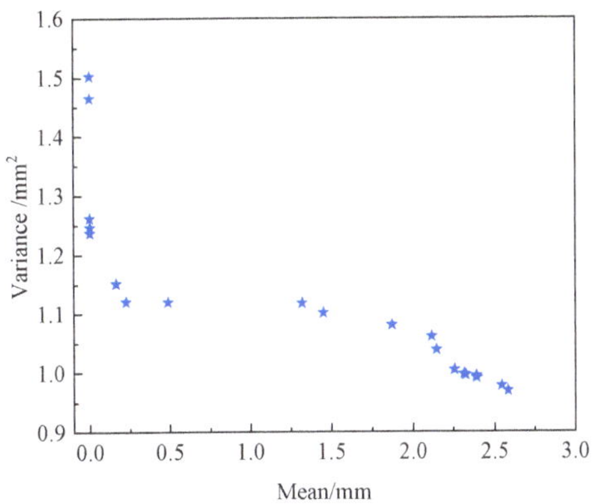

Figure 13. The Pareto optimal solution set of the propellant transport process.

Table 3. The results of robust optimization of propellant transport.

Serial Number	R_1	R_2	R_t	R_s	R_r	L_m
1	12.6548	26.5823	83.0432	83.0384	12.4935	0.089
2	12.6544	26.5583	82.1473	82.1504	12.4933	0.043
3	12.6539	26.5476	83.0571	83.0413	12.4935	0.037
4	12.6489	26.5716	83.0147	83.0274	12.4945	0.049
5	12.6435	26.5804	82.7831	82.5731	12.4946	0.015
6	12.6601	26.5749	82.4397	82.3176	12.4945	0.085
7	12.6573	26.5614	83.0312	83.1047	12.4935	0.016
8	12.6683	26.5743	82.8734	82.7496	12.4935	0.0014
9	12.6415	26.5827	83.0187	83.0674	12.4945	0.0023
10	12.6573	26.5479	83.0243	83.0249	12.4934	0.0018

Five hundred sets of samples are taken within the error range of the initial values (nominal values) in Table 1, and these are input into the chain transmission mechanism dynamic model to obtain 500 sets of the conveyed object arrival values. A statistical analysis of the conveyed object arrival values at the initial values is conducted to derive the probability density function (as shown by the curve on the right side of Figure 14). Then, the optimized design parameters (with the average of the 10 optimized results in Table 3 taken as the optimal solution) replace the nominal values in Table 1 for resampling 500 sets. These are then input into the chain transmission mechanism dynamic model to calculate 500 sets of the conveyed object arrival values. A statistical analysis of the optimized conveyed object arrival values is conducted to derive the new probability density function (as shown by the curve on the left side of Figure 14).

From Figure 14, it can be seen that both before and after optimization, the conveyed object arrival positions are approximately normally distributed. The optimized conveyed object arrival position ensures that the target value is reached, and the variance of the conveyed object arrival position is reduced from the original 3.1515 to 1.1615. The optimized variance is 63.14% smaller than the initial variance, indicating a significant improvement in the consistency of the conveyed object arrival. By rounding the design parameters in Table 3 and comparing them with the initial design parameters, it can be observed that based on the nominal values, R_1 and R_2 can be slightly reduced, and R_r and L_m can be basically unchanged, while R_t and R_s can be slightly increased. These adjustments can serve as a reference in the design process.

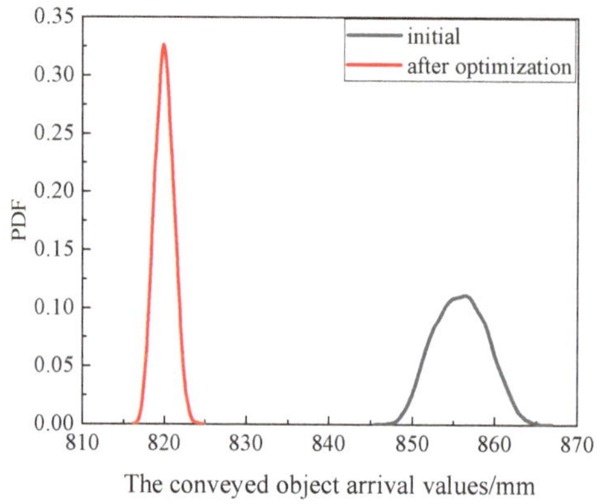

Figure 14. PDF of the modular charge displacement before and after optimization.

5. Conclusions

This paper addresses the robust optimization design problem of chain transmission mechanisms considering time-varying load uncertainties. A robust optimization design method for chain transmission mechanisms based on K-L expansion and PCE methods is proposed. The motion mechanism of multi-contact modes in chain transmission systems is revealed, and a dynamic model of the chain transmission mechanism is established and validated through experiments. The time-varying random loads are dimensionally reduced using the K-L expansion, and then, based on the PCE method, a fast-mapping relationship between uncertainty parameters and output responses is established. Finally, with the objective of minimizing the mean and standard deviation of the conveyed object positioning accuracy, and with position and velocity at the destination as constraints, a robust optimization design model for the position consistency of the chain drive conveyor is established. The multi-objective particle swarm optimization algorithm is applied for solving the model, and the following conclusions can be obtained:

(1) The K-L expansion method can effectively reduce the dimensionality of time-varying random loads, solving the high-dimensional modeling difficulties by the PCE method.
(2) A surrogate model for the chain transmission mechanism is established based on the PCE method, which greatly improves computational efficiency and saves computational costs while ensuring calculation accuracy.
(3) The robust optimization design method can effectively improve the transmission accuracy of the chain transmission mechanism. Compared to the initial design, the variance of the transported object is reduced by 63.14%.
(4) The optimized results prove the effectiveness of the proposed method in improving the consistency of the conveyed object positioning. This method and its results provide theoretical guidance for the design of chain transmission mechanisms.

This article also has certain limitations. Due to the complexity of the chain transmission mechanism model, certain simplifications and assumptions were made during the dynamic modeling process, which had a slight effect on the accuracy of the dynamic model. In future work, the method proposed in this paper can be extended to the design process of other chain transmission mechanisms to improve motion robustness. Furthermore, to enhance the service reliability of chain transmission mechanisms, further research can focus on time-varying robust optimization design based on reliability. This will better promote the application and development of chain transmission systems in complex environments, providing stronger theoretical support and technical assurance for the engineering design of chain transmission mechanisms.

Author Contributions: T.L.: software, methodology, validation, investigation, writing—original draft, writing—review and editing. Y.L.: validation, resources. P.L.: software, writing—original draft. X.D.: funding acquisition, writing—review and editing. All authors have read and agreed to the published version of the manuscript.

Funding: This work was supported by the National Natural Science Foundation of China (52405275), the Scientific Research Foundation for High-level Talents of Nanjing Institute of Technology (YKJ202104), and the Natural Science Foundation of the Higher Education Institutions of Jiangsu Province, China (24KJB460017). The authors also gratefully thank the editors and reviewers of this manuscript.

Data Availability Statement: Data are contained within the article.

Conflicts of Interest: The authors declare that they have no conflicts of interest in this work. We declare that we do not have any commercial or associative interest that represents a conflict of interest in connection with the work submitted.

Nomenclature

n_t	the teeth number of the sprocket	R_s	the original radius of the sprocket
α	the pitch angle of the sprocket	$\boldsymbol{r}_s$	position vector of the sprocket's center in the global coordinate system
$\boldsymbol{r}_{rc}$	the position vector of the roller in the global coordinate system	$\boldsymbol{s}_{rc}$	the vector from the sprocket center to the roller center
$O-XY$	global coordinate system	$O_s-X_sY_s$	sprocket coordinate system
$O_s-\xi_{sn_i}\eta_{sn_i}$	local coordinate system of the tooth groove	A_{sn_i}	the coordinate transformation matrix
R_1	the radius of the circular arcs in the respective segments *ab* and *cd*	R_2	the radius of the circular arcs in the respective segment *bc*
R_r	the radius of the roller	$\boldsymbol{e}_{ab},\boldsymbol{e}_{bc},\boldsymbol{e}_{cd}$	the eccentric vector between the center of the roller and the center of the bottom arc
$\boldsymbol{s}_{O_{ab}},\boldsymbol{s}_{O_{bc}},\boldsymbol{s}_{O_{cd}}$	the vector from the center of the bottom arc to the center of the sprocket	$\theta_{ab},\theta_{bc},\theta_{cd}$	angle of arc segment *ab*, *bc*, and *cd*
θ_1	the angle between $\boldsymbol{s}_{O_{ab}a}$ and $\boldsymbol{e}_{ab}$	θ_2	the angle between $\boldsymbol{e}_{bc}$ and $-\boldsymbol{s}_{O_{bc}}$
$\delta_{ab},\delta_{bc},\delta_{cd}$	penetration between the roller and the sprocket tooth groove	$\boldsymbol{r}_Q,\boldsymbol{r}_P$	the position vector from the origin of the global coordinate system to points *Q and P*
$\boldsymbol{r}_{PQ}$	the position vector from point *P* to point *Q*	δ_p	penetration between *P* and *Q*
F_n	the normal collision force	K	the stiffness coefficient of the two contacting bodies
h	the penetration depth of the contact bodies	n	the contact index
C	the contact damping coefficient	$\dot{h}$	the relative collision velocity at the contact point
K_p	the equivalent stiffness of the planar contact	F_t	the tangential friction force
c_f	the coefficient of sliding friction	c_d	the dynamic correction factor
v_t	the relative tangential velocity	t	the motion time of the transmission process
$q,\dot{q},\ddot{q}$	the relative displacement, relative velocity, and relative acceleration of the rotational joints in the chain transmission mechanism	$\boldsymbol{\Phi}(q,t)$	constraint equation of the chain transmission mechanism
$\boldsymbol{\Phi}_q$	the derivative of $\boldsymbol{\Phi}(q,t)$ with respect to q	λ	the Lagrange multiplier
$\boldsymbol{M}(q,t)$	the mass matrix of the chain transmission mechanism	$C(q,\dot{q},t)$	the generalized damping matrix of the chain transmission mechanism
Q	the external force acting on the chain transmission mechanism	i_q	the current of the chain transmission mechanism
T_e	the motor torque	K_T	the motor torque constant
k	transmission ratio		

Appendix A

Appendix A.1 Kinematics Equation of the Mechanism

In order to establish the kinematics equation of the mechanism with the Newton Euler method represented by the relative coordinate method, it is necessary to derive the kinematics relationship between adjacent rigid bodies. $O_rx_ry_rz_r$ and $O_cx_cy_cz_c$ are used as examples to establish the motion relationship between two adjacent rigid bodies, as shown in Figure A1. $O_0x_0y_0z_0$ is the global coordinate system, $p_{rO}\xi_{rO}\eta_{rO}\zeta_{rO}$ and $p_{cI}\xi_{cI}\eta_{cI}\zeta_{cI}$ are the coordinate systems where hinge J_c is connected to the coordination cylinder piston (represented by B_r) and the coordination hydraulic cylinder (represented by B_c), respectively.

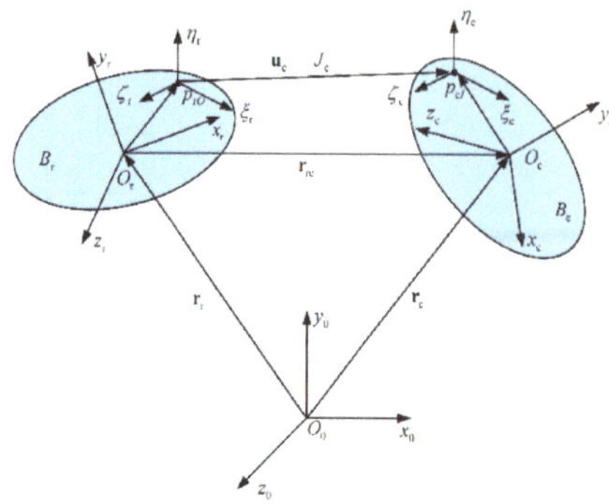

Figure A1. The position relation of adjacent rigid bodies.

Define the velocity and acceleration vectors of $O_rx_ry_rz_r$ and $O_cx_cy_cz_c$ relative to $O_0x_0y_0z_0$ as

$$\mathbf{S}_r = \begin{bmatrix} \boldsymbol{\omega}_r \\ \boldsymbol{v}_r \end{bmatrix},\ \mathbf{S}_c = \begin{bmatrix} \boldsymbol{\omega}_c \\ \boldsymbol{v}_c \end{bmatrix},\ \dot{\mathbf{S}}_r = \begin{bmatrix} \boldsymbol{\varepsilon}_r \\ \boldsymbol{a}_r \end{bmatrix},\ \dot{\mathbf{S}}_c = \begin{bmatrix} \boldsymbol{\varepsilon}_c \\ \boldsymbol{a}_c \end{bmatrix} \tag{A1}$$

where ω, ε, v, and $\boldsymbol{a}$ represent the angular velocity, angular acceleration, linear velocity, and linear acceleration of the joined body coordinate system, respectively; subscripts r and c indicate the serial number of the bodies; and the relative velocity and relative acceleration vectors between the hinge coordinate systems $p_{cI}\xi_{cI}\eta_{cI}\zeta_{cI}$ and $p_{rO}\xi_{rO}\eta_{rO}\zeta_{rO}$ are $\dot{q}_c$ and $\ddot{q}_c$, respectively.

According to the relative coordinate method, the relationship between the motion vector of the body coordinate system and the relative motion vector of the hinge coordinate system can be recursively obtained.

$$\mathbf{S}_c = \boldsymbol{T}_c\mathbf{S}_r + \boldsymbol{D}_c\dot{\boldsymbol{q}}_c \tag{A2}$$

$$\dot{\mathbf{S}}_c = \boldsymbol{T}_c\dot{\mathbf{S}}_r + \dot{\boldsymbol{T}}_c\mathbf{S}_r + \boldsymbol{D}_c\ddot{\boldsymbol{q}}_c + \dot{\boldsymbol{D}}_c\dot{\boldsymbol{q}}_c \tag{A3}$$

where $\boldsymbol{T}_c$, $\dot{\boldsymbol{T}}_c$, $\boldsymbol{D}_c$, and $\dot{\boldsymbol{D}}_c$ are the corresponding recursive relationship matrix obtained by the relative coordinate method. According to Equations (A5) and (A6), the comprehensive motion relationship equation can be obtained by integrating the three bodies of the coordination mechanism as follows:

$$\mathbf{S} = \boldsymbol{H}\dot{\boldsymbol{q}} \tag{A4}$$

$$\dot{\mathbf{S}} = \boldsymbol{H}\ddot{\boldsymbol{q}} + \dot{\boldsymbol{H}}\dot{\boldsymbol{q}} \tag{A5}$$

where $\mathbf{S} = \begin{bmatrix} \mathbf{S}_1 & \mathbf{S}_2 & \mathbf{S}_3 \end{bmatrix}^{\mathrm{T}}$; $\dot{\mathbf{S}} = \begin{bmatrix} \dot{\mathbf{S}}_1 & \dot{\mathbf{S}}_2 & \dot{\mathbf{S}}_3 \end{bmatrix}^{\mathrm{T}}$; $\dot{q} = \begin{bmatrix} \dot{q}_1 & \dot{q}_2 & \dot{q}_3 \end{bmatrix}^{\mathrm{T}}$; $\ddot{q} = \begin{bmatrix} \ddot{q}_1 & \ddot{q}_2 & \ddot{q}_3 \end{bmatrix}^{\mathrm{T}}$; and $\boldsymbol{H}$ is the corresponding coefficient matrix.

Appendix A.2 Dynamics Equation of the Mechanism

For any rigid body $B_i (i = \mathrm{a}, \mathrm{r}, \mathrm{c})$, assuming c_i is the vector of any point in the body relative to the connected coordinate system $O_ix_iy_iz_i$, the acceleration at this point can be obtained as follows:

$$\boldsymbol{a}_{c_i} = \boldsymbol{a}_i + \boldsymbol{\varepsilon}_i \times \boldsymbol{c}_i + \boldsymbol{\omega}_i \times (\boldsymbol{\omega}_i \times \boldsymbol{c}_i) \tag{A6}$$

where $\boldsymbol{a}_i$ is the acceleration of O_i relative to $O_0x_0y_0z_0$.

According to the principle of virtual power, the virtual power equation of the body can be obtained as follows:

$$\int_V \delta \boldsymbol{v}_{c_i}^{\mathrm{T}} (\rho \boldsymbol{a}_{c_i} + \boldsymbol{F}_i^{\mathrm{a}} + \boldsymbol{F}_i^{\mathrm{n}}) \mathrm{d}V = \mathbf{0} \tag{A7}$$

where $\boldsymbol{F}_i^{\mathrm{a}}$ and $\boldsymbol{F}_i^{\mathrm{n}}$ are the generalized external loads and ideal hinge constraint loads on the body, respectively; ρ is the density of the body; $\delta \boldsymbol{v}_{c_i}$ is the virtual velocity as

$$\delta \boldsymbol{v}_{c_i} = \delta \boldsymbol{v}_i + \delta(\boldsymbol{\omega}_i \times \boldsymbol{c}_i) \tag{A8}$$

Substitute Equations (A1), (A2), and (A6) into Equation (A7), and apply the virtual power equation of a single body to the coordinate system, and then organize them into a compact form as follows:

$$\delta \boldsymbol{S}^{\mathrm{T}} \boldsymbol{R} \dot{\boldsymbol{S}} + \delta \boldsymbol{S}^{\mathrm{T}} \boldsymbol{W} \boldsymbol{R} \boldsymbol{E} \boldsymbol{S} = \delta \boldsymbol{S}^{\mathrm{T}} \boldsymbol{Q}^{\mathrm{a}} + \delta \boldsymbol{S}^{\mathrm{T}} \boldsymbol{Q}^{\mathrm{n}} \tag{A9}$$

where $\boldsymbol{R}$, $\boldsymbol{W}$, $\boldsymbol{E}$, $\boldsymbol{Q}^{\mathrm{a}}$, and $\boldsymbol{Q}^{\mathrm{n}}$ are the integration forms of each item in Equation (A7) in the coordinate system.

Combining the kinematics equation, substitute Equations (A4) and (A5) into Equation (A9) and obtain

$$\begin{aligned} &\delta \dot{\boldsymbol{q}}^{\mathrm{T}} \boldsymbol{H}^{\mathrm{T}} \boldsymbol{R} \boldsymbol{H} \ddot{\boldsymbol{q}} + \delta \dot{\boldsymbol{q}}^{\mathrm{T}} \boldsymbol{H}^{\mathrm{T}} \left(\boldsymbol{R} \dot{\boldsymbol{H}} + \boldsymbol{W} \boldsymbol{R} \boldsymbol{E} \boldsymbol{H} \right) \dot{\boldsymbol{q}} \\ &= \delta \dot{\boldsymbol{q}}^{\mathrm{T}} \boldsymbol{H}^{\mathrm{T}} \boldsymbol{Q}^{\mathrm{a}} + \delta \dot{\boldsymbol{q}}^{\mathrm{T}} \boldsymbol{H}^{\mathrm{T}} \boldsymbol{Q}^{\mathrm{n}} \end{aligned} \tag{A10}$$

Due to the constraint force and torque not doing work, their virtual powers are zero; write Equation (A10) in compact form as follows:

$$\delta \dot{\boldsymbol{q}}^{\mathrm{T}} \boldsymbol{M} \ddot{\boldsymbol{q}} + \delta \dot{\boldsymbol{q}}^{\mathrm{T}} \boldsymbol{C} \dot{\boldsymbol{q}} = \delta \dot{\boldsymbol{q}}^{\mathrm{T}} \boldsymbol{F} \tag{A11}$$

where $\boldsymbol{M} = \boldsymbol{H}^{\mathrm{T}} \boldsymbol{R} \boldsymbol{H}$, $\boldsymbol{C} = \boldsymbol{H}^{\mathrm{T}} \left(\boldsymbol{R} \dot{\boldsymbol{H}} + \boldsymbol{W} \boldsymbol{R} \boldsymbol{E} \boldsymbol{H} \right)$,$\boldsymbol{Q} = \boldsymbol{H}^{\mathrm{T}} \boldsymbol{Q}^{\mathrm{a}}$.

According to the Variational principle, the dynamics equation of the coordination mechanism can be obtained as follows:

$$\boldsymbol{M} \ddot{\boldsymbol{q}} + \boldsymbol{C} \dot{\boldsymbol{q}} = \boldsymbol{Q} \tag{A12}$$

Appendix A.3 Dynamic Equations with Constraint Equations

During the movement of the mechanism, there are additional constraints. When using the principle of relative motion to deal with the closed loop problem, it is necessary to disconnect a hinge and add a virtual body at the position where the hinge is disconnected, so that the motion of the virtual body is consistent with that of the actual body.

The diagrammatic sketch of the virtual body is simplified to Figure A2a, which is composed of three bodies, three rotary hinges (hinge 1, 3, 4), and one mobile hinge (hinge 2). When conducting kinematics analysis, first disconnect hinge 4, as shown in Figure A2b, and then add the virtual body, as shown in Figure A2c. Thus, the motion state of $O_{\mathrm{a}}' - x_{\mathrm{a}}' y_{\mathrm{a}}' z_{\mathrm{a}}'$ and $O_{\mathrm{a}} - x_{\mathrm{a}} y_{\mathrm{a}} z_{\mathrm{a}}$ is the same, and the kinematics relationship is written as

$$\mathbf{S}_{\mathrm{a}} = \mathbf{S}_{\mathrm{a}}' \tag{A13}$$

Calculate the first derivative of time for Equation (13) and incorporate the relationship between velocity and acceleration; it can be obtained as follows:

$$\left(\mathbf{H}_{\mathrm{a}} - \mathbf{H}_{\mathrm{a}}' \right) \ddot{q} = \left(\dot{\mathbf{H}}_{\mathrm{a}}' - \dot{\mathbf{H}}_{\mathrm{a}} \right) \dot{q} \tag{A14}$$

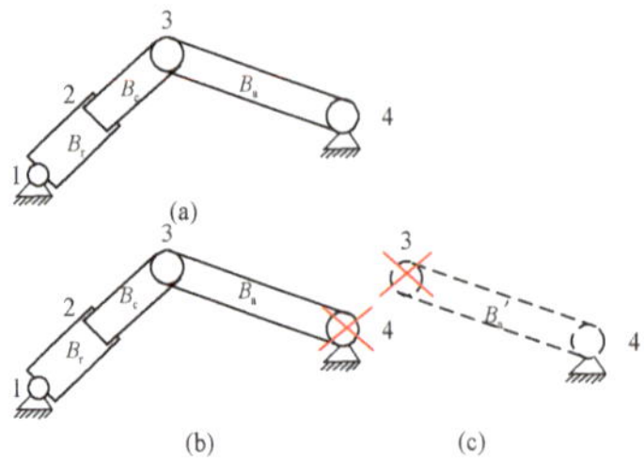

Figure A2. Schematic diagram of virtual body constraints. (**a**) the Closed loop mechanism; (**b**) disconnect hinge 4; (**c**) add the virtual body.

Equation (A14) represents the additional constraint relationship of the coordination process. Due to the existence of additional constraints, for the convenience of derivation, the constraint relationships of displacement, velocity, and acceleration are expressed as follows:

$$\mathbf{\Phi}(q,t) = 0 \tag{A15}$$

$$\mathbf{\Phi}_q(q,t)\dot{q} + \mathbf{\Phi}_t(q,t) = 0 \tag{A16}$$

$$\mathbf{\Phi}_q(q,t)\ddot{q} + \mathbf{\Phi}_{qq}(q,t)\dot{q}^2 + 2\mathbf{\Phi}_q(q,t)\dot{q} + \mathbf{\Phi}_{tt}(q,t) = 0 \tag{A17}$$

where $\mathbf{\Phi}(q,t)$ represents the displacement constraint relationship of the coordination mechanism regarding displacement q and time t; $\mathbf{\Phi}_q(q,t)$ and $\mathbf{\Phi}_t(q,t)$ are the derivative of $\mathbf{\Phi}(q,t)$ over q and t, respectively; $\mathbf{\Phi}_{qq}(q,t)$ and $\mathbf{\Phi}_{tt}(q,t)$ are the second derivative of $\mathbf{\Phi}(q,t)$ over q and t, respectively.

Assuming $\dot{q}_1^*$ and $\dot{q}_2^*$ are the two sets of velocities of the coordination mechanism at the same position and time, they can be obtained as follows:

$$\mathbf{\Phi}_q(q,t)\dot{q}_1^* + \mathbf{\Phi}_t(q,t) = 0 \tag{A18}$$

$$\mathbf{\Phi}_q(q,t)\dot{q}_2^* + \mathbf{\Phi}_t(q,t) = 0 \tag{A19}$$

Subtract Equations (A18) and (A19) and then obtain

$$\mathbf{\Phi}_q(q,t)\delta(\dot{q}) = 0 \tag{A20}$$

where $\delta(\dot{q}) = \dot{q}_2^* - \dot{q}_1^*$ is the virtual velocity of the coordination mechanism.

In order to consider constraints in the dynamic equation of the coordination mechanism, Lagrange multipliers are introduced and combined to obtain the dynamic equation of the coordination mechanism as follows:

$$\begin{cases} M(q,t)\ddot{q} + C(q,\dot{q},t)\dot{q} + \mathbf{\Phi}_q^{\mathrm{T}}\lambda = Q \\ \mathbf{\Phi}(q,t) = \mathbf{0} \end{cases} \tag{A21}$$

References

1. Li, Y.; Qian, L.; Chen, G.; Huang, W. Multiple clearance robustness optimization of a chain ramming machine based on a data-driven model. *Nonlinear Dyn.* **2023**, *111*, 13807–13828. [CrossRef]
2. Zhuang, M.; Li, G.; Ding, K.; Xu, G. Optimized design of mechanical chain drive based on a wireless sensor network data algorithm. *J. Sens.* **2021**, *2021*, 2901624. [CrossRef]
3. Chai, M.; Yuan, Y.; Zhao, W. An improved particle swarm optimization algorithm for dynamic analysis of chain drive based on multidisciplinary design optimization. *Adv. Mech. Eng.* **2019**, *11*, 1687814019829611. [CrossRef]
4. Jiang, S.B.; Huang, S.; Zeng, Q.L.; Wang, C.L.; Gao, K.D.; Zhang, Y.Q. Dynamic properties of chain drive system considering multiple impact factors. *Int. J. Simul. Model* **2022**, *21*, 284–295. [CrossRef]

5. Hong, L.; Li, X.J. Structural optimization of a chain transmission with constraints. *Appl. Mech. Mater.* **2014**, *687*, 523–526. [CrossRef]
6. Da, L.; Ying, L.; Zheng, G.; Sun, M.; Liang, Z. Design of mechanical chain drive based on particle swarm optimization. In Proceedings of the 2022 International Conference on Intelligent Transportation, Big Data & Smart City (ICITBS), Hengyang, China, 26–27 March 2022; IEEE: New York, NY, USA, 2022; pp. 97–101.
7. Cheng, Y.; Guo, H.; Fu, Z.; Wan, N.; Li, L.; Wang, Y. Design and analysis of the Gemini chain system in dual clutch transmission of automobile. *Chin. J. Mech. Eng.* **2015**, *28*, 38–45. [CrossRef]
8. Wang, Y.; Du, J.; Yang, T.; Zhou, J.; Wang, B.; Shi, Y.; Bai, J. Robust Optimization Design of a Blended Wing-Body Drone Considering Influence of Propulsion System. *Aerosp. Sci. Technol.* **2024**, *156*, 109751. [CrossRef]
9. Zhang, W.; Qiang, W.; Fanzhi, Z.; Chao, Y. A novel robust aerodynamic optimization technique coupled with adjoint solvers and polynomial chaos expansion. *Chin. J. Aeronaut.* **2022**, *35*, 35–55. [CrossRef]
10. Najlawi, B.; Nejlaoui, M.; Affi, Z.; Romdhane, L. Multi-objective robust design optimization of a sewing mechanism under uncertainties. *J. Intell. Manuf.* **2019**, *30*, 783–794. [CrossRef]
11. Zhang, Y.; Chen, L.; Chen, G.; Liu, T. Data-Driven Modeling and Robust Optimization of the Ammunition Ramming Mechanism Based on Artificial Neural Network. In Proceedings of the International Conference on Mechanical System Dynamics, Beijing, China, 1–5 September 2023; Springer Nature: Singapore, 2023; pp. 691–705.
12. Gao, H.; Sun, G.; Liu, Z.; Sun, C.; Li, N.; Ding, L.; Yu, H.; Deng, Z. Tension distribution algorithm based on graphics with high computational efficiency and robust optimization for two-redundant cable-driven parallel robots. *Mech. Mach. Theory* **2022**, *172*, 104739. [CrossRef]
13. Guo, F.; Sun, Z.; Zhang, S.; Cao, R.; Li, H. Optimal design and reliability analysis of a compliant stroke amplification mechanism. *Mech. Mach. Theory* **2022**, *171*, 104748. [CrossRef]
14. Wu, W.; Rao, S.S. Uncertainty analysis and allocation of joint tolerances in robot manipulators based on interval analysis. *Reliab. Eng. Syst. Saf.* **2007**, *92*, 54–64. [CrossRef]
15. Du, X.; Venigella, P.K.; Liu, D. Robust mechanism synthesis with random and interval variables. *Mech. Mach. Theory* **2009**, *44*, 1321–1337. [CrossRef]
16. Krüger, J.C.; Kranz, M.; Schmidt, T.; Seifried, R.; Kriegesmann, B. An efficient and non-intrusive approach for robust design optimization with the first-order second-moment method. *Comput. Methods Appl. Mech. Eng.* **2023**, *414*, 116136. [CrossRef]
17. Meng, Z.; Wu, Y.; Wang, X.; Ren, S.; Yu, B. Robust topology optimization methodology for continuum structures under probabilistic and fuzzy uncertainties. *Int. J. Numer. Methods Eng.* **2021**, *122*, 2095–2111. [CrossRef]
18. Kriegesmann, B.; Lüdeker, J.K. Robust compliance topology optimization using the first-order second-moment method. *Struct. Multidiscip. Optim.* **2019**, *60*, 269–286. [CrossRef]
19. Cheng, J.; Lu, W.; Liu, Z.; Wu, D.; Gao, W.; Tan, J. Robust optimization of engineering structures involving hybrid probabilistic and interval uncertainties. *Struct. Multidiscip. Optim.* **2021**, *63*, 1327–1349. [CrossRef]
20. Lee, S.H.; Chen, W.; Kwak, B.M. Robust design with arbitrary distributions using Gauss-type quadrature formula. *Struct. Multidiscip. Optim.* **2009**, *39*, 227–243. [CrossRef]
21. Chatterjee, T.; Chowdhury, R.; Ramu, P. Decoupling uncertainty quantification from robust design optimization. *Struct. Multidiscip. Optim.* **2019**, *59*, 1969–1990. [CrossRef]
22. Lee, D.; Rahman, S. Robust design optimization under dependent random variables by a generalized polynomial chaos expansion. *Struct. Multidiscip. Optim.* **2021**, *63*, 2425–2457. [CrossRef]
23. Zafar, T.; Zhang, Y.; Wang, Z. An efficient Kriging based method for time-dependent reliability based robust design optimization via evolutionary algorithm. *Comput. Methods Appl. Mech. Eng.* **2020**, *372*, 113386. [CrossRef]
24. Jiang, Z.; Wu, J.; Huang, F.; Lv, Y.; Wan, L. A novel adaptive Kriging method: Time-dependent reliability-based robust design optimization and case study. *Comput. Ind. Eng.* **2021**, *162*, 107692. [CrossRef]
25. Cheng, J.; Wang, R.; Liu, Z.; Tan, J. Robust equilibrium optimization of structural dynamic characteristics considering different working conditions. *Int. J. Mech. Sci.* **2021**, *210*, 106741. [CrossRef]
26. Pereira, C.; Ambrósio, J.; Ramalho, A. Dynamics of chain drives using a generalized revolute clearance joint formulation. *Mech. Mach. Theory* **2015**, *92*, 64–85. [CrossRef]
27. Pedersen, S.L. Model of contact between rollers and sprockets in chain-drive systems. *Arch. Appl. Mech.* **2005**, *74*, 489–508. [CrossRef]
28. Ma, J.; Bai, M.; Wang, J.; Dong, S.; Jie, H.; Hu, B.; Yin, L. A novel variable restitution coefficient model for sphere–substrate elastoplastic contact/impact process. *Mech. Mach. Theory* **2024**, *202*, 105773. [CrossRef]
29. Ma, J.; Wang, J.; Peng, J.; Yin, L.; Dong, S.; Tang, J. Data-driven modeling for complex contacting phenomena via improved neural networks considering link switches. *Mech. Mach. Theory* **2024**, *191*, 105521. [CrossRef]
30. Movahedi-Lankarani, H. Canonical Equations of Motion and Estimation of Parameters in the Analysis of Impact Problems. Ph.D. Thesis, University of Arizona, Tucson, AZ, USA, 1988.

31. Liu, T.; Niu, Y.; Chen, G. A general methodology for simulation model validation and model updating of the coordination mechanism with uncertainty. *J. Mech. Sci. Technol.* **2023**, *37*, 6271–6286. [CrossRef]
32. Bae, D.S.; Han, J.M.; Choi, J.H. An implementation method for constrained flexible multibody dynamics using a virtual body and joint. *Multibody Syst. Dyn.* **2000**, *4*, 297–315. [CrossRef]
33. Tong, M.N.; Zhao, Y.G.; Zhao, Z. Simulating strongly non-Gaussian and non-stationary processes using Karhunen–Loève expansion and L-moments-based Hermite polynomial model. *Mech. Syst. Signal Process.* **2021**, *160*, 107953. [CrossRef]
34. Blatman, G.; Sudret, B. Adaptive sparse polynomial chaos expansions based on Least Angle Regression. *J. Comput. Phys.* **2011**, *230*, 2345–2367. [CrossRef]
35. Chen, G.S. The Study on the Dynamic of the Projectile-Barrel Coupled System and the Corresponding Key Parameters. Ph.D. Thesis, University of Science and Technology, Nanjing, China, 2016.
36. Efron, B.; Hastie, T.; Johnstone, I.; Tibshirani, R. Least Angle Regression. *Annels Stat.* **2004**, *32*, 407–499. [CrossRef]

machines

MDPI

Review

Power Transmission Mechanism and Tribological Performance of Modern Bicycle Drivetrains—A Review

Yook Wah Liew [1], Owen Matthews [2], Dzung Viet Dao [1] and Huaizhong Li [1,*]

[1] School of Engineering and Built Environment, Gold Coast Campus, Griffith University, Southport, QLD 4222, Australia; yookwah.liew@griffithuni.edu.au (Y.W.L.); d.dao@griffith.edu.au (D.V.D.)
[2] Ride Mechanic Pty Ltd., P.O. Box 5191, Daisy Hill, QLD 4127, Australia; owen@ridemechanic.com.au
* Correspondence: h.li@griffith.edu.au; Tel.: +61-(7)-5552-8252

Abstract: Bicycles are one of the most sustainable forms of transportation and sports available today, known for their environmental friendliness, cost-effectiveness, lightweight design, compactness, and health benefits. The efficiency and power transmission of bicycle drivetrains have emerged as crucial concerns for engineers, bicycle manufacturers, and both professional and amateur cyclists. However, research and publications related to bicycle drivetrain systems and their tribological performance are notably limited. There is a lack of systematic reviews on technological progress and recent research works in this field. This paper aims to redress this imbalance by presenting a comprehensive literature review of power transmission and tribology in bicycle drivetrains through assessing an extensive body of theoretical and practical work encompassing bicycle drivetrains and roller chain drive mechanisms and performance. This review comprises an exploration of bicycle drivetrain mechanisms and components, an examination of subjects related to power transmission mechanics and efficiency, and a thorough analysis of tribological factors in bicycle drivetrains, including friction, wear, and lubrication. A particular focus has been put on the performance of roller chain drives. This review consolidates research findings related to power transmission within the bicycle drivetrain systems and outlines some future perspectives in relevant research. Through this review, we aim to shed light on the existing knowledge gaps within bicycle drivetrain research and offer constructive recommendations for advancements in this field.

Keywords: bicycle; drivetrain; roller chain; friction; wear; tribology

1. Introduction

Bicycles, powered by human energy, have undergone substantial evolution since their invention in the early 19th century, and they have become a popular form of transportation, recreation, and sports around the world. The utilisation of bicycles offers numerous advantages, such being environmentally friendly and pollution-free, good for the rider's health, low cost, lightweight, and enjoyable. A recent study found that cycling can be a viable transport option for short- and medium-length trips for many individuals and trip purposes in a modern city [1], which is considered the most successful option to relieve traffic jams [2]. Chen et al. [3] complied a global dataset for bicycle ownership and use from 1962 to 2015 and reported that the global production of bicycles had a compound annual growth rate of 3.4% in this period, increasing from 20.7 million units in 1962 to 123.3 million units in 2015. This is higher than that of global car production (3.0% from 14.0 million units to 68.6 million units) in the same period [3]. The aggregated amount of global bicycle production in this period is 4.65 billion units, which is 2.4 times the aggregated amount of

https://doi.org/10.3390/machines13010066

global car production [3]. The global production of bicycles and e-bikes reached 193 million units in 2021, growing 11 percent from the previous year [4]. In comparison, the production of global motor vehicles was at around 85.4 million units in 2022, denoting an increase of 5.7% from 2021. According to a market analysis report by the Grand View Research organisation, the global bicycle market size was valued at USD 64,625.7 million in 2022. The compound annual growth rate from 2023 to 2030 was expected to be at 9.7% [5].

Various types of bicycles have been invented, including mountain bikes, road bikes, hybrid bikes, cruise bikes, track bikes, etc. [6]. The advent of electric bicycles (e-bikes) [7], the establishment of public bike-sharing systems [8], and the growing interest in cycling activities have further propelled the popularity of bicycles, leading to an increase in the number of cyclists [9]. In addition to being a popular recreational activity, cycling can also be pursued as a competitive sport. Bicycle races are popular worldwide, which include several categories such as road bicycle racing, mountain bike racing, and track cycling. Bicycles have great potential to contribute to a more sustainable, healthy, and equitable society.

Despite their great economic and social significance, research and publications related to bicycles, especially their drivetrain systems, have been notably limited. It has been reported that cycling races are sometimes won by time differences of fractions of a second or a few centimetres [10]. Transmission inefficiency due to chain and bearing losses can be a critical factor contributing to the loss of a competition [6]. Therefore, the design and manufacture of bicycles with a higher power efficiency are becoming increasingly of interest. This paper aims to redress this imbalance by undertaking a comprehensive literature review of the bicycle drivetrain system, with a specific focus on its power transmission mechanisms and tribological performance. This emphasis is driven by the paramount importance of power transmission to riders, bicycle designers, and engineers, coupled with the effectiveness of tribology as a tool in comprehending and enhancing the vehicle's drivetrain system [11]. It has been reported that competitive track cycling races are won by milliseconds [12], and the power transmission efficiency is an important factor in performance. To ensure that this review was able to capture all the significant contributions in bicycle drivetrain power transmission and tribology, a systematic quantitative literature review technique was used [13], which is a derivation of the classic systematic literature review approach.

This paper is structured as follows. Section 2 introduces the various mechanisms and fundamental components found in bicycle drivetrain systems. Section 3 explores the power transmission mechanics and efficiency within the drivetrain, covering power loss mechanisms, factors influencing efficiency, the contact mechanics between chain and sprockets, load assessment, and the kinematic and dynamic behaviour of the system. Section 4 delves into the tribological behaviour of bicycle drivetrain systems, with a focus on wear and lubrication. Conclusions are provided in the final section, with current trends and future perspectives in bicycle drivetrain research outlined.

2. Mechanisms and Components in Bicycle Drivetrains

The drivetrain system of bicycles transfers power from the rider's pedals to the rear wheel, propelling the bike forward. Traditionally, single-speed drivetrains have been widely used in classic, casual, and track bikes. As shown in Figure 1a, this type of drivetrain comprises a single chainring and a single cog connected by a chain, thus providing only a fixed gear ratio [14]. Single-speed drivetrains offer the advantages of affordability, reduced weight, minimal maintenance, and overall reliability. However, they do not allow for gear changes and thus multispeed ratios, which can be limiting in certain situations such as tackling steep hills [15].

In contrast, most modern bicycles more often employ multispeed drivetrains that enable riders to shift between different gear ratios and adjust their pedalling effort to various road conditions and personal fitness levels [16]. Multispeed drivetrains consist of various types of chains, sprockets, and additional gear-shifting mechanisms. They can be further classified into 1× drivetrains and 2/3× drivetrains based on the number of front sprockets or chainrings present, as in Figure 1b,c. A 1× drivetrain consists of a single front chainring/sprocket and a rear cassette with multiple cogs, while 2/3× drivetrains comprise two or three front chainrings/sprockets and a rear cassette with multiple cogs. Gravel and mountain bikes generally employ 1× drivetrains, while road or hybrid bikes commonly utilise 2/3× drivetrains.

(a)

(b)

(c)

(d)

Figure 1. Various bicycle drivetrains: (**a**) fixed gear—single speed bicycles; (**b**) 1× drivetrain—multispeed bicycles; (**c**) 2/3× drivetrain—multispeed bicycles (courtesy of Dawson, T., https://vintagebicycle.wordpress.com/page/2/, accessed on 28 October 2024); [17]; (**d**) drivetrain with power-assisted module—e-bikes.

2.1. Roller Chain Drives

Most bicycles utilise the roller chain drive as their power transmission mechanism, with few exceptions like indoor exercise cycling machines or unconventional bikes.

Although alternative transmission mechanisms like gears, hydraulics, strings, or shafts do exist, the chain drive system remains the most common due to its numerous advantages, including cost-effectiveness, compactness, efficiency, and customisability [18].

A typical roller chain drive consists of components such as the crank arms, pedals, bottom bracket, chainring, chain, cog, or cassette, and the closed-loop roller chain comprises a series of interconnected links with rollers and pins [19]. When the driving sprocket rotates, the chain is pulled along, causing the driven sprocket to rotate and transfer power from the driving shaft to the driven shaft [20]. A normal roller chain drive can achieve an efficiency of around 97–99% if it is working properly. Figure 2 illustrates the common components or features found in a bicycle drivetrain.

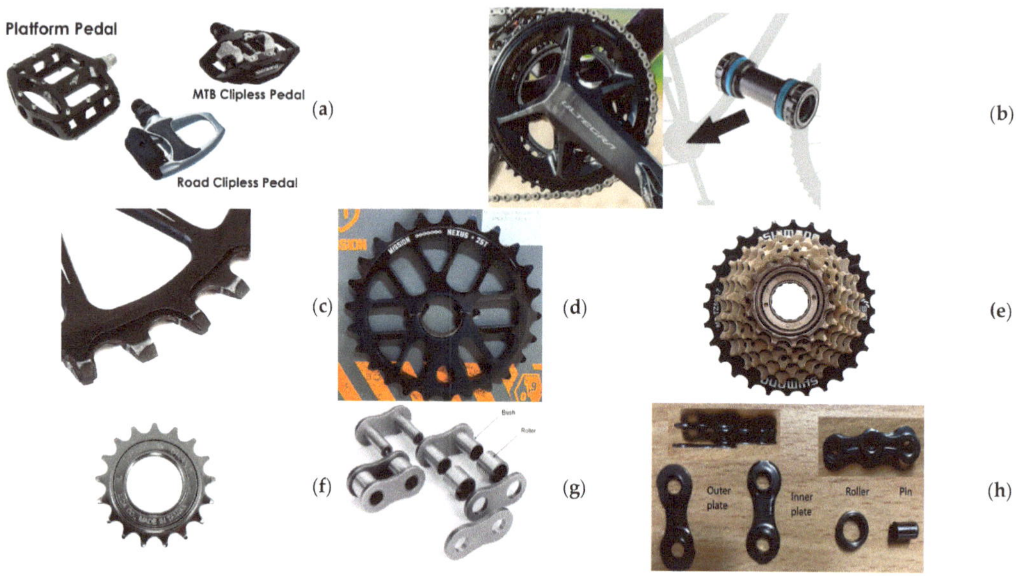

Figure 2. Component present in bicycle drivetrain's roller chain drive mechanism. (**a**) Pedal; (**b**) crank arm and bottom bracket; (**c**) wide narrow chainring—single-speed/1× drivetrain; (**d**) uniform chainring—multispeed drivetrain; (**e**) sprocket/cog—single-speed drivetrain; (**f**) cassette—multispeed drivetrain; (**g**) roller chain with bush—single-speed drivetrain; (**h**) bushingless chain—multispeed drivetrain.

2.1.1. Cranksets and Sprockets

The bicycle crankset converts the reciprocating motion of the rider's legs into rotational motion, driving the sprocket and chains to achieve power transmission in the bicycle. It consists of several components, including the crank arms, pedals, bottom bracket or shaft, and bearings, as shown in Figure 2.

The sprocket is a toothed wheel component that engages with the chain links to perform motion and power transmission. The sprockets can be categorised into front chainring and rear cog (or cassette for multispeed bikes), as in Figure 2c–f. The front chainring is the driving sprocket component that is attached to the crankset. Its function is to move and rotate the chain when the rider applies force to the pedals. Chainrings come in a variety of sizes and shapes, with the number of teeth ranging from 20T to 53T. Chainrings generally have evenly spaced teeth, but some chainrings are in wide–narrow teeth design [21], which contain two groups of teeth that have different thicknesses. This design ensures better engagement between the chain and sprockets, preventing chain slips-off during cycling activity [21]. Some chainrings even incorporate optional hook features along the rear flank to provide a better guide of the chain [22]. Wide–narrow teeth chainrings are commonly used in fixed gear and 1× bicycle drivetrains. There are also some specific designs like oval chainrings or compact chainrings.

The driven sprocket In a bicycle drivetrain Is typically attached to the rear wheel and can rotate or propel the bicycle forward when it is driven by the connected chain. The rear sprocket can be a single sprocket, known as a cog or freewheel, or multiple sprockets, known as a cassette (Figure 2f). The cassette is a gear block component that consists of several sprockets with different sizes and numbers of teeth. It allows the bicycle drivetrain to achieve different gear ratios and is present in most modern multispeed bikes. Both cogs and cassettes come in a range of sizes, which are typically measured in terms of the number of teeth on the sprocket. The gear ratio of the drivetrain is determined by the relationship between the number of teeth on the front chainring and the number of teeth on the rear cog or cassette. This indicates how many times the rear wheel turns in one rotation of the cranks. A larger cog with more teeth results in a lower gear ratio, providing more torque and lower rotating speed, thus making pedalling easier. On the other hand, a smaller cog with fewer teeth yields a higher gear ratio, enabling faster pedalling and greater speed [23].

2.1.2. Chains

The two main types of chains in bicycles are the standard bush roller chain used in traditional bikes and the bushingless roller chain used in modern bikes with derailleur systems [24], as shown in Figure 2g,h. A standard roller chain with bushing consists of outer plates/links, pins, rollers, and inner plates with sleeves (bushings). The inner pins are covered by multiple layers of free-to-rotate components like bushings and rollers [6]. The bushings are press-fitted into each link to reduce wear on the chain's pins and plates. These bushings help the chain maintain its shape and provide an extended service life. Standard bush roller chains are often used in more economical bicycles, where durability and longevity are more important considerations. However, bush roller chains can be heavier and less efficient due to the added friction of the bushings. That is why in modern high-performance bicycles, standard roller chains have been replaced by bushingless chains.

The bushingless chain lacks the bushing component in its inner plate. The pins are designed to rotate directly on the inner plates. It is specifically designed to fit the gear-shifting features (derailleur) found in modern bicycles. This design reduces the weight of the chain and reduces the friction between the components by eliminating the bushings. In comparison to a standard roller chain, a bushingless chain is narrower, lighter, and more flexible. Higher-speed bicycle chains tend to have smaller plate thicknesses and inner and outer widths [25]. As a result, bushingless roller chains are typically more efficient than bush roller chains. In addition, the bushingless chain also has a small design change in its inner plate to provide better alignment and engagement with sprocket teeth, which can greatly eliminate the chance of the chain slipping off [26].

2.1.3. Chain Drive Configuration

Several studies have investigated the chain drive setup and configuration. Cho et al. [27] examined the optimal gear ratio for multispeed bicycles and proposed that reducing the number of gears without a loss of physiological efficiency would result in easier gear-shifting, as a fewer number of gears which are optimal can provide a performance equivalent to the multispeed gear system.

Van Soest [28] investigated the impact of bicycle chainring shape and geometry on overall performance with a focus on non-circular chainrings and developed algorithms and models that represent the gear ratio through the functions of crank angle and effective radius. It was concluded that chainring shapes have no significant impact on overall efficiency. Computer-aided engineering (CAE) tools and techniques have been used in the analysis of and improvement in chain drive geometry, such as the 3D modelling of roller chain drives [29,30].

2.2. *Gear-Shifting Mechanisms*

In multispeed bicycles, the gear-shifting mechanisms are designed to enable riders to adjust the gear ratio of their bicycles, allowing them to maintain an efficient pedalling cadence and adapt to different terrains and riding conditions. The two main types of gear-shifting components are derailleurs (external) and internal gear hubs.

2.2.1. Derailleurs

The derailleur system was developed in the early 20th century [31]. It introduced the ability to easily move the roller chain between different chainrings or gears on the cassette block, allowing for a wider range of gear ratios. In modern multispeed bicycle drivetrains, the derailleur system typically includes a shifter, a derailleur, and a cassette or freewheel. The shifter serves as the control mechanism that riders use to adjust the gear ratio through a cable that runs to the derailleur. Derailleurs come in various types and designs, including front and rear derailleurs, as shown in Figure 3. It should be noted that the derailleur system introduces additional friction between the side plates of the roller chain, leading to energy loss at the articulation points due to higher bearing pressure. It also introduces extra tension in the return side of the drivetrain system. In experimental research on bicycle derailleur systems [32], the stability region in the derailleur system was reported, and a performance index called the stability ratio was introduced for better shift lever design in the rear derailleur.

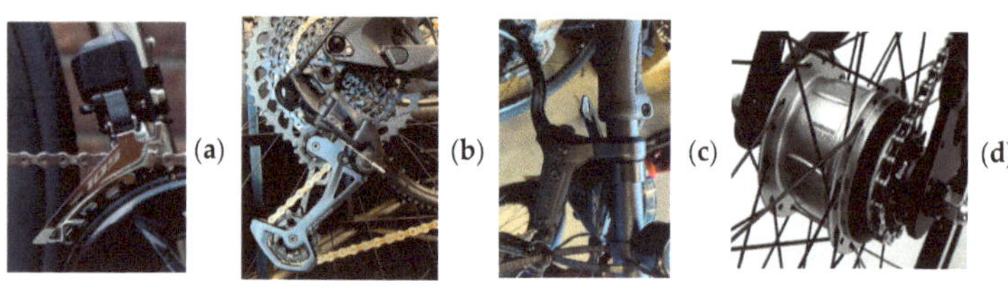

Figure 3. Main component of gear-shifting mechanism. (**a**) Front derailleur—multispeed 2/3× drivetrain only; (**b**) rear derailleur—multispeed drivetrain; (**c**) shifter—multispeed bike; (**d**) internal hub gear—multispeed bike.

2.2.2. Gear Hubs and Gearboxes

The internal gear hub system (also known as hub gear), as shown in Figure 3d, is located within the rear wheel drive. It consists of various small mechanical elements like Pinions, gear rings, clutches, etc. The design is very compact and almost everything is sealed within a protective shell, usually with an epicyclic gear configuration [21]. The desired gear ratio is achieved through the interaction between the rear sprocket and the different positions of gears inside the internal hub. The main advantages of a planetary gear mechanism include low space occupancy, a high speed reduction ratio, and power transmission capabilities from the interaction of its gears of different sizes. It also allows for a better load distribution, which can further reduce wear and improve overall system durability [33]. On the other hand, hub gears are usually more expensive, heavier, and less efficient. Three-speed hubs were popular, but the number of gears offered in epicyclic hubs gradually increased. Hub gears can be found with two, three, five, seven, eight, nine, eleven, and fourteen speeds [33]. The 14-speed Rohloff 500/14 (shown in Figure 4a), made in Germany and launched in 1998, has been considered the king of hub gears due to its excellent performance that approaches the efficiency and range of a modern derailleur and provides more evenly spaced ratios, better reliability, and longer service life [33].

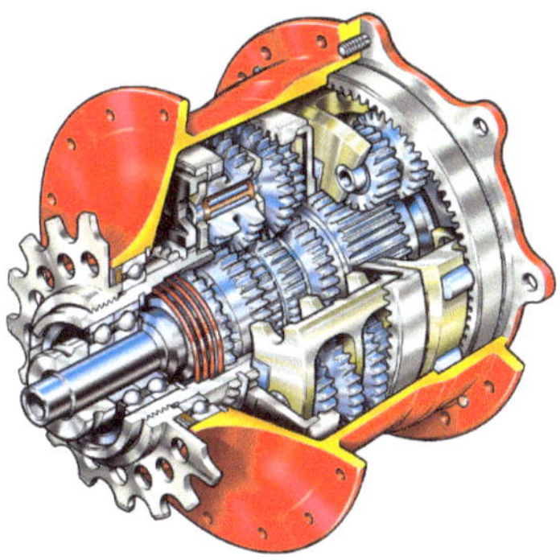

(**a**)

(**b**)

Figure 4. (**a**) Internal schematic of a 14-speed gear box Rohloff Speedhub 500/14 (courtesy of https://commons.wikimedia.org/wiki/File:Speed1c.png, accessed on 10 January 2025); (**b**) Pinion P1.18 gearbox interior (courtesy of https://commons.wikimedia.org/wiki/File:Pinion_P1.18_interior.jpg, accessed on 10 January 2025).

Another type of gear system for power transmission in bicycles is the gearbox. The gearbox is usually integrated into the frame near the crank, and it may be used in addition to or instead of derailleur gears or a hub gear [34]. A bicycle gearbox offers low-maintenance drivetrain and enables gears to be instantly shifted without the need to pedal. Two decades ago, Honda developed some top-secret bicycle gearbox drivetrains for downhill racing and won the world championships and five World Cup rounds [35]. Unfortunately, there are very limited information about the designs outside of their patent (US7503862B2, filed 2005) [36]. Cesar from UNNO bikes holds a recent patent (ES2684528A1, filed 2017) [37] for a similar gearbox, in which there is no derailleur; instead, it is the cassette that slides back and forth [35]. Shimano also filed their 13-speed gearbox patent (US 2019/0,011,037 A1) [38] that uses a form of derailleur. The Pinion gearbox is a relative newcomer. The gear is located in the bottom bracket position and requires frames to be made especially for this location [6]. Pinion gearboxes use spur gearing with two sub-units connected in sequence. Power is transmitted through only two sets of cogs. A photo of the Pinion P1.18 gearbox interior is shown in Figure 4b. The gearbox can be shifted through full-range gears all at once, shift part-way, or one at a time, whether stationary or while riding. The operation of this system is superbly explained at pinion.eu/en/technology/ [6].

In most cases, cyclists have to manually change gear ratio depending on resistance. The so-called autobike or automatically geared bicycle, whether with a derailleur or internally geared hub, is supposed to shift gears without intervention from the rider. The automatic speed changer can be divided into a step speed changer, including the fixed-axle gear type and the planetary gear type according to the gear train mechanism, and a continuously variable speed changer [39]. There have been some efforts in this pursuit, and some new patents have been filed in recent years, such as the US patent Automatic Transmission System for a Bicycle filed by Haven Mercer in 2023 (US 2023/0373592 A1) [40], and the patent "CVT automatic variator transmission for a bicycle" filed by inventor Peyman Asadi in 2019 (Patent number: 11885415; US20220213948A1) [41]. Nivex developed automatic transmission, which consists of a specially designed GPS unit and proprietary AI system that also control the derailleurs. The system interfaces with the power metre to detect whether a rider is attacking or taking it easy, to select the optimum gear for each situation, and it works with wireless derailleurs, such as SRAM eTap or the Shimano Di2 [42]. Shimano developed some new hardware and technologies towards automatic transmission, such as the new XT Di2, LinkGlide, the EP8-01 e-bike motor, and Auto Shift and Free Shift [43]. The Auto Shift, as the software, is a new mode that automatically shifts gears by employing speed, torque, and cadence sensors along with an algorithm to determine what gear you should be in and requiring hardware such as Di2 working with the derailleur and EP motor to execute a shift [43]. The development of automatic transmissions for bicycles is still at a very early stage and far more challenging than automatic transmissions for cars. So far, the peer reviewed literature is very limited, and more research efforts are expected.

2.3. Belt Drive Transmission

In lieu of the conventional steel chain-based drivetrains for bicycle transmission, there are other types of chainless drivetrains, such as those with toothed belts or with cardan shaft [44]. In the early days of cycling, flat belts were occasionally employed in the drive system of bicycles, but little progress was made due to the dependence on friction [33]. Since the 1960s, toothed belts have become more durable and efficient. Bridgestone launched its Picnica, a folding model bicycle, in the 1980s and claimed it as the world's first series-produced belt drive bicycle [33]. The Gates Corporation, a market leader in drive belts, designed and made the belt drives used on the Strida from the 1980s onwards. It was not until 2007, when Gates developed the modern Carbon Drive system for bicycles, that the belt drive gained widespread popularity [33].

A belt drivetrain system consists of a belt that wraps around two pulleys or rotating shafts. Figure 5a shows a bicycle belt drive with horizontal fork ends and tugs for providing belt tension [45]. The belt drives for bicycles provide the benefits of being low maintenance, lubrication-free, lightweight, having a smooth operation, and enjoying a long life [45]. Modern transmission belts may consist of multiple layers, such as a protective top and bottom layer, with a flexible, high-resilience composite material in the middle. Gates's Carbon Drive belt has strands of carbon fibres and teeth of nylon-jacketed polyurethane [33]. It has been reported that the belt drive can last upwards of 30,000 km/19,000 mi, which offers 3–4 times the service life of a chain [46]. Other belt drive manufacturers include Advanced Belt Drive, Continental AG, and Veer Cycle, to name a few [45].

(a) (b)

Figure 5. (**a**) A bicycle belt drive with horizontal fork ends and tugs for providing belt tension (courtesy of https://commons.wikimedia.org/wiki/File:Bicycle_belt_drive_1.JPG, accessed on 10 January 2025); (**b**) bicycle belt drive with internal-geared multispeed rear hub (courtesy of https://commons.wikimedia.org/wiki/File:Belt-drive_internal-geared_multi-speed_rear_hub.JPG, accessed on 10 January 2025).

However, there are several drawbacks to the belt drive, such as the large space requirement due to the belt's width, material sensitivity to weather conditions, relatively high cost, and lower efficiency compared to a traditional roller chain drive [44]. As belts are one-piece and cannot be broken, the bicycle frame will need to be designed with a 'belt splitter' or an opening in the rear triangle. The belt's width and pre-tension must be carefully chosen, calculated, and adjusted [6]. A tensioner needs to be added to a drive to apply tension to the belt. Since belts cannot run at an angle, they are not compatible with derailleur gears, which means that belts are only compatible with internally geared hubs (Figure 5b), Pinion gearboxes, or single-speed gears. A detailed comparison of the pros and cons of belt drive bicycles can be found in [47]. It should be noted that peer reviewed journal articles on bicycle belt drives are very limited. It is difficult to find papers in the literature specific to their tribological behaviour.

2.4. Drivetrain with Power-Assisted Module for E-Bikes

A major variation in the realm of bicycles is the introduction of e-bikes and power-assisted bikes. These innovations address some of the limitations of the traditional human-powered bicycles while retaining most of their advantages. E-bikes incorporate a motor or engine into their drivetrain to assist the cyclist's pedalling action, as shown in Figure 1d. Other components, such as the chain–sprocket system, wheels, and frame, are largely the same as those of traditional bicycles, thus allowing existing research to remain relevant [48]. E-bikes have been steadily gaining popularity worldwide, with a notable increase in research focus over the years [49]. Hung and Lim [50] presented a comprehensive review on the development of e-bikes. Studies on e-bike performance and dynamic analysis, such as those by Arango, Lopez and Ceren [51], and Stackhouse and Dong [52], further highlight the growing interest and advancements in this field.

3. Power Transmission Mechanics and Efficiency

3.1. Mechanics of Power Loss and Factors Affecting Efficiency

The power transmission efficiency in bicycling can be defined as the power output at the driving wheel divided by the mechanical power input from the human body [6]. Although it is believed that the roller chain drive can achieve around 97–99% of power transmission efficiency [53], many factors can affect the drivetrain performance, such as friction in the chain drive and bearings, impact losses, and slip loss [6]. Existing research also stated that the efficiency of the chain drivetrain should be between 95% and 98.5% [54]. According to Sgamma et al. [55], the energy loss in a roller chain system can be attributed to several factors, including the frictional forces between the pins and bushings in the links as they articulate onto and off the sprockets, the impact forces generated when the chain rollers engage with the sprocket teeth, and the vibrations that result from these interactions. In a roller chain drive, friction is primarily due to the sliding actions between solid components [56], such as between a fixed pin and bushing. The chain is composed of many small components, and internal collision between components generates high amounts of friction and reduces overall transmission efficiency. There is also friction between a roller and the sprocket teeth due to meshing, and between the side plate and the side of the sprocket tooth [6]. When the chain sits on sprockets that are not completely in line, lateral offset occurs, introducing additional stress and friction to the side plates, further reducing performance [57]. If the chain is dirty or worn, some rolling friction may occur due to the presence of small particles.

Additionally, dynamic characteristics such as resonance, transverse vibration, and damping forces generated by the chain in motion can also contribute to power loss [53]. In a recent work, Lanaspeze et al. [58] showed that the power loss contributed by the meshing motion and the roller motion is in similar magnitude for a wide range of gear ratios and loading conditions in a simple two sprocket drive; however, it was also reported that power loss in the chain drive can be reduced by selecting larger gear ratios in race tandem bicycles [59]. Various ways of measuring chain drive efficiency, friction, and wear were compared in a recent review paper [60]. It was interesting to note that almost no research has investigated the effects of chain contamination on efficiency.

Impulse and vibration can occur due to meshing or interlock between the chain and sprockets, especially when the chain operates close to its natural frequency. Polygonal action is another issue that can arise when the chain wraps around a small diameter sprocket, resulting in periodic fluctuation and transverse displacement that reduces the system's performance [61]. Other sources of energy loss include shock loads that occur during shaft revolution, inertia loads resulting from sudden momentary jams in the chains, centrifugal tension when the chain travels over the curved part between sprockets, additional loads when the chain is in a hanging or catenary state, as well as noise and heat generated during operation [62].

3.2. Contact Mechanics Between the Chain and Sprockets

Several researchers have investigated the contact mechanics between the chain and sprockets in order to better understand the drivetrain's friction and kinematics. Pedersen et al. [63] measured the pseudo-penetration of a roller in the sprocket tooth and presented a more realistic circular tooth profile consisting of seven contact areas, enabling a detailed analysis of the contact vector and forces at different areas of the sprocket [64]. Ambrosio et al. [65] developed a planar roller chain drive model based on the cylindrical contact and real tooth profiles and studied the contact forces and trajectory of pins and bushings using the multibody dynamic approach. Further, it was reported [66–68] that the chain drive system also has other contact points such as the chain-to-chain guide and chain-to-tensioner, which may be present in some customised bikes to help hold the chain in place.

Meshing noise, vibration, and energy loss in chain drive transmission can occur during the meshing process. Binder and Mize [69] investigated roller chain drive strand vibration during operation. They modelled tight strand chains as tension strings with loaded masses at each roller centre to determine the natural frequency. Wang et al. [70,71] examined the stability of chain drive systems under periodic sprocket oscillations and proposed a numerical model with an impulse function based on their couple effects. Liu et al. [72] incorporated both longitudinal and transverse chain motions in a meshing impulse calculation and found that chain resonant and meshing frequency can affect the impact intensity or magnitude. Liu's thesis [73] presented mathematical models for meshing and illustrated the relationship between meshing, impact, and noise and conducted several experiments to determine the meshing noise. Zheng et al. [74] studied the meshing noise of chain drives and found that the noise that radiated from both the drive and driven sprocket was small compared to the roller noise. The acoustical energy loss from meshing noise can be neglected in most cases.

3.3. Analysis of Forces

A pioneering analysis of chain drive forces was conducted by Binder and Covert [75], which focused on the impact velocity and energy transfer between the roller chain drive and the sprocket. They developed analytical equations to illustrate the roller–sprocket interaction, with the effects of friction ignored, and conducted laboratory tests to analyse the correlation between impact velocity and factors such as sprocket wear, roller failure, heat, and noise. Their research suggested that the impact energy, determined by sprocket speed, is closely related to roller breakage in the chain. This publication, along with a follow-up work by Binder [76], laid a foundation for understanding the mechanics of chain drives.

Naji and Marshek [77] analysed the geometric progressive load distribution in the roller chain drive around the sprocket teeth region (Figure 6a) and determined the effects of pitch difference, friction, and centrifugal forces on it. The interaction between the chain rollers and sprocket teeth at various positions was formulated, and the impact of various factors such as chain components, sprocket types, pitch size, and wrap or articulation angle on friction and centrifugal forces was examined, which in turn all influence the overall tension and loading on the sprocket teeth. Key findings reveal that the tooth load on a driven sprocket is higher than that of a driver sprocket due to the friction between the sprocket teeth and the chain rollers, and that the roller teeth of the sprocket are more susceptible to wear than the pin teeth. Additionally, a larger pitch size tends to increase the tension within the chain links. Naji and Marshek [78] also conducted an experimental study to determine the sprocket and roller chain load distributions, which was found to be independent of the elastic properties of the chain and the sprocket, and the lubrication as well. This study confirmed that the load distribution for the chain on a driver sprocket differed from that for a chain on a driven sprocket due to the different direction of the friction force. It was found that a larger pitch of the sprocket tooth can amplify the tension in the chain link. Collectively, these analyses and findings significantly advanced the understanding of the mechanics of roller chain drives.

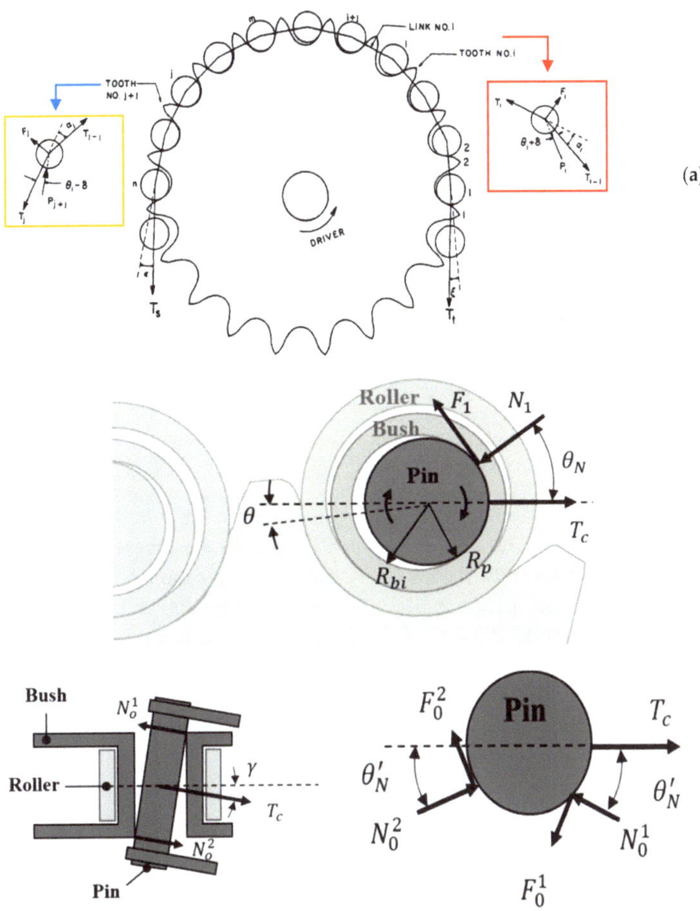

(a) (b) (c)

Figure 6. *Cont.*

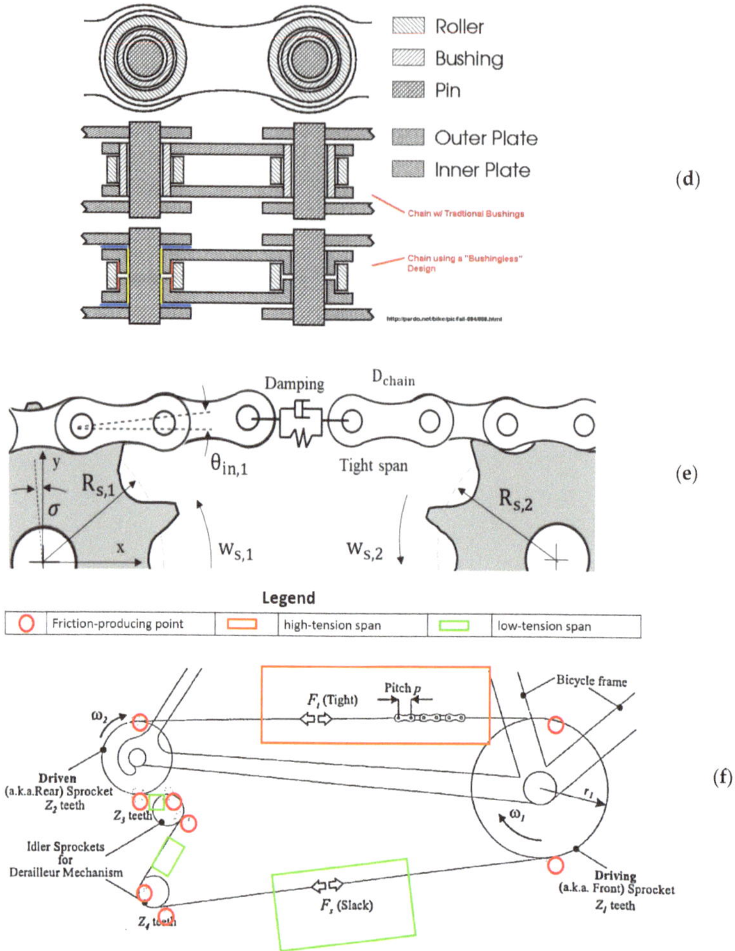

Figure 6. Schematic diagram for bicycle drivetrain's force, friction, or energy analyses. (**a**) Force analysis [77]; (**b**) pin articulation [53]; (**c**) lateral offset [77]; (**d**) friction (sliding) area [79]; (**e**) damping in chain span [77]; (**f**) overall tension distribution [57].

Building upon the research findings of Naji and Marshek, Johnson and co-workers carried out further investigations during the 1990s. This included a more detailed model of the quasi-static contact phenomenon from the first principles by Kim and Johnson [80], which considered the external loading conditions, the curvature difference between the roller and the seating curve, the net error in the chain pitch, and the Coulomb friction in chain motion. Conwell and Johnson [81] experimentally investigated the dynamic behaviour of roller chain drives. They found that the tension in a chain link increases very rapidly as the link exits the driven sprocket and decreases rapidly when the link enters the drive sprocket.

Troedsson and Vedmar [82–84] conducted a comprehensive analysis of the entire chain drive system, focusing on the static and dynamic load distribution, and with the effects of elasticity to the link considered. They examined the forces in both the tight (top) and slack (bottom) regions of the chain, as well as in the driving and driven sprocket sections. Their load distribution model incorporates variables such as inertia, damping force, and moment in addition to the frictional force component to accurately represent the dynamic behaviour in chain drive operation. Kidd, Loch, and Reuben [57] published an important study on tribology in bicycle drivetrains. Through a comprehensive force analysis, they classified the bicycle drivetrain into different regions and defined all friction-producing points, as shown in Figure 6. The work formulated a power loss equation for these engagement points and indicated the effects of lateral behaviour and chain offset.

Lodge and Burgess [85] proposed an alternative approach for evaluating the forces and efficiency of roller chains with a novel chain efficiency model that considers only the geometry of the chain link on sprocket teeth and the associated force components around articulation points such as the pin, bushing, and roller. It was shown that frictional losses are a significant source of energy loss in chain transmission. The energy loss equations during pin and bushing articulation were developed using Coulomb's law of friction (sliding), though it was also shown that at low torque, losses due to impact, adhesion, and/or vibration become more significant. A report in Friction Facts [79] identified the friction-producing mechanism and sliding surface for both bush and bushingless bicycle chains, which can be used to further develop an efficiency model specifically for bicycle chains based on the work of Lodge and Burgess [85]. Although it was traditionally assumed that small sprockets are the best solution for whole life cost and performance, Lodge and Burgess [86,87] disproved this through an experimental investigation. They introduced a chart for the selection of optimum chain and sprocket size and demonstrated that significant energy savings can be achieved by using it. Their findings support the view held by some researchers that using larger sprockets, despite the increase in mass and inertia, can enhance transmission efficiency due to reduced frictional losses. Binder and Covert [75] showed that when the number of teeth in the sprocket was increased, the impact energy decreased, and the number of failures was also reduced. Hollingworth and Hills [88] attributed the reduction in power loss to the smaller angles of articulation imposed on the chain when using sprockets with a larger number of teeth.

Spicer et al. [89] investigated the efficiency of bicycle chain drives both experimentally and theoretically. Their findings revealed that the primary factors affecting efficiency are sprocket size and chain tension, with friction surprisingly accounting for only a small portion of overall losses. In a subsequent study, Spicer [90] analysed the non-linear elastic behaviour in bicycle chain drives and its impact on transmission efficiency and developed mathematical models of power loss, taking into account the effects of tensions, pin–bushing interaction, chain extension, and other properties like Young's modulus. It was concluded that power loss decreases with an increase in roller chain tension during operation. Zhang and Tak [91] proposed a transmission efficiency model for roller chain drive based on sliding friction and damping force, which accounts for the effects of lateral offset and other operating parameters. Sgamma et al. [55] developed a phenomenological model for chain transmissions efficiency and introduced a new parameter called chain tension efficiency to model the distribution of losses within the system. It was observed that the efficiency increases nearly linearly with the number of teeth of both sprockets. In addition, the effectiveness of finite element analysis (FEA) as a tool in determining the forces, stresses, and displacements of the roller chain drive under various tension and loading conditions was demonstrated by Nikhil et al. [92], as well as Dhage and Diwate [93]. The integration of FEA with the multibody system (MBS) technique has been explored to enhance the analysis of the mechanics and dynamics of roller chain drives [94].

3.4. Kinematic and Dynamic Behaviour in Bicycle Drivetrains

Various studies of the kinematics and dynamics of roller chain drives have been conducted, aiming to develop numerical models based on force and motion analysis. Lai et al. [95] analysed the motion of the rear derailleur shifting process in bicycle chains and presented a design methodology for the tooth profile and changer mechanism. The jumping-over-teeth phenomenon and the impact on the overall chain drive performance were discussed by Wang, Zheng, and Zhang [96]. Ma and Chiou [97] studied the condition of the chain link during the shifting process and proposed an optimal path model for drivetrain up-shifting. The path is formed by several bent chain links that can sustain the yaw and roll for the sprocket tooth chamber, and the path can shorten the required up-shifting distance, decrease the phase angle, promote efficiency, and reduce movement yield by the derailleur tappet pressure.

Pedersen [98] developed mathematical models to predict the chain drive's kinematic, force, damping, and other parameters based on a multibody system dynamic analysis approach. Wu et al. [99] performed a kinematic analysis for an eight-speed bicycle transmission hub employing a fundamental circuit method. It included diagrams that illustrate the power flow paths of the transmission hub. Omar [100,101] modelled the bicycle chain with bushings using multibody dynamics based on spatial algebra. Non-linear kinematic and dynamic equations of motion between different bodies are derived based on the spatial Cartesian coordinates and joint variables in the roller chain link. This approach can be applied for the investigation of ride and shift quality, contact forces, wear prediction, and noise emission.

4. Tribological Performance in Bicycle Drivetrains

4.1. Wear

4.1.1. Mechanics of Wear in Chain Drives

Mechanical wear occurs due to the relative motion between surfaces in contact. Components like chains and sprockets, which experience elevated friction during operation, are particularly prone to wear. This wear has the potential to cause a deterioration in functionality, ultimately leading to material failure. There are different types of wear, including abrasive, sliding, rolling, impact, corrosion, and erosion. In roller chain drives, the primary forms of wear are abrasive and sliding [102]. Abrasive wear occurs when materials rub against each other, causing the softer material to break off due to cutting, ploughing, or surface fracturing of a harder material. During sliding, where there is relative movement between the pin and bush in a roller chain, particularly when passing over sprockets, joint abrasion occurs, with wear concentrated mainly in the contact zone between the pin and bush [103]. The smoothness of the material influences this wear, and polishing the object's surface can effectively reduce it. Sliding or adhesive wear occurs when objects are in contact, and their surface asperities collide, forming adhesion that leads to material transfer [104]. Adhesive wear reduces if metal oxide films are present on the material surface, as they can minimise or prevent real contact between materials [105]. Various regions of the chain drive, including the chain pin–bush, bush–roller, roller–sprocket teeth, and side plates, are subject to contact and wear. Figure 7 illustrates comparisons of some new and worn components of roller chain drives and sprockets.

Elongation is a clear indication of roller chain wear. Wear-induced pitch elongation of the chain not only causes significant variations in static load but also generates additional dynamic load on the drive and driven sprockets, leading to the obvious vibration of the whole chain [106]. Once the chain elongation exceeds 0.5–0.75% in the elongation checker tool, it needs to be replaced before it further damages other drivetrain components [107]. Regarding the sprocket, wear mainly occurs in the gear teeth region, resulting in chipping or material removal. When the outer surface of the gear teeth becomes worn or rough, this affects the contact between the chain and sprocket, resulting in poor fit and accelerated chain wear, reduced transmission of motion, and reduced power efficiency. Although many cyclists assume that the sprockets in the chain drive system should be as small as possible, a study by Burgess [56] showed that large sprockets are best for cycling as a larger chain set has the advantage of reducing chain forces and thus internal friction losses, which also leads to lower wear rates. It was claimed that doubling sprocket size will have the effect of doubling the wear life [56].

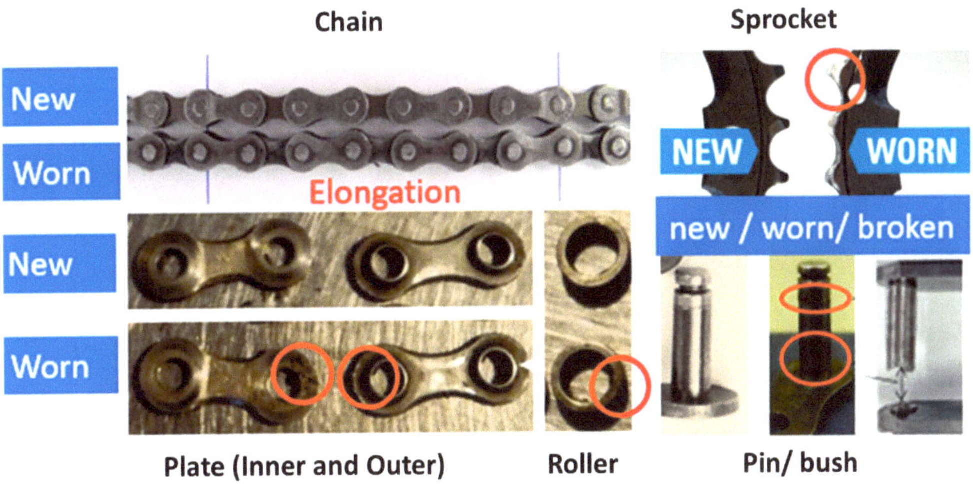

Figure 7. Wear phenomenon in roller chain components and sprocket, including elongation of chain due to wear and new and worn chain plates, rollers, pins, and sprockets. The red circles indicate noticeable worn locations present in the chain drive components.

4.1.2. Detection of Chain Drive Wear

Wear in roller chains can be detected using various approaches. Measuring elongation with chain gauges or checker tools is the method commonly used by cyclists. Becker et al. [108] introduced two chain wear test rigs for friction and wear investigations, including a chain joint tribometer for single-chain joint analysis, as shown in Figure 8 [109], which successfully identified the run-in wear period and the state of steady wear in an initial test. This test rig enables wear and friction measurement and investigation on all parts inside a chain joint like pin and bush [109] and has been used to investigate the influence of triboactive coating on the joint wear of grease-lubricated roller chains [103]. Zhou introduced a mathematical method as a tool to facilitate the test rig design for chain wear testing [110]. Rosenkranz et al. [111] employed a specially modified, commercial ball-on-disc tribometer to investigate the friction and wear behaviour of laser-patterned chain links and found that cross-like patterns can reduce friction and wear noticeably. Saito, Noda, and Sano [112] developed a testing machine that can simulate realistic wear conditions on roller chain links in a short amount of time using limited resources and energy in generating a worn roller chain for wear testing and evaluation experiments. The wear resistance of roller chain components can be enhanced through material selection, optimised manufacturing processes, heat treatment, and surface coatings [103]. It should be noted that while these studies are not specific to bicycle drivetrains, the mechanisms they explore are relatively generic and can be applied to bicycle drivetrains.

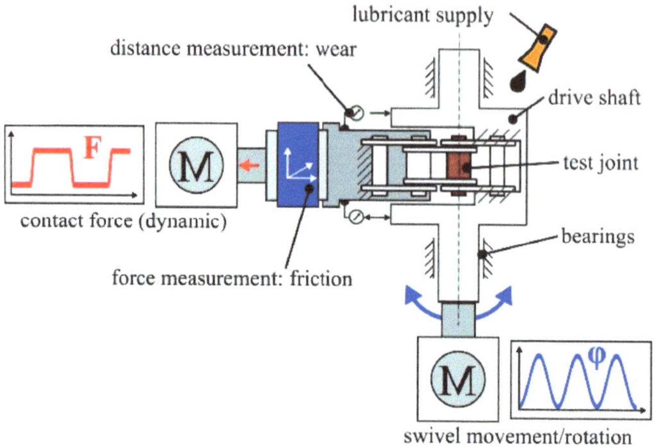

Figure 8. Schematic representation of the chain joint tribometer [109]. In this setup, high-resolution eddy current sensors are used to continuously detect the distance increase between the tensile-side adapter and the swivel shaft surface so as to measure the elongation of the chain segment and thus the wear progress in the contact area between the test pin and bush. The friction occurring in the contact is measured through a multi-component force and torque sensor [109].

4.2. Lubrication

4.2.1. Types of Chain Drive Lubrication

Lubrication is the use of a lubricant to reduce friction and wear between two contact surfaces. Proper application of lubricant can increase the transmission smoothness and efficiency of the chain drive by up to 5% [113,114]. There are many types of lubricants designed for different uses and varied opinions of what is the best lubricant. Lubricants are primarily categorised based on their phase, i.e., oil-based liquids (natural, mineral, or synthetic oil), non-oil liquids, cohesive types (such as waxes, greases, or pastes), and solid dry lubricants [115]. Performance and efficiency are the most crucial criteria when selecting lubrication.

It is agreed that the type of riding and location best determines the type of chain lubrication, as different types of lubricants are suited for different situations [116]. Some bike lubricants have features that help minimise wear, corrosion, oxidation, water resistance, or grime protection. In dry and dusty areas, dry lubricants (thin or light lubricants in a volatile carrier) would do better as they attract and hold less dust and grit than oils that remain moist. Wax-based lubes are also popular that are applied and left to dry. On the other hand, in wet and humid areas with a lot of precipitation, a thicker lubricant is commonly used that will adhere to the chain [116]. For the lubrication of bicycle chain transmission, water resistance and viscosity play an important role in lubricant parameter selection [117]. It was suggested that the mixed grade lubricant performs better in the rainy operating environment through an experimental investigation by Shen et al. [117].

Most roller chains and bicycle drivetrains use synthetic or mineral oil lubricants with additives [118]. Compared to grease or wax, liquid synthetic oil has low viscosity and a high penetration rate. It can easily slip into the tight gaps between the chain's components and form a protection film on the metal surface. Additives are used to maintain lubricant performance under various cycling environments. One advantage of liquid lubricant is that it can be directly applied to the system while the chain drive is in operation, whereas grease or solid lubricants require stopping the chain movement before applying. Liquid lubricants also take less time to take effect compared to dry lubricants, wax, or grease products. As a result, many bicycle riders prefer using liquid lubricants for their drivetrain and sometimes use dry or grease lubricants in the preparation stage or after cycling. Liquid lubricants can be applied in various ways, including dripping from a bottle/nozzle; constant bathing or spraying; partially dipping in a pool of lubricant; fully soaking/submerging the parts; operating within a closed container full of lubricant, etc. [119]. However, only the bottle dripping method is suitable for use while riding a bicycle.

4.2.2. Effects of Chain Drive Lubrication

Due to the extremely narrow spacing between chain components, with little to no separation between opposing surfaces, lubrication in roller chain drive applications typically operates under the boundary lubrication regime, which is featured with a high friction coefficient and thin lubricating film, although all lubrication regimes are encountered in elite cycling. Actually, the lubrication regime can be affected by chain tension, alignment and velocity, sprocket size, lubricants, and contamination, and it also influences the coefficient of friction [60].

There have been some experimental studies [120–122] investigating the efficiency and effects of lubricants on chain drives and bicycle drivetrains. The key findings are as follows:

- Dry lubricant, such as paraffin wax, offers superior performance in terms of efficiency, longevity, and resistance to water and dirt. However, the application method is fully immersive and time-consuming.
- Liquid lubricants with additives such as PTFE or wax generally exhibit better performance.
- Renewable lubricants derived from bio-derived oils can achieve good friction reduction but may lack resistance to oxidation and heat.
- Wax and grease have a longer lasting time than liquid products but are difficult to clean afterwards.
- Wet lubricants and weather conditions can attract contaminants, leading to increased friction in the chain. Thus, resetting the lubrication is recommended if cycling off-road.
- High temperatures can reduce oil viscosity, resulting in a smaller lubrication film interacting with contact surface asperities. Clean oil can improve performance, but the presence of dust, debris, or particles can introduce additional friction to the system.
- The lubrication method significantly impacts efficiency. Frequent lubrication to the drivetrain during cycling, such as drip or spray application, is recommended. While immersive soaking offers better results, it is impractical for typical cycling conditions.

5. Conclusions and Future Perspectives

This paper presents a comprehensive literature review of bicycle drivetrain systems, with a primary emphasis on the power transmission and tribological performance of roller chain drives. It discussed various aspects of bicycle drivetrains, exploring the diversity in their core components and mechanisms. This review consolidates research findings pertaining to power transmission within the bicycle drivetrain systems, encompassing discussion on power loss mechanisms, efficiency factors, contact mechanics, force analysis, as well as the kinematic and dynamic behaviour of the bicycle roller chain drive mechanisms. Furthermore, different facets of friction, wear, and lubrication are examined, outlining their relevance to the bicycle drivetrain systems. The findings of this work can inspire further academic research on advancing our scientific understanding of the drivetrain systems, create new tools for performance simulation and product development for higher efficiency lower costs to benefit cyclists and the wider community, and promote collaboration between researchers and industry to tackle technical challenges for product innovations.

While the demand for bicycles remains high and the drivetrain remains a fundamental element of any bicycle, there exists immense potential in this field of bicycle drivetrain technology. Some future perspectives in relevant research can be outlined as follows to underscore potential avenues for exploration and development.

- Modern bicycle drivetrains have widely utilised bushingless chains and the wide–narrow teeth chainrings, which deviate from the conventional design of roller chain drive systems. However, there is very little research that specifically delves into these types of chain drives. Investigating the performance and characteristics of bushingless chains can provide valuable insights into their physical properties, efficiencies, and tribological behaviours. Achievements on understanding the friction force, wear resistance, dynamic characteristics, and development of corresponding simulation models can facilitate industry to improve products.
- There is a critical need for research and innovations in friction reduction within the drivetrain to enhance tribological performance. This can be achieved through innovative lubrication techniques, surface treatments, and exploring the potential of novel materials and coatings to minimise wear and increase efficiency. The integration of advanced materials, such as light alloys and ceramics, can also play a pivotal role in developing more durable and lightweight drivetrain components.
- The trend in modelling and simulation for bicycle drivetrain research involves a more profound understanding of the power transmission mechanisms and an increasing reliance on advanced computational tools. It allows for a detailed analysis and optimisation of various aspects of the drivetrain system. Accurate characterisation of the empirical parameters, including friction coefficients, is crucial. Further research is needed to explore the impact of key components in the chain drive, such as pin, bushing, and roller dimensions, on power loss and wear rates. This investigation will aid in designing more efficient and durable chain drives.
- The electric bicycle industry has experienced significant development in recent years. The adoption of electronic drivetrains, including electronic shifting systems and advanced motor controllers, is on the rise. Innovations such as internal gear hubs, belt drives, and automatic transmission are gaining popularity. There is a need to make them lightweight, more reliable, and affordable. This evolving landscape presents new challenges and opportunities that warrant special attention in drivetrain research.

Author Contributions: Conceptualisation, Y.W.L., O.M., D.V.D. and H.L.; data curation, Y.W.L. and H.L.; writing—original draft preparation, Y.W.L.; writing—review and editing, Y.W.L. and H.L., supervision, O.M., D.V.D. and H.L. All authors have read and agreed to the published version of the manuscript.

Funding: Y.L. receives a PhD scholarship funded by GRIFFITH UNIVERSITY and RIDE MECHANICS.

Acknowledgments: During the preparation of this work, the authors used ChatGPT to improve the language and readability of the article. After using this tool/service, the authors reviewed and edited the content as needed and take full responsibility for the content of the publication.

Conflicts of Interest: Author O.M. was employed by the company Ride Mechanic Pty Ltd. The remaining authors declare that the research was conducted in the absence of any commercial or financial relationships that could be construed as a potential conflict of interest.

References

1. Tabascio, A.; Tiznado-Aitken, I.; Harris, D.; Farber, S. Assessing the potential of cycling growth in Toronto, Canada. *Int. J. Sustain. Transp.* **2023**, *17*, 1370–1383. [CrossRef]
2. Oliveira, F.; Nery, D.; Costa, D.G.; Silva, I. A Survey of Technologies and Recent Developments for Sustainable Smart Cycling. *Sustainability* **2021**, *13*, 3422. [CrossRef]
3. Chen, W.; Carstensen, T.A.; Wang, R.; Derrible, S.; Rueda, D.R.; Nieuwenhuijsen, M.J.; Liu, G. Historical patterns and sustainability implications of worldwide bicycle ownership and use. *Commun. Earth Environ.* **2022**, *3*, 171. [CrossRef]

4. Thomas, F. 2022 World Bicycle Industry Report: An Industry Addicted to N + 1. a-dbikes.com. 2024. Available online: https://a-dbikes.com/blogs/news/analysis-of-world-bicycle-industry-association-report-2022?srsltid=AfmBOopdwK0F6VwbsJTcTIYJlrFfi_YKhHirQEaOZtCtEgQ8H9DYyouc&utm_source=chatgpt.com (accessed on 14 January 2025).
5. Grand View Research. Bicycle Market Analysis Report. 2023. Available online: https://www.grandviewresearch.com/industry-analysis/bicycle-market#:~:text=ReportOverview,anticipatedtopropelmarketgrowth (accessed on 1 February 2023).
6. Wilson, D.G.; Schmidt, T. *Bicycling Science*, 4th ed.; Massachusetts Institute of Technology: Cambridge, MA, USA, 2020.
7. Rérat, P. The rise of the e-bike: Towards an extension of the practice of cycling? *Mobilities* **2021**, *16*, 423–439. [CrossRef]
8. Teixeira, J.F.; Silva, C.; Moura, F. Empirical evidence on the impacts of bikesharing: A literature review. *Transp. Rev.* **2021**, *41*, 329–351. [CrossRef]
9. Popan, C. Introduction Tips of the cycling iceberg. In *Bicycle Utopias: Imagining Fast and Slow Cycling Futures*; Taylor & Francis Group: Abingdon, UK, 2018. [CrossRef]
10. Malizia, F.; Blocken, B. Cyclist aerodynamics through time: Better, faster, stronger. *J. Wind. Eng. Ind. Aerodyn.* **2021**, *214*, 104673. [CrossRef]
11. Holmberg, K.; Erdemir, A. Influence of tribology on global energy consumption, costs and emissions. *Friction* **2017**, *5*, 263–284. [CrossRef]
12. Underwood, L.; Jermy, M. Determining optimal pacing strategy for the track cycling individual pursuit event with a fixed energy mathematical model. *Sports Eng.* **2014**, *17*, 183–196. [CrossRef]
13. Pickering, C.; Byrne, J. The benefits of publishing systematic quantitative literature reviews for PhD candidates and other early-career researchers. *High. Educ. Res. Dev.* **2014**, *33*, 534–548. [CrossRef]
14. Lovejoy, K.; Handy, S. Developments in bicycle equipment and its role in promoting cycling as a travel mode. In *City Cycling*; MIT Press: Cambridge, MA, USA, 2012; pp. 75–104.
15. Stead Cycles. Single Speed vs Geared Bikes. 2023. Available online: https://www.steadcycles.com.au/singlespeed-vs-gearedbikes/ (accessed on 23 March 2023).
16. Thakur, T.S.; Babu, P.M. Evolution of Bicycles and their utility as Fitness aids—A Review. *IOSR J. Nurs. Health Sci.* **2019**, *8*, 19–23.
17. Dawson, T. H.R.Morris 1968, with Beautifully Carved Lugwork. Vintage Bicycle Blog. 2020. Available online: https://vintagebicycle.wordpress.com/page/2/ (accessed on 5 March 2024).
18. Malppan, G.J.; Sunny, T. A Review on Design Developments in Bicycle. *Int. Res. J. Eng. Technol.* **2015**, *2*, 1794–1799. Available online: https://www.researchgate.net/profile/Tom-Sunny-2/publication/280698135_A_Review_on_Design_Developments_in_Bicycle/links/602ef0a292851c4ed5804753/A-Review-on-Design-Developments-in-Bicycle.pdf (accessed on 28 October 2024).
19. Urunkar, R.U.; Deshpande, P.P. Study of Drive Mechanisms of Bicycle, Tricycle or Like Vehicles to Optimize Operating Performance-A Review. *J. Eng. Res. Appl.* **2014**, *4*, 214–219. Available online: https://www.ijera.com/papers/Vol4_issue1/Version%202/AC4102214219.pdf (accessed on 28 October 2024).
20. Stratview Research. Belts, Chains and Gears: How Power Transmission Works. EETech Media. 2022. Available online: https://eepower.com/technical-articles/belts-chains-and-gears-how-power-transmission-works/# (accessed on 24 March 2023).
21. SRAM. *Spare Parts Catalog*; SRAM: Chicago, IL, USA, 2023. [CrossRef]
22. Reiter, M.; Florczyk, R.; Braedt, H. Chainring. U.S. Patent US 11,110,991 B2, 7 September 2021. Available online: https://patents.justia.com/patent/11110991 (accessed on 10 January 2025).
23. Livesey, A. *Bicycle Engineering and Technology*; Routledge, Taylor & Francis Group: Abingdon, UK, 2021.
24. Berto, F.J. The Sunset of Suntour. In Proceedings of the 9th International Cycling History Conference, Ottawa, ON, Canada, 19–21 August 1998; Volume 9, pp. 116–140. Available online: https://www.mechanischehirngespinnste.de/wp-content/uploads/2021/07/sunset_of_suntour.pdf (accessed on 28 October 2024).
25. Wagner, M.; Koch, T.; Kuehne, I.; Frey, A. Analysis of Static Stress in a Bicycle Chain Plate. In Proceedings of the 2015 COMSOL Conference, Grenoble, France, 14–16 October 2015; pp. 1–6.
26. Dupoyet, G.J.M. The Transmission Chain. Published Online 1981. Available online: https://ridemechanic.com.au/wp-content/uploads/2019/08/Sedis_US4265134A.pdf (accessed on 28 October 2024).
27. Cho, C.K.; Yun, M.H.; Yoon, C.S.; Lee, M.W. An ergonomic study on the optimal gear ratio for a multi-speed bicycle. *Int. J. Ind. Ergon.* **1999**, *23*, 95–100. [CrossRef]
28. Van Soest, A.J. From bicycle chain ring shape to gear ratio : Algorithm and examples. *J. Biomech.* **2014**, *47*, 281–283. [CrossRef] [PubMed]
29. Mo, Y.; Xu, X. Solidworks chain drive design and 3D modeling techniques. *Appl. Mech. Mater.* **2012**, *215–216*, 1146–1149. [CrossRef]
30. Wang, L. Simulation of chain drive based on MSC. Visual nastran 4D. *Appl. Mech. Mater.* **2012**, *101–102*, 267–270. [CrossRef]
31. Suciu, P. When Were Bicycle Gears Invented? Bikeradar. 2012. Available online: https://www.bikeradar.com/features/when-were-bicycle-gears-invented/#:~:text=In1895FrenchmanJeanLoubeyre,thePolyceler(multispeed).&text=Thederailleurwasdevelopedin,allowedgearingtobeused (accessed on 4 May 2023).

32. Lin, T.Y.; Tseng, C.H.; Fong, Z.H. An experimental approach characterizing rear bicycle derailleur systems Part II: The stability region and its applications. *Int. J. Veh. Des.* **1998**, *19*, 371–384.
33. Hadland, T.; Lessing, H. *Bicycle Design: An Illustrated History*; The MIT Press: Cambridge, MA, USA, 2014.
34. Wikipedia Contributors. Gearbox Bicycle. Wikipedia, The Free Encyclopedia. 2024. Available online: https://en.wikipedia.org/w/index.php?title=Gearbox_bicycle&oldid=1225189249 (accessed on 11 January 2025).
35. Denham, A. Inside The Revolutionary Honda Bicycle Gearboxes (All 3 Generations). cyclingabout.com. 2023. Available online: https://www.cyclingabout.com/inside-the-revolutionary-honda-bicycle-gearboxes/ (accessed on 11 January 2025).
36. Kaga, H.; Kimura, T. Transmission with Internal Protective Shield and Bicycle Incorporating Same. U.S. Patent US7503862B2, 17 March 2009. Available online: https://patents.google.com/patent/US7503862B2/en (accessed on 10 January 2025).
37. Sistema de Transmisión para Bicicleta, by César ROJO VIDAL. Spain patent ES2684528A1, 30 March 2017. Available online: https://patents.google.com/patent/ES2684528A1/es (accessed on 10 January 2025).
38. Iwai, T.; Taniguchi, M. Sliding Component and Bicycle Internal Transmission Device. U.S. Patent US2019/0011037 A1, 10 July 2017. Available online: https://patentimages.storage.googleapis.com/ac/c6/46/60b59fb9800461/US20190011037A1.pdf (accessed on 10 January 2025).
39. Hsieh, L.c. The Kinematic Design of Automatic Gear Transmissions for Bicycles. *Eng. Lett.* **2008**, *16*. Available online: https://www.engineeringletters.com/issues_v16/issue_3/EL_16_3_17.pdf (accessed on 10 January 2025).
40. Mercer, H. Automatic Transmission System for a Bicycle. U.S. Patent US20230373592A1, 23 November 2023. Available online: https://patents.google.com/patent/US20230373592A1/en (accessed on 10 January 2025).
41. Asadi, P. Cvt Automatic Variator Transmission for a Bicycle. U.S. Patent US20220213948A1, 7 July 2022. Available online: https://patents.google.com/patent/US20220213948A1/en (accessed on 10 January 2025).
42. Heine, J. Nivex Automatic Transmission. renehersecycles.com. 2023. Available online: https://www.renehersecycles.com/nivex-automatic-transmission/#:~:text=ThenewNivexautomatictransmission,optimumgearforeachsituation (accessed on 11 January 2025).
43. Roggeman, P.; Shimano Auto Shifts Into the Future. nsmb.com. 2023. Available online: https://nsmb.com/articles/shimano-auto-shifts-into-the-future/#:~:text=Specs*1x12-speedwith10-51tCassetteOptions.,Di2RightShiftSwitchandI-SPECEV (accessed on 11 January 2025).
44. Levarda, E. Bicycle transmissions. *IOP Conf. Ser. Mater. Sci. Eng.* **2018**, *444*, 052013. [CrossRef]
45. Wikipedia Contributors. Belt-Driven Bicycle. Wikipedia, The Free Encyclopedia. 2024. Available online: https://en.wikipedia.org/w/index.php?title=Belt-driven_bicycle&oldid=1227938963 (accessed on 10 January 2025).
46. Denham, A.; Gates Carbon Belt Drive: Everything You Ever Need to Know. cyclingabout.com. 2019. Available online: https://www.cyclingabout.com/carbon-belt-drive-everything-you-ever-need-to-know/ (accessed on 10 January 2025).
47. Norman, P. What Are Belt Drive Bikes and Why Are They a Popular Alternative to a Bike with a Chain? bikeradar.com. 2024. Available online: https://www.bikeradar.com/advice/buyers-guides/belt-drive-bicycles/ (accessed on 10 January 2025).
48. Abagnale, C.; Strano, S.; Cardone, M.; Iodice, P.; Terzo, M.; Vorraro, G. Design and Development of a Test Rig for E-bike Performance Evaluation. *SAE Tech. Pap.* **2015**, *2015*, 1–7. [CrossRef]
49. Deleenheer, W.; Jáneš, L.; Jayakumar, A. Development of an Electric Bicycle for a Sharing System in Prague. *Acta Polytech. CTU Proc.* **2017**, *12*, 24. [CrossRef]
50. Hung, N.B.; Lim, O. A review of history, development, design and research of electric bicycles. *Appl. Energy* **2020**, *260*, 114323. [CrossRef]
51. Arango, I.; Lopez, C.; Ceren, A. Improving the autonomy of a mid-drive motor electric bicycle based on system efficiency maps and its performance. *World Electr. Veh. J.* **2021**, *12*, 59. [CrossRef]
52. Stackhouse, R.V.; Dong, Z. Development of electric bicycle performance testing techniques and adaptable electric bicycle power system. In Proceedings of the2007 International Design Engineering Technical Conferences and Computers and Information in Engineering Conference, Las Vegas, NV, USA, 4–7 September 2007; Volume 3, pp. 1113–1124. [CrossRef]
53. Zhang, S.P.; Tak, T.O. Efficiency evaluation of electric bicycle power transmission systems. *Sustainability* **2021**, *13*, 10988. [CrossRef]
54. Lukes, R.; Hart, J.; Haake, S. An analytical model for track cycling. *Proc. Inst. Mech. Eng. Part P J. Sports Eng. Technol.* **2012**, *226*, 143–151. [CrossRef]
55. Sgamma, M.; Bucchi, F.; Frendo, F. A phenomenological model for chain transmissions efficiency. *IOP Conf. Ser. Mater. Sci. Eng.* **2021**, *1038*, 012060. [CrossRef]
56. Burgess, S.C. Improving cycling performance with large sprockets. *Sports Eng.* **1999**, *1*, 107–113. [CrossRef]
57. Kidd, M.D.; Loch, N.E.; Reuben, R.L. *Bicycle Chain Efficiency*; Heriot-Watt University: Edinburgh, UK, 2000. [CrossRef]
58. Lanaspeze, G.; Guilbert, B.; Manin, L.; Ville, F. Preliminary modelling of power losses in roller chain drive: Application to single speed cycling. *Mech. Ind.* **2022**, *23*, 217–226. [CrossRef]
59. Smit, A.; van der Zwaard, S.; Janssen, I.; Janssen, T.W.J. Power loss of the chain drive in a race tandem bicycle. *Sports Eng.* **2023**, *26*, 1–8. [CrossRef]

60. Aubert, R.; Roizard, X.; Grappe, F.; Lallemand, F. Tribological devices in cycling: A review. *Proc. Inst. Mech. Eng. Part P J. Sports Eng. Technol.* **2023**. [CrossRef]
61. Bouillon, G., Tordion, G.V. On polygonal action in roller chain drives. *J. Manuf. Sci. Eng. Trans. ASME* **1965**, *87*, 243–251. [CrossRef]
62. Bhoite, T.D.; Pawar, P.M.; Gaikwad, B.D. Fea Based Study of Effect of Radial Variation of Outer Link in A Typical Roller Chain Link Assembly. *Int. J. Mech. Ind. Eng.* **2012**, *1*, 69–74. [CrossRef]
63. Pedersen, S.L.; Hansen, J.M.; Ambrósio, J.A.C. A roller chain drive model including contact with guide-bars. *Multibody Syst. Dyn.* **2004**, *12*, 285–301. [CrossRef]
64. Pedersen, S.L. Model of contact between rollers and sprockets in chain-drive systems. *Arch. Appl. Mech.* **2005**, *74*, 489–508. [CrossRef]
65. Ambrosio, J.; Malça, C.; Ramalho, A. Planar roller chain drive dynamics using a cylindrical contact force model. *Mech. Based Des. Struct. Mach.* **2016**, *44*, 109–122. [CrossRef]
66. Papuc, R.; Velicu, R. Tribological Study of Guide-Chain Contact. *Ann. ORADEA Univ. Fascicle Manag. Technol. Eng.* **2013**, *XXII*, 257–260. [CrossRef]
67. Choi, W.; Johnson, G.E. Transverse vibration of roller chain drive with tensioner. In Proceedings of the International Design Engineering Technical Conferences and Computers and Information in Engineering Conference, Albuquerque, NM, USA, 19–22 September 1993; Volume 11795, pp. 19–28. [CrossRef]
68. Dwyer-Joyce, R.S.; Lewis, R.; Ward, A.; Patterson, E.A. Determination of impact stresses in an automotive chain drive component. *SAE Tech. Pap.* **2006**, *2006*, 1–9. [CrossRef]
69. Binder, R.C.; Mize, G.G. Strand vibrations in a roller-chain drive. *J. Frankl. Inst.* **1949**, *247*, 25–32. [CrossRef]
70. Wang, K.W. On the Stability of Chain Drive Systems Under Periodic Sprocket Oscillations. *J. Vib. Acoust.* **1992**, *114*, 119–126. [CrossRef]
71. Wang, K.W.; Liu, S.P.; Hayek, S.I.; Chen, F.H.K. On the impact intensity of vibrating axialiy moving roller chains. *J. Vib. Acoust. Trans. ASME* **1992**, *114*, 397–403. [CrossRef]
72. Liu, S.P.; Wang, K.W.; Hayek, S.I.; Trethewey, M.W.; Chen, F.H.K. A global-local integrated study of roller chain meshing dynamics. *J. Sound Vib.* **1997**, *203*, 41–62. [CrossRef]
73. Liu, S.P. *Impact Dynamics of Chain Drive System*; The Pennsylvania State University: University Park, PA, USA, 1998. Available online: https://www.proquest.com/openview/360c1d041908eff3765b1429475e48ad/1?pq-origsite=gscholar&cbl=18750&diss=y (accessed on 8 January 2025).
74. Zheng, H.; Wang, Y.Y.; Liu, G.R.; Lam, K.; Quek, K.; Ito, T.; Noguchi, Y. Efficient modelling and prediction of meshing noise from chain drives. *J. Sound Vib.* **2001**, *245*, 133–150. [CrossRef]
75. Binder, R.C.; Covert, W.V. Impact between chain roller and sprocket in a chain drive. *J. Frankl. Inst.* **1948**, *245*, 319–329. [CrossRef]
76. Binder, R.C. *Mechanics of the Roller Chain Drive*; Prentice-Hall: Englewood Cliffs, NJ, USA, 1956.
77. Naji, M.R.; Marshek, K.M. The effects of the pitch difference on the load distribution of a roller chain drive. *Mech. Mach. Theory* **1989**, *24*, 351–362. [CrossRef]
78. Naji, M.R.; Marshek, K.M. Experimental determination of the roller chain load distribution. *J. Mech. Des. Trans. ASME* **1983**, *105*, 331–338. [CrossRef]
79. Friction Facts. Friction-Producing Mechanisms of a Bicycle Chain—Effectiveness of Chain Lubricants for Reducing Frictional Losses. 2014. Available online: https://zerofrictioncycling.com.au/wp-content/uploads/2020/06/Friction-Producing-Mechanisms.pdf (accessed on 5 March 2024).
80. Kim, M.S.; Johnson, G.E. Mechanics of Roller Chain-Sprocket Contact: A General Modelling Stragegy. In Proceedings of the International Design Engineering Technical Conferences and Computers and Information in Engineering Conference, Scottsdale, AZ, USA, 13–16 September 1992; American Society of Mechanical Engineers: New York, NY, USA, 1992; Volume 9372, pp. 689–695. [CrossRef]
81. Conwell, J.C.; Johnson, G.E. Experimental investigation of link tension and roller-sprocket impact in roller chain drives. *Mech. Mach. Theory* **1996**, *31*, 533–544. [CrossRef]
82. Troedsson, I.; Vedmar, L. A method to determine the static load distribution in a chain drive. *J. Mech. Des. Trans. ASME* **1999**, *121*, 402–408. [CrossRef]
83. Troedsson, I.; Vedmar, L. A dynamic analysis of the oscillations in a chain drive. *J. Mech. Des. Trans. ASME* **2001**, *123*, 395–401. [CrossRef]
84. Troedsson, I.; Vedmar, L. A method to determine the dynamic load distribution in a chain drive. *Proc. Inst. Mech. Eng. Part C J. Mech. Eng. Sci.* **2001**, *215*, 569–579. [CrossRef]
85. Lodge, C.J.; Burgess, S.C. A model of the tension and transmission efficiency of a bush roller chain. *Proc. Inst. Mech. Eng. Part C J. Mech. Eng. Sci.* **2002**, *216*, 385–394. [CrossRef]

86. Lodge, C.J.; Burgess, S.C. An investigation into the selection of optimum chain and sprocket size. *J. Eng. Des.* **2004**, *15*, 563–580. [CrossRef]
87. Burgess, S.; Lodge, C. Optimisation of the chain drive system on sports motorcycles. *Sports Eng.* **2004**, *7*, 65–73. [CrossRef]
88. Hollingworth, N.E.; Hills, D.A. Theoretical efficiency of a cranked link chain drive. *Proc. Inst. Mech. Eng. Part C J. Mech. Eng. Sci.* **1986**, *200*, 375–377. [CrossRef]
89. Spicer, J.B.; Richardson, C.J.K.K.; Ehrlich, M.J.; Bernstein, J.R.; Fukuda, M.; Terada, M. Effects of frictional loss on bicycle chain drive efficiency. *J. Mech. Des. Trans. ASME* **2001**, *123*, 598–605. [CrossRef]
90. Spicer, J.B. Effects of the Nonlinear Elastic Behavior of Bicycle Chain on Transmission Efficiency. *J. Appl. Mech.* **2013**, *80*, 021005. [CrossRef]
91. Zhang, S.P.; Tak, T.O. Efficiency estimation of roller chain power transmission system. *Appl. Sci.* **2020**, *10*, 7729. [CrossRef]
92. Nikhil, M.; Pisal, S.; Khot, J. Structural analysis of motorcycle chain by using CAE software. *Int. J. Adv. Res. Sci. Eng. IJARSE* **2015**, *8354*, 540–546.
93. Dhage, B.H.; Diwate, A.D. Finite Element Analysis and Experimentation of Carbon Fiber Chain Drive. *GRD J. Eng.* **2017**, *2*, 24–33. Available online: https://www.grdjournals.com/uploads/article/GRDJE/V02/I12/0008/GRDJEV02I120008.pdf (accessed on 8 January 2025).
94. Suwannahong, W.; Suvanjumrat, C. An integrating finite element method and multibody simulation for drive systems analysis. *Eng. J.* **2017**, *21*, 221–234. [CrossRef]
95. Lai, W.H.; Sung, C.K.; Wang, J.B. Motion analysis of a bicycle rear derailleur during the shifting process. *Mech. Mach. Theory* **1998**, *33*, 365–378. [CrossRef]
96. Wang, Y.; Zheng, Z.; Zhang, G. A study on jumping-over-teeth phenomenon in roller chain drive. *J. Mech. Des. Trans. ASME* **1990**, *112*, 569–574. [CrossRef]
97. Ma, Y.Z.; Chiou, S.J. An Optimal Model on Contour of Up-Shifting Tooth for Derailleur System of Bicycle. *J. Mech.* **2017**, *33*, 759–767. [CrossRef]
98. Pedersen, S.L. *Simulation and Analysis of Roller Chain Drive Systems*; Technical University of Denmark: Kongens Lyngby, Denmark, 2004. Available online: https://core.ac.uk/download/pdf/13738147.pdf (accessed on 8 January 2025).
99. Wu, Y.C.; Ren, P.W.; Chen, L.A. Kinematic analysis of an 8-speed bicycle transmission hub. *Appl. Mech. Mater.* **2014**, *479–480*, 234–238. [CrossRef]
100. Omar, M.A. Chain drive simulation using spatial multibody dynamics. *Adv. Mech. Eng.* **2014**, *6*, 378030. [CrossRef]
101. Omar, M. Multibody dynamics formulation for modeling and simulation of roller chain using spatial operator. *MATEC Web Conf.* **2016**, *51*, 01003. [CrossRef]
102. Budinski, K.G. Types of Friction and Friction Testing. In *Guide to Friction, Wear, and Erosion Testing*; ASTM International: West Conshohocken, PA, USA, 2007; p. 17.
103. Rank, M.; Oehler, M.; Koch, O.; Bobzin, K.; Kalscheuer, C.; Möbius, M.P. Investigation of the Influence of Triboactive CrAlMoN Coating on the Joint Wear of Grease-Lubricated Roller Chains. *Tribol. Trans.* **2023**, *66*, 1105–1116. [CrossRef]
104. Tichy, J.A.; Meyer, D.M. Review of solid mechanics in tribology. *Int. J. Solids Struct.* **2000**, *37*, 391–400. [CrossRef]
105. Benabdallah, H.S.; Aguilar, D.A. Acoustic emission and its relationship with friction and wear for sliding contact. *Tribol. Trans.* **2008**, *51*, 738–747. [CrossRef]
106. Zhao, J.T.; Wang, S.Z.; Wang, Z.X. The effects of the wear elongation on the load of a long-distance transmission chain. *Appl. Mech. Mater.* **2014**, *456*, 60–64. [CrossRef]
107. David, R.; Oscar, H. Bicycle Chain Wear Explained | How to Know when to Replace a Bike Chain. Bikeradar. Available online: https://www.bikeradar.com/advice/workshop/how-to-know-when-its-time-to-replace-your-bicycle-chain (accessed on 8 January 2025).
108. Becker, A.; Meffert, D.; Sauer, B. Friction and wear investigations on single chain joints. *Forsch. Im Ingenieurwesen/Eng. Res.* **2019**, *83*, 53–63. [CrossRef]
109. Meffert, D.; Oehler, M.; Sauer, B. Precise friction measurement in drive chains using a chain joint tribometer. *Tribol. Online* **2021**, *16*, 151–158. [CrossRef]
110. Zhou, Y. Mathematical Modeling of Chain Drive Geometries for a Durability Test Rig. *Exp. Tech.* **2015**, *40*, 1137–1146. [CrossRef]
111. Rosenkranz, A.; Krupp, F.; Reinert, L.; Mücklich, F.; Sauer, B. Tribological performance of laser-patterned chain links—Influence of pattern geometry and periodicity. *Wear* **2017**, *370–371*, 51–58. [CrossRef]
112. Saito, R.; Noda, N.A.; Sano, Y. Newly developed wear testing machine having sufficient reproducibility useful for investigating roller chains. *ISIJ Int.* **2020**, *60*, 2255–2265. [CrossRef]
113. Fawcett, J.N.; Nicol, S.W. Influence of Lubrication on Tooth-Roller Impacts in Chain Drives. *Inst. Mech. Eng. Proc.* **1977**, *191*, 271–275. [CrossRef]
114. Palazzetti, R.; Yan, X.T. Study on lubrication effect on motorbike chain transmissions. *Ind. Lubr. Tribol.* **2016**, *68*, 561–568. [CrossRef]

115. Bloch, H.P.; Bannister, K.E. *Practical Lubrication for Industrial Facilities*, 3rd ed.; River Publishers: Aalborg, Denmark, 2017. [CrossRef]
116. Downs, T. *The Bicycling Guide to Complete Bicycle Maintenance & Repair*, 6th ed.; Rodale Press: Emmaus, PA, USA, 2010.
117. Shen, C.J.; Abdul Rahim, N.; Abdul Rahman, M.R. An Investigation of Bicycle Chain Lubrication Performance in Rainy Condition. *ZULFAQAR J. Def. Sci. Eng. Technol.* **2022**, *5*, 171–179.
118. Tung, S.C.; McMillan, M.L. Automotive tribology overview of current advances and challenges for the future. *Tribol. Int.* **2004**, *37*, 517–536. [CrossRef]
119. Neale, M.J. Lubrication of components. In *Lubrication and Reliability Handbook*; Elsevier Science & Technology: Amsterdam, The Netherlands, 2000.
120. Kozlov, K.E.; Egorov, A.V.; Belogusev, V.N. Experimental Evaluation of Chain Transmissions Lubricants Quality Using a New Method Based on Additional Inertia Moment Use. *Procedia Eng.* **2017**, *206*, 617–623. [CrossRef]
121. Tandler, R.; Bohn, N.; Gabbert, U.; Woschke, E. Experimental investigations of the internal friction in automotive bush chain drive systems. *Tribol. Int.* **2019**, *140*, 105871. [CrossRef]
122. Michelsen, B. *The Performance of Renewable Bicycle Chain Lubricants*; University of Delaware: Newark, DE, USA, 2015. Available online: https://udspace.udel.edu/handle/19716/17093 (accessed on 28 October 2024).

machines

MDPI

Article

Research on Active–Passive Training Control Strategies for Upper Limb Rehabilitation Robot

Yongming Yang

Public Experiment Center, University of Shanghai for Science and Technology, Shanghai 200093, China; jackyang@usst.edu.cn

Abstract: Due to accidents, upper limb movement disorders have become increasingly common. Training can help restore muscle strength and rebuild neurological function. However, the existing single mode has limitations in adapting to the training needs of different rehabilitation stages. Therefore, this paper conducts research on active–passive training control strategies for an upper limb rehabilitation robot. It establishes an upper limb kinematic model based on the Lagrange method and builds a man–machine integration dynamics model for upper limb rehabilitation in MATLAB (R2016a)/Simulink. A design active controller, passive controller, and switching controller based on PI and feedforward compensation strategies are proposed to improve training control accuracy. The output moment of the system during active training is planned to ensure the safety and stability of the training process. By utilizing neural networks to train sample data during rehabilitation training, the fuzzy rules and membership functions in fuzzy intention recognition algorithm are optimized to improve the accuracy of intention recognition during training. By adopting the independently developed experimental platform for the upper limb rehabilitation robot, active–passive training, intention recognition, and training mode switching are achieved. The results show that the active and passive training processes are smooth, the training intention recognition is accurate, and the switching between active and passive training modes is steady. This verifies the feasibility and effectiveness of the established mathematical model in upper limb rehabilitation training.

Keywords: active–passive training; intention recognition; adaptive neural fuzzy network; upper limb rehabilitation robot; involute planetary gear reducer

1. Introduction

Through systematic limb exercises and joint activities, the functional strength of muscles can be effectively improved, promoting the recovery process of individuals in rehabilitation [1].

With the widespread application of rehabilitation robots, problems such as single mode and a low level of intelligence have emerged. Scholars are researching a new generation of intelligent rehabilitation robots with diversified modes, characterized by the deep integration of man–machine interaction [2–5]. MIT developed the MIT-Manus rehabilitation robot in the 1990s [6]. HOCOMAAG, a Swiss company, is committed to the research and development of high-performance rehabilitation and training robots [7]. Domestic companies such as DaAi and Step are committed to the research and development of rehabilitation robots with mature technology and stable performance.

Rehabilitation robots are divided into auxiliary and training types. The auxiliary type is mainly aimed at rehabilitation of individuals with limb paralysis and an inability to recover by utilizing rehabilitation robots to assist with certain training movements; this training type is mainly aimed at rehabilitation of individuals who have suffered limb injuries but can still recover through training [8]. The most widely used training type rehabilitation robot uses active–passive training, but there are problems such as difficulty in mode switching and a non-compliant training process [9].

Early rehabilitation robots often used passive training. Mason, M.T., et al. [10] proposed a force–position hybrid control strategy that can simultaneously control both the force and position. Hogan, N. [11], proposed an impedance control strategy that can accurately track the desired trajectory of a robot when it is subjected to external interference, allowing the robot to adjust the interaction force with the external environment in a timely manner. Cao, J.H., et al. [12] designed a MIMO synovial controller for trajectory control and flexible control of a lower limb exoskeleton equipped with two pneumatic artificial muscles. Zhang, Y.K. [13], designed an upper limb pose maintenance rehabilitation training robot, established a pose maintenance algorithm by using an impedance-compliant control model, and conducted kinematic and dynamic simulations.

In addition to advanced driving and control strategies [14], rehabilitation robots also need to accurately recognize the training intentions of individuals in rehabilitation during the training process. Fu, R.R., et al. [15] used EEG signals to extract motion intention recognition signals, achieving the identification of pre-movement potentials for left and right hands. Zhai, M.Q. [16], used sEMG surface electromyography signals to extract the characteristics generated during human motion. They adopted an adaptive-energy microproduct-based acquisition method to collect muscle strength electrical signals and then classified the collected signals for motion intention recognition. Fan, Y.J. [17], developed a rehabilitation robot based on the interaction force and sEMG, which integrates signals such as the sEMG, joint rotation angle, and man–machine interaction force for motion intention recognition. In clinical trials at the hospital, the robot accurately identified the motion intention of individuals in rehabilitation. Farina, D., et al. [18] studied the characteristics of electromyographic signals generated during continuous movement of the wrist joint with one to two degrees of freedom. They trained the electromyographic motion mapping relationship using methods such as linear regression and neural networks and successfully applied it to electromyographic control of prosthetic wrist joints. Liu, Z.J., et al. [19] applied inertial sensors to intelligent prostheses and classified the collected swing phase temporal motion data using a support vector machine classifier. This effectively identified motion states such as going upstairs, going downstairs, and walking on flat ground, with an accuracy rate of 95.8%. Li, G.N., et al. [20] designed a terminal traction rehabilitation robot, which can effectively recognize the movement intention of individuals in rehabilitation in real time by registering the man–machine interaction force and its change rate under different motion states and using a fuzzy algorithm to classify its features. Tran, H.T., et al. [21] used Gaussian process regression to study the relationship between the interaction forces and dynamic factors between the human body and external bone system and achieved human posture prediction. Wang, C.X., et al. [22] designed an upper limb motion recognition system using three-dimensional acceleration sensors. The feature vectors were obtained through wavelet transform and classified using support vector machines, which can effectively recognize various types of upper limb movements.

The purpose of this paper is to conduct research on active–passive training control strategies for an upper limb rehabilitation robot. It establishes an upper limb kinematic model based on the Lagrange method and builds a man–machine integration dynamics model for upper limb rehabilitation in MATLAB (R2016a)/Simulink. By adopting the independently developed experimental platform for the upper limb rehabilitation robot, active–passive training, intention recognition, and training mode switching are achieved. The results show that the active and passive training processes are smooth, the training intention recognition is accurate, and the switching between active and passive training modes is steady. This verifies the feasibility and effectiveness of the established mathematical model in upper limb rehabilitation training.

2. Man–Machine Model for Upper Limb Rehabilitation

2.1. Upper Limb Dynamics Equation

In order to analyze the relationship between the position and force in upper limb rehabilitation training, the upper limb is regarded as movement in a plane, with shoulder movements including bending and extension and elbow movements including elbow bending and extension. In this plane, the upper limb can be approximated as a linkage mechanism. Due to the contact between the palm and operating arm, the distance between them is negligible. Consider the wrist joint, palm, and operating arm as a single point, and the shoulder joint as a fixed fulcrum, to establish the upper limb plane model shown in Figure 1. l_1 represents the length of the upper arm; l_2 represents the length of the forearm; A represents the shoulder joint; B represents the elbow joint; C represents the wrist joint, palm, and manipulator arm; D is the center of mass of the upper arm; and E is the center of mass of the forearm. The angle between the upper arm and the vertical axis, y_1, is q_1, and the angle between the forearm and the vertical axis, y_2, is q_2.

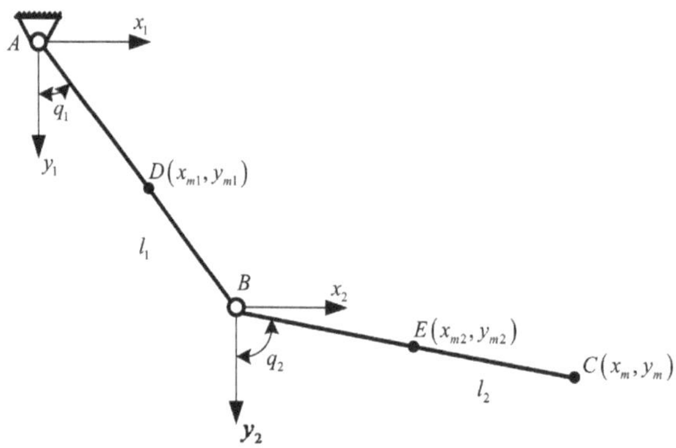

Figure 1. Upper limb plane model.

Based on system dynamics, the generalized force equation for the upper limb can be derived as Equation (1).

$$\begin{cases} Q_1 = T_j - m_1 g l_{m1} sinq_1 - (m_2 g + F) l_1 sinq_1 \\ Q_2 = T_z - m_2 g l_{m2} sinq_2 - F l_2 sinq_2 \end{cases} \quad (1)$$

where T_j is the joint resultant moment of the shoulder joint; T_z is the resultant moment of the elbow joint; F is the force exerted on the end pivot point; m_1, m_2 are the mass of the forearm and upper arm; and l_{m1}, l_{m2} are the center position of the mass of the forearm and upper arm, respectively.

Based on the Lagrange equation, the upper limb dynamics equation can be derived as Equation (2).

$$\begin{cases} J_{11}\ddot{q}_1 + J_{12}\ddot{q}_2 + m_2 l_1 l_{m1} sin(q_1 - q_2)\dot{q}_2^2 = Q_2 \\ J_{21}\ddot{q}_1 + J_{22}\ddot{q}_2 + m_2 l_1 l_{m1} sin(q_2 - q_1)\dot{q}_1^2 = Q_2 \end{cases} \quad (2)$$

where $J_{11} = m_1 l_{m1}^2 + J_{m1} + m_2 l_1^2$; $J_{12} = J_{21} = 2m_2 l_1 l_{m2} \cos(q_2 - q_1)$; $J_{22} = m_2 l_{m2}^2 + J_{m2}$.

2.2. Upper Limb Man–Machine Integration Model

The upper limb man–machine integration model is established by programing in MATLAB(R2016a)/Simulink to analyze the load characteristics and motion laws of the robot during rehabilitation training and is shown in Figure 2.

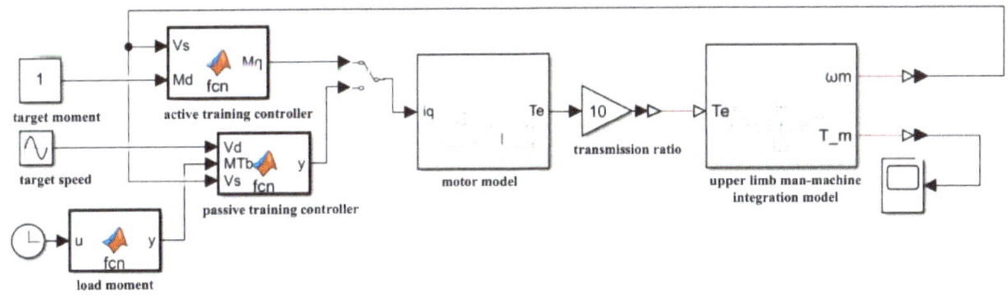

Figure 2. Upper limb man–machine integration model.

For active training, a moment open-loop control system is adopted. The training resistance moment is preset in the active controller, and the system comprehensively considers factors such as the actual training speed and external interference moment to output the corresponding driving current to the motor model, thereby driving the upper limb man–machine integration model to move. For passive training, a speed-based closed-loop control system with speed as the feedback variable is adopted. The preset training speed is input into the passive controller, and the controller calculates the appropriate current for the motor model based on the real-time feedback speed and interference moment, thereby driving the upper limb man–machine integration model to move according to the predetermined speed trajectory.

Figure 3 depicts a current closed-loop control motor model utilizing PI control. This model calculates the moment output by the motor based on its input current value. This moment represents the resistance moment applied by the motor to the individual in rehabilitation during training.

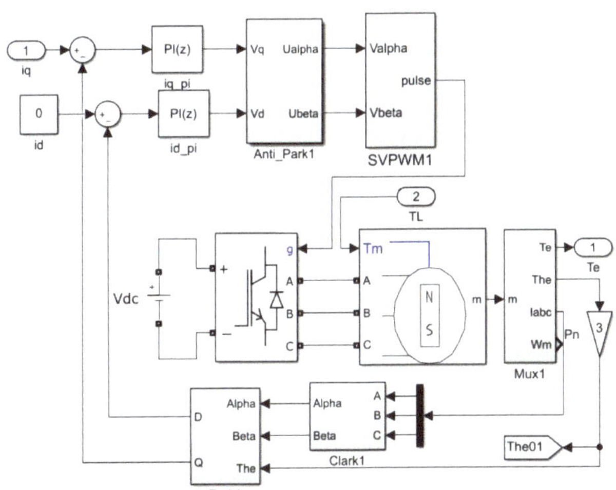

Figure 3. Motor model.

Figure 4 depicts the upper limb man–machine integration dynamics model. Based on the motor's output speed and load moment, the moment at the motor's output end can be determined under varying speed conditions. The moment in the stable state represents the load moment that the rehabilitation robot needs to overcome, arising from the upper limbs and operating arm.

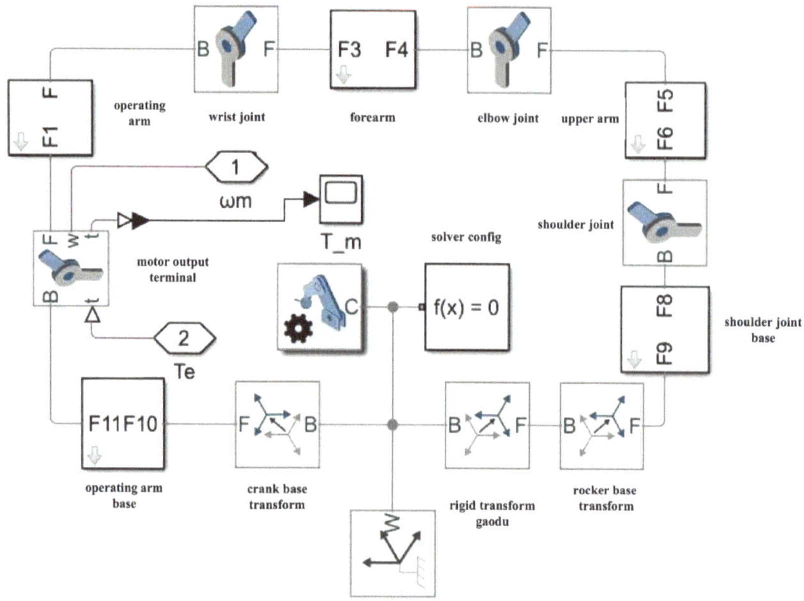

Figure 4. Upper limb man–machine integration dynamics model.

3. Control Strategies for Upper Limb Rehabilitation Training

3.1. Active–Passive Training Control Scheme

Figure 5 depicts the active–passive training control scheme, divided into the decision level and execution level. The decision level includes mode setting, intent recognition, switching controller, and security protection. The execution level includes the active controller and passive controller.

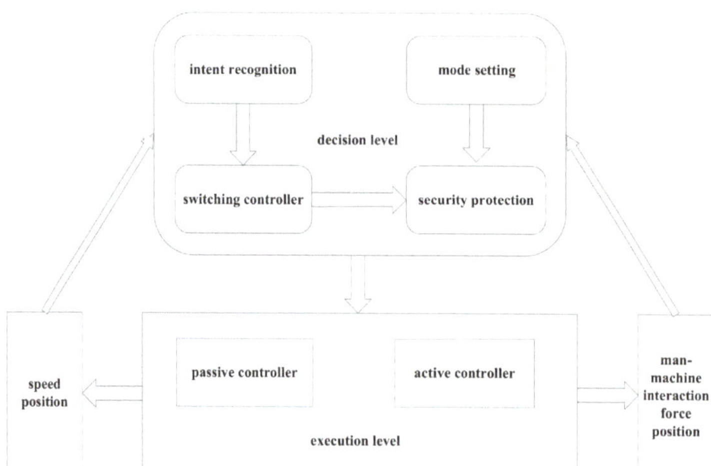

Figure 5. Active–passive training control scheme.

3.1.1. Active Controller

When conducting active training, the rehabilitation robot needs to provide a constant resistance moment to cooperate with the individual in rehabilitation for impedance training.

In the actual training process, factors such as the inherent friction moment of the system and the gravity moment generated by the operating arm itself can affect the accuracy of the moment output by the controller. Therefore, this paper adopts moment open-loop control of the motor to achieve active training. Figure 6 shows the established active controller.

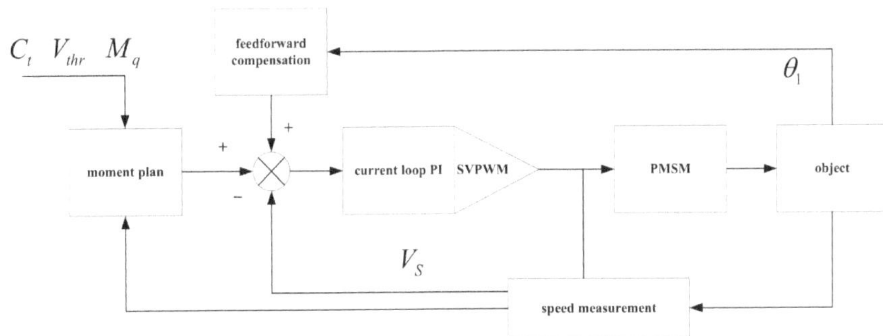

Figure 6. Active controller.

The moment required to compensate for the feedforward compensation link set in the active controller can be derived as Equation (3).

$$M_{fz} = M_f + m_s g l_{m3} \cos\theta_1 \tag{3}$$

Due to the output of a constant moment by the motor without load, rapid rotation of the manipulator arm may occur, which can cause harm to the individual in rehabilitation. Therefore, it is necessary to plan the output moment of the motor. The motion intention is defined as λ_A. This applies when extending the elbow $\lambda_A = 1$, bending the elbow $\lambda_A = -1$, and stopping the movement $\lambda_A = 0$. From this, Equation (4) can be derived.

$$\lambda_A = \begin{cases} 1 & V_S \geq V_{thr} \\ 0 & -V_{thr} < V_S < V_{thr} \\ -1 & V_S \leq -V_{thr} \end{cases} \tag{4}$$

where V_{thr} is the speed threshold; V_S is the current training speed.

If the moment value is controlled and output according to the inherent curve, it is easy for individuals in rehabilitation to become injured when they experience spasms. Therefore, the output of rising and falling phases of the moment value is correlated with the current training speed. The planned output curve of a moment can be derived as Equation (5).

$$M_q = \lambda_A C_t |V_{thr}| \quad M_q < M_d \tag{5}$$

where M_q is the planned output moment; C_t is the moment coefficient; and M_d is the set training resistance moment.

According to Equation (6), the larger the moment coefficient is, the faster the moment value rises or falls. When the moment value reaches the set value, the increase in speed will not affect the change in moment value.

$$M_q = \lambda_A M_d \tag{6}$$

By combining Equations (5) and (6), the equation for the moment planning output value can be derived as Equation (7).

$$M_q = \begin{cases} \lambda_A C_t |V_{thr}| & \lambda_A C_t |V_{thr}| < M_d \\ \lambda_A M_d & \lambda_A C_t |V_{thr}| \geq M_d \end{cases} \tag{7}$$

3.1.2. Passive Controller

When conducting passive training, the individual in rehabilitation's upper limb does not actively exert force, and the robot drives their upper limb to perform reciprocating movements at a set speed. Due to the presence of system friction and dynamic loads generated by the upper limb and manipulator arm in the rehabilitation robot, these factors have a significant impact on the stability of the control system. Therefore, a speed-based closed-loop control motor is used to achieve passive training. Figure 7 shows the established passive controller.

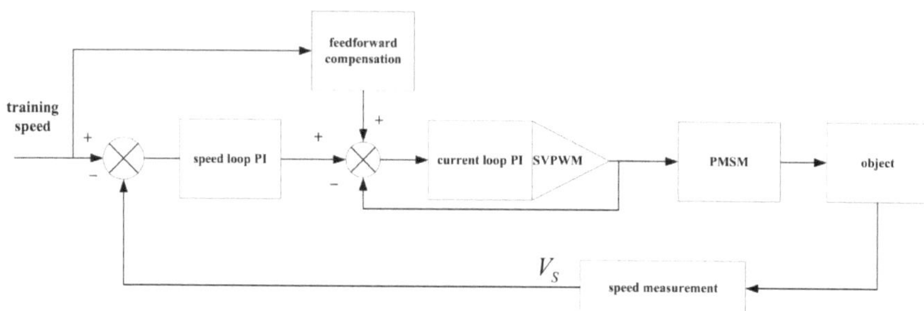

Figure 7. Passive controller.

The moment required to compensate for the feedforward compensation link in the passive controller can be derived as Equation (8).

$$M_{fb} = M_f + M_{Tb} \tag{8}$$

where M_{fb} is the moment that needs to be compensated; M_{Tb} is the load moment obtained from the passive training simulation.

3.1.3. Switching Controller

In the process of active–passive rehabilitation training, it mainly includes active intention, active-to-passive intention, passive intention, and passive-to-active intention. The switching between adjacent motion intentions is the mutual switching between moment open-loop and velocity closed-loop. If the motion intention switching is directly performed, the sudden change in the controlled variable will cause vibration, impact, and other phenomena in the rehabilitation system, resulting in secondary injury to the individual in rehabilitation. Therefore, an active–passive training switching controller as shown in Figure 8 is established, and its input is the individual in rehabilitation's motion intention and current training position. Using a cosine function to smoothly transition the switching amount [23] increases the safety of the rehabilitation robot. The expression for the cosine function is Equation (9).

$$y = \frac{1 + \cos(\frac{\pi x}{z})}{2} \quad (9)$$

where z is the switching time.

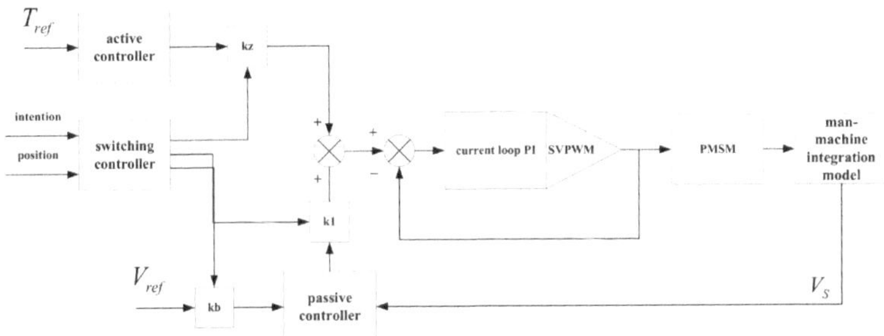

Figure 8. Active–passive training switching controller.

Here, T_{ref} is the target training moment; V_{ref} is the target training speed; k_z and k_b are the proportional coefficients of the active and passive controllers, respectively, ranging from 0 to 1; and k_1 is the passive controller switch, with a value of 0 or 1.

When conducting active training, $k_z = 1$, $k_b = k_1 = 0$. If the switching controller receives a command to switch from active to passive, k_z will decrease to 0 according to the cosine function. When the individual in rehabilitation moves the operating arm to the starting position of the training, $k_b = k_1 = 1$.

When conducting passive training, $k_z = 0$, $k_b = k_1 = 1$. If the switching controller receives a command to switch from passive to active, k_b will decrease to 0 according to the cosine function. When $k_b = 0$, $k_1 = 0$, $k_z = 1$.

3.2. Active–Passive Training Intention Recognition Based on Adaptive Neural Fuzzy Network

In the process of mixed active–passive training, individual in rehabilitation have five training intentions, active intention, active-to-passive intention, passive intention, passive-to-active intention, and spasms. In order to improve the accuracy of training intention recognition, help individuals in rehabilitation switch training modes more easily according to their own intentions during training, and enhance the safety of the training, the rehabilitation training intention recognition model based on the Adaptive Network Fuzzy Inference System (ANFIS) is established and shown in Figure 9. It is a system based on the Takagi Sugeno model and neural network, which have both the learning mechanism of a neural network and the inference ability of a fuzzy system [24]. ANFIS utilizes the least squares method and backpropagation method of a BP neural network to update the parameters of the algorithm and automatically generates If–Then rules, which avoids the shortcomings of traditional methods that overly rely on expert experience to design fuzzy inference rules, improves the control accuracy, and has a self-learning ability. In addition, ANFIS has the characteristics of small errors, convergence speed block, and requiring fewer training samples, making it very suitable for the identification and control of complex and variable systems [25]. By adjusting the parameters, the output of the ANFIS model gradually approaches the expected output, thereby achieving high-precision recognition of the individual in rehabilitation's training intention.

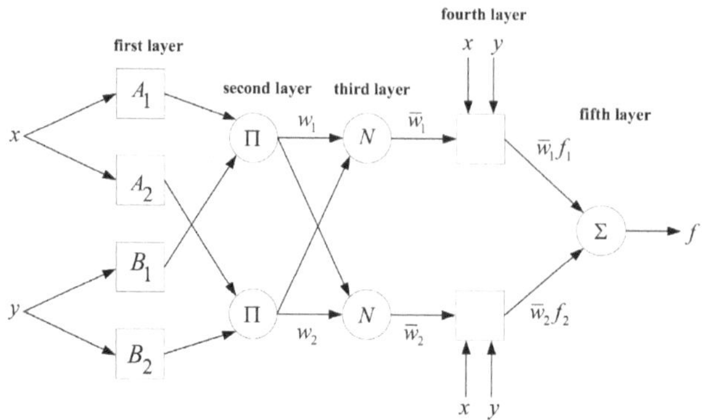

Figure 9. ANFIS schematic diagram.

The ANFIS structure consists of five layers of networks. Taking x_i, y_i as input and f as output, the rule of ANFIS is expressed as Equation (10).

$$\begin{cases} \text{IF x is } A_1 \text{ and } B_1, \text{ then } f_1 = p_1 + q_1 + r_1 \\ \text{IF x is } A_2 \text{ and } B_2, \text{ then}hen\ f_2 = p_2 + q_2 + r_2 \end{cases} \tag{10}$$

where A_1 and B_1 is a fuzzy set; f_i is the output; and p_i, q_i, and r_i are adaptive parameters.

The first layer is the membership layer, which maps the input data to the corresponding fuzzy set through membership functions. The node function of this layer is expressed as Equation (11).

$$\begin{cases} O_i^1 = \mu_{Ai}(x), \quad i = 1,2 \\ O_i^1 = \mu_{Bi}(x), \quad i = 1,2 \end{cases} \tag{11}$$

where O_i^1 is the membership function, which is the degree of A_i, B_i's mapping to x, y.

The function of the second layer is to calculate the fitness of the input to each rule and determine the corresponding output and weight for each rule, which is expressed as Equation (12).

$$O_i^2 = w_i = \mu_{A_i}(x)\mu_{B_i}(y), \quad i = 1,2 \tag{12}$$

The third layer normalizes the fitness of each rule calculated in the second layer and calculates the credibility of the i-th rule for the i-th node, which is expressed as Equation (13).

$$O_i^3 = \overline{w}_i = \frac{w_i}{w_1 + w_2} \tag{13}$$

The fourth layer calculates the output of each rule, expressed as Equation (14).

$$O_i^4 = w_i f_i = w_i(a_i x + b_i y + c_i) \tag{14}$$

where the parameter set $\{a_i, b_i, c_i\}$ is the input parameters of the subsequent layer.

The fifth layer outputs the final calculation result, expressed as Equation (15).

$$O_i^5 = f = \sum \overline{w}_i f_i = \frac{\sum_i w_i f_i}{\sum_i w_i} \tag{15}$$

4. Simulation Analysis

4.1. Analysis of Characteristics of Active–Passive Training

Due to the fact that active training is based on the fixed resistance moment provided by the rehabilitation robot, the increase and decrease in resistance moment are determined by the rehabilitation robot. Therefore, this paper only analyzes the characteristics of the man–machine interaction moment during passive training.

Assuming an adult male with a weight of 65 Kg and a height of 175 cm, the simulation analysis is conducted based on his upper limb parameters. Set the training amplitude to 90 and the angular frequencies to 0.5 rad/s, 0.7 rad/s, and 0.9 rad/s for sinusoidal velocity signals and finally obtain the passive training man–machine interaction moment characteristic curve as shown in Figure 10. Here, Tb1, Tb2, and Tb3 correspond to the characteristics of the man–machine interaction moment at angular frequencies of 0.5 rad/s, 0.7 rad/s, and 0.9 rad/s, respectively.

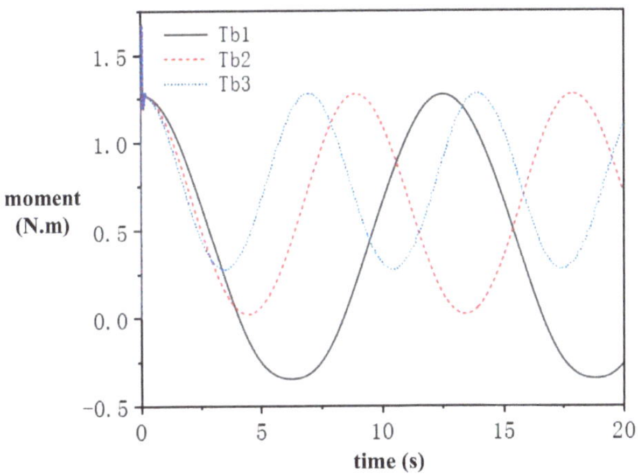

Figure 10. Characteristic curve of man–machine interaction moment under passive training.

According to Figure 10, the driving moment provided by the rehabilitation robot during passive training undergoes periodic changes. As the angular frequency of the sine speed signal increases, the minimum required driving moment will also increase. When the initial position of the manipulator arm is not at an angle of 90° to the horizontal reference line, the robot needs to overcome the heavy moment generated by the upper limbs and manipulator arm due to gravity. Therefore, during passive training, feedforward compensation for the heavy moment can improve the control accuracy of the controller, accurately determine the force exerted by the individual in rehabilitation during training, and accurately determine the individual in rehabilitation's training intention.

4.2. Simulation of ANFIS

According to the intention recognition schemes of active and passive training, the ANFIS system is of the dual-input single-output type. The inputs of the neural fuzzy controller during active training are the speed and speed change rate, while the inputs of the neural fuzzy controller during passive training are the moment and moment change rate. The outputs in both modes are 1, 2, and 3, corresponding to passive intention, active intention, and muscle spasms, respectively. We used the MATLAB-ANFIS editor to train the intention data during the rehabilitation process and then optimized the parameters of fuzzy rules and membership functions in two modes. This paper takes intention recognition in passive training mode as an example, and its neural fuzzy network structure is shown in Figure 11.

Passive training is performed based on the written control algorithm program. The 32 sets of training data are listed and shown in Table 1.

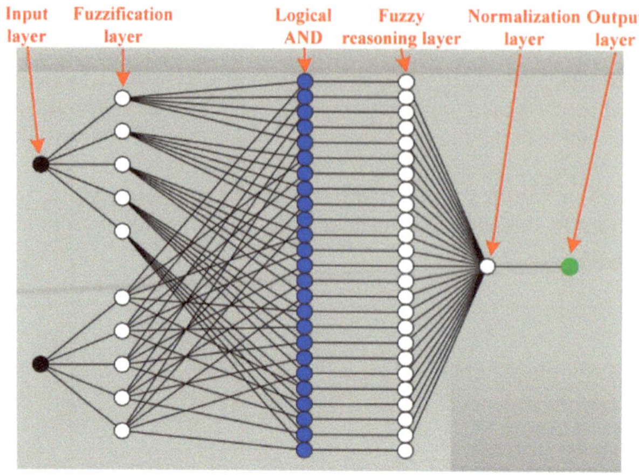

Figure 11. Neural fuzzy network structure.

Table 1. Passive training intention recognition data.

Serial Number	Moment (N·m)	Moment Change Rate	Intention	Serial Number	Moment (N·m)	Moment Change Rate	Intention
1	−2.89919	−0.18061	1	17	5.8574	2.73288	2
2	−2.90065	0.14449	1	18	7.17456	1.33904	2
3	−2.67257	2.39283	1	19	−1.98654	−6.70764	2
4	0.74798	1.78711	1	20	−2.21987	−6.11473	2
5	1.55429	0.69505	1	21	−3.79773	0.58544	2
6	1.74214	1.065	1	22	−6.1614	−2.37539	2
7	1.9033	0.9093	1	23	12.00933	14.67709	3
8	2.12509	−0.98777	1	24	13.39354	9.46294	3
9	1.91203	−1.38281	1	25	14.14827	2.49995	3
10	−1.21186	−1.51965	1	26	14.08117	−4.24755	3
11	−1.31405	−1.75009	1	27	13.35658	−8.25594	3
12	−6.79149	5.95313	2	28	−6.72165	−13.3704	3
13	−4.8158	8.45648	2	29	−8.57654	−12.7002	3
14	−0.97069	8.67197	2	30	−11.8130	−1.22569	3
15	1.5103	9.14033	2	31	−12.0104	0.35375	3
16	4.72376	4.05078	2	32	11.0025	8.51758	3

The intention recognition sample data during passive training were trained 100 times, and the output error after training was 0.142, as shown in Figure 12, which meets the rehabilitation training requirements.

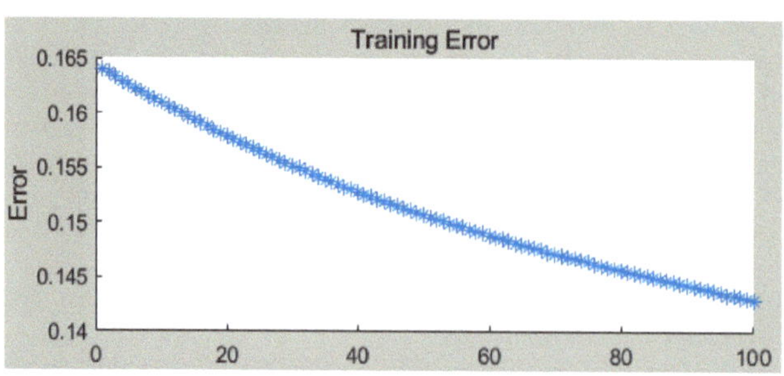

Figure 12. Neural network training error.

The membership functions under the active training mode optimized by the neural network are shown in Figure 13.

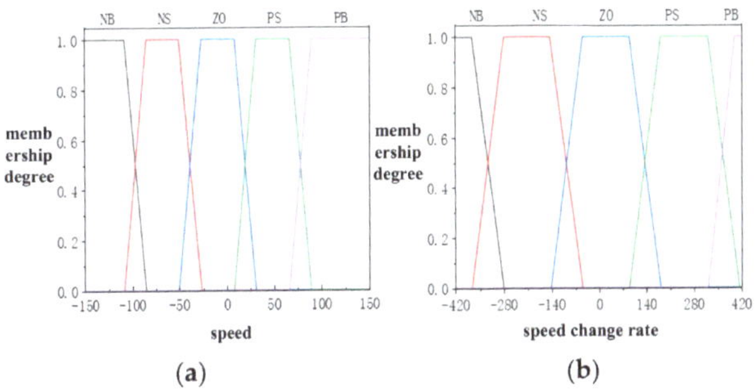

Figure 13. Membership function during active training: (**a**) speed membership function and (**b**) speed change rate membership function.

The membership functions under the passive training mode optimized by the neural network are shown in Figure 14.

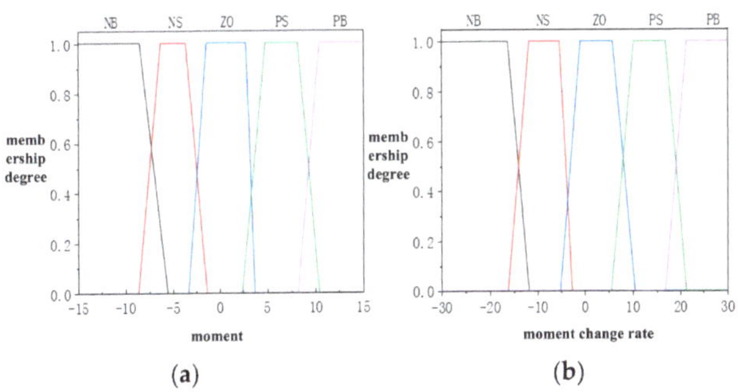

Figure 14. Membership function during passive training: (**a**) moment membership function and (**b**) moment change rate membership function.

By using a neural fuzzy network model to train the fuzzy rules and membership functions in the controller, a fuzzy intention recognition controller that is suitable for active and passive training can be obtained.

5. Results and Discussion

5.1. Equipment Selection and Platform Construction

The mechanical structure diagram of the upper limb rehabilitation robot is shown in Figure 15, and the independently developed experimental platform for the upper limb rehabilitation robot is shown in Figure 16.

The independently developed experimental platform for the upper limb rehabilitation robot is divided into hardware and software parts. The hardware part includes an industrial computer, Siemens 1TF7 permanent magnet synchronous servo motor, SIMOTION motion controller, SINAMICS S120 motor drive module, involute planetary gear reducer, German ETH DRDL series moment sensor, German Beifu EtherCAT data input and output module, coupling, and operating arm. The software part uses Twincat3.1 from Beifu for algorithm development and Siemens Scout for motor configuration. The rated torque (T) of the Siemens 1FT7 motor is 2 N·m, the rated speed (n) is 3000 rpm, the rated power (P) is 0.9 KW, and the rated current (I) is 2.1 A. The reduction ratio of the involute planetary gear reducer is 1:10. The DRDL series torque sensor from ETH Germany requires a 12 V DC voltage supply, with a torque measurement range of ±20 N·m, an output signal of ±10 V voltage signal, and a measurement accuracy of 0.2%. The acquisition rate of the Twincat3.1 oscilloscope is set as 10 ms. Both the speed control and torque control use the built-in PID of the Siemens controller, which can meet the control requirements and adapt to ensure better reproducibility.

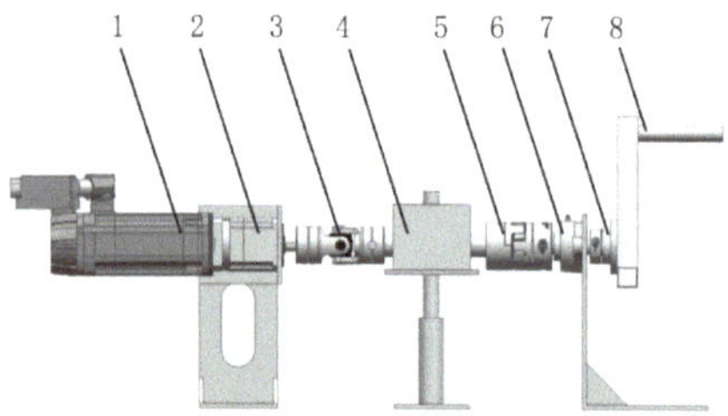

1.servo motor 2.planetary gear reducer
3.coupling 4.moment sensor 5.coupling
6.bearing seat 7.guide shaft support 8.operating arm

Figure 15. Mechanical structure diagram of the upper limb rehabilitation robot.

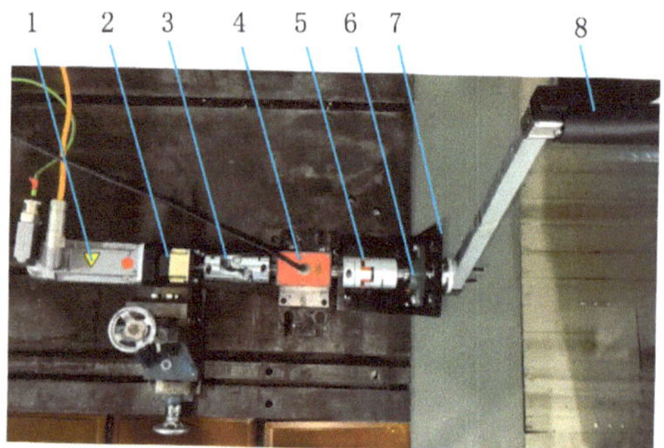

1.servo motor 2.planetary gear reducer
3.coupling 4.moment sensor 5.coupling
6.bearing seat 7.guide shaft support 8.operating arm

Figure 16. Independently developed upper limb rehabilitation robot experimental platform.

5.2. Upper Limb Rehabilitation Experiments

5.2.1. Active Training Experiment

When conducting active training, an individual in rehabilitation first drives the operating arm to move. The moment planner calculates the required output resistance moment value based on the real-time monitoring data of the training speed until the target resistance moment is reached. Active–passive training is suitable for the early and middle stages of rehabilitation training, when the muscle strength of the rehabilitation is still weak. We selected healthy adult males for our active–passive training experiments to test the control effectiveness of the active controller, passive controller, and intent recognition algorithm. The experimental process of the active–passive training is shown in Figure 17.

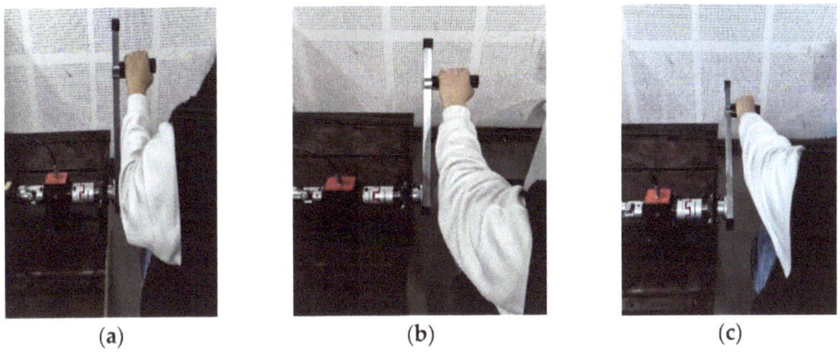

Figure 17. Active–passive training process: (a) contracted state, (b) intermediate state, (c) extended state.

Given that individuals in rehabilitation have varying perceptions of resistance during different training stages, the main objective of this experiment was to investigate the response speed and stable output performance of the motor moment under different resistance coefficient settings. In the experiment, the training resistance moment value was set to 5 N·m, and based on this, two levels of resistance coefficients of 0.25 and 0.45 were selected for testing. The values monitored by the moment sensor and speed sensor are shown in Figures 18 and 19. According to the figs, the response speed and stable output of the motor moment are perfect, which contributes to a good rehabilitation training effect.

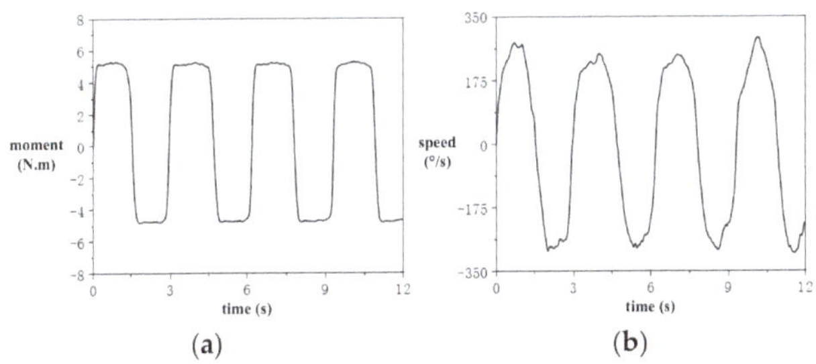

Figure 18. Training results (the training coefficient is set as 0.25): (**a**) training resistance moment and (**b**) training speed.

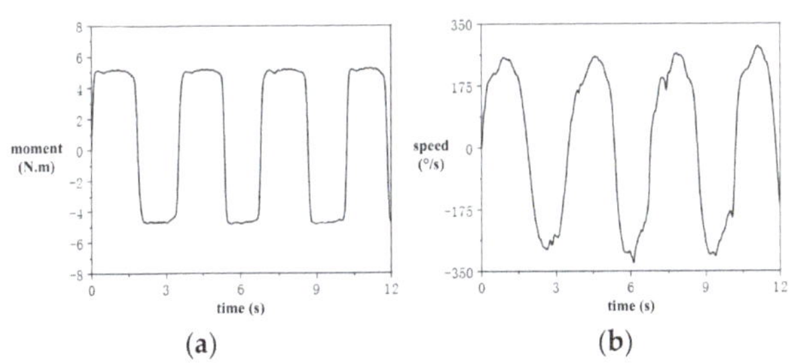

Figure 19. Training results (the training coefficient is set as 0.45): (**a**) training resistance moment and (**b**) training speed.

5.2.2. Passive Training Experiment

The analysis of the dynamic model of the upper limb man–machine integration shows that the load moment generated by individuals in rehabilitation at different training positions has a periodic variation characteristic, and this dynamic load has a significant impact on the stability of the passive training. The purpose of this experiment is to verify the variation law of the load moment in the actual training environment.

Figure 20 shows the moment curves of the passive training experiment conducted under the conditions of an amplitude of 90 and angular frequencies of 0.5 rad/s, 0.7 rad/s, and 0.9 rad/s. During the experiment, the individuals in rehabilitation kept their upper limbs relaxed without exerting any force. From the results, it can be seen that there is a certain difference between the load moment and the simulation results in the actual passive training process, but the overall pattern is similar to the simulation results. Therefore, taking the load moment as feedforward compensation is feasible.

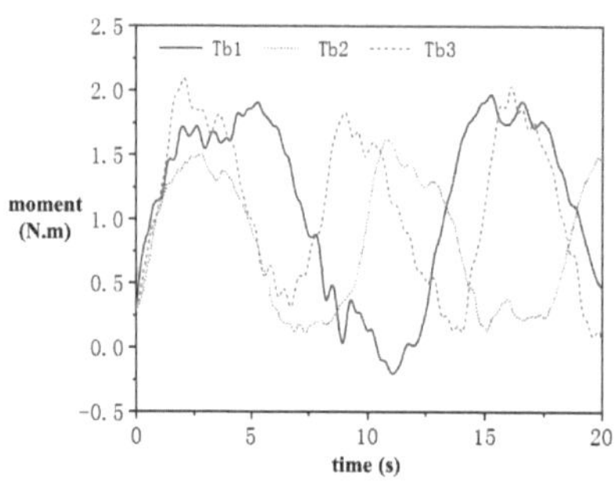

Figure 20. Passive training experimental data.

5.2.3. Training Intention Recognition Experiment

The occurrence of muscle spasms and unexpected movements during active training can lead to sudden changes in the speed, torque, and their rate of change, which can be used for judging muscle spasms and unexpected movements. In addition, unexpected movements can be prevented using software and mechanical hardware. In terms of software, if an unexpected movement occurs, the rehabilitation robot can be stopped immediately, and in terms of machinery, limit devices can be added. Due to the need for simultaneous validation of the training intention recognition and mode switching, this paper simulates active, passive, and spasm in both active and passive training modes. The collected data are then fed into a trained fuzzy intention recognizer to observe the accuracy of the pattern recognition.

Intention recognition is performed during active training. The intention recognition data shown in Figures 21–24 were obtained in active training mode, with resistance coefficients set as 0.2 and 0.5, respectively, and corresponding resistance moments set as 2 N·m and 5 N·m. In the experiment, active training, fatigue training, and spasms were simulated. According to the principle of maximum similarity, it can be seen that in both cases, the intention recognizer can accurately recognize the training intention. When simulating a state of fatigue, the recognition results may fluctuate due to the increased rate of speed change during commutation.

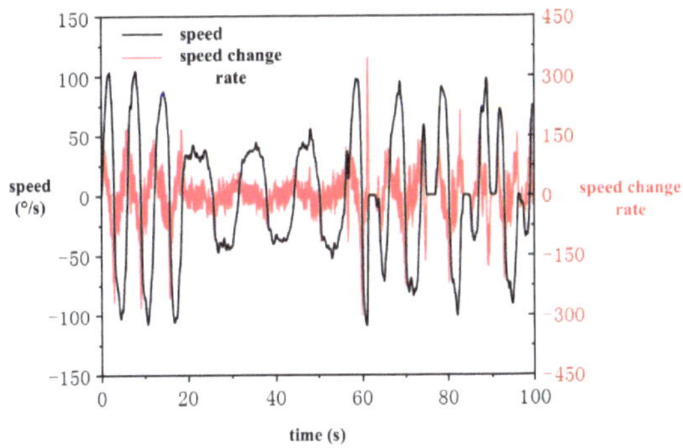

Figure 21. Experimental results of speed and speed change rate (C_t = 0.2, M_d = 2 N·m).

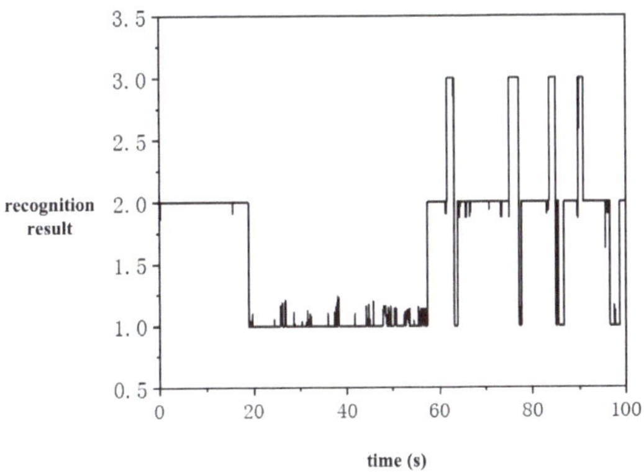

Figure 22. Intention recognition results (C_t = 0.2, M_d = 2 N·m).

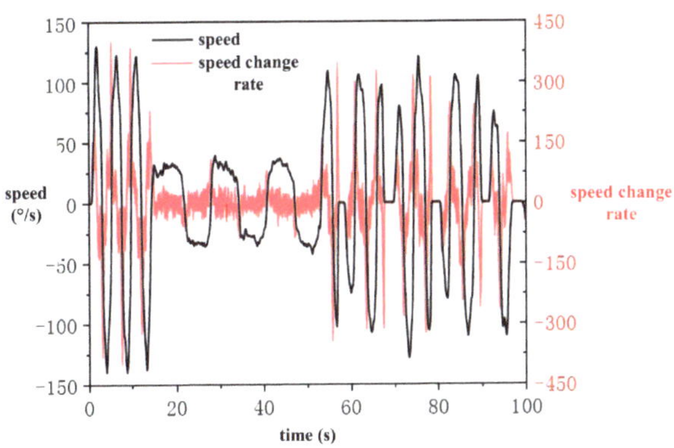

Figure 23. Experimental results of speed and speed change rate (C_t = 0.5, M_d = 5 N·m).

The intention recognition process shown in Figures 25–28 was performed in passive training mode with velocity patterns of 30sin ($\pi/5 \times$ t) and 60sin ($\pi/5 \times$ t), respectively. Firstly, passive training was performed for a period of time, then force was actively exerted during the passive training process, and finally, the occurrence of spasms was simulated during passive training. According to the principle of maximum similarity, it can be seen that in both cases, the intention recognizer can accurately recognize the individual in rehabilitation's movement intention. When the individual in rehabilitation actively applies force, the output of the active training intention is the number 2. If the recognition sensitivity is too high, it will be detrimental to training. When spasms occur, they can be quickly recognized, with an output intention result of the number 3. The faster the recognition of spasms occurs, the more effective it is at protecting the individual in rehabilitation.

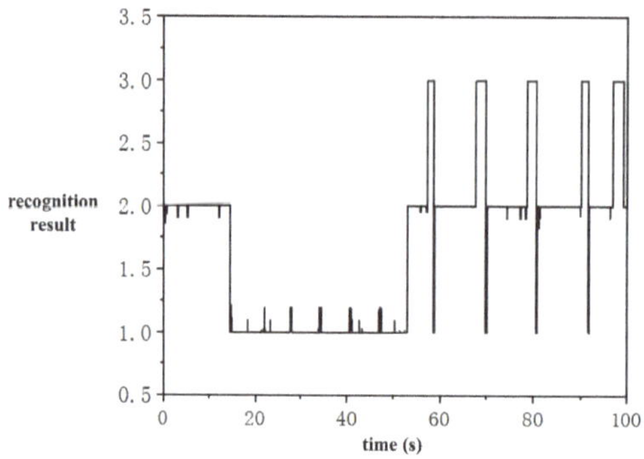

Figure 24. Intention recognition results (C_t = 0.5, M_d = 5 N·m).

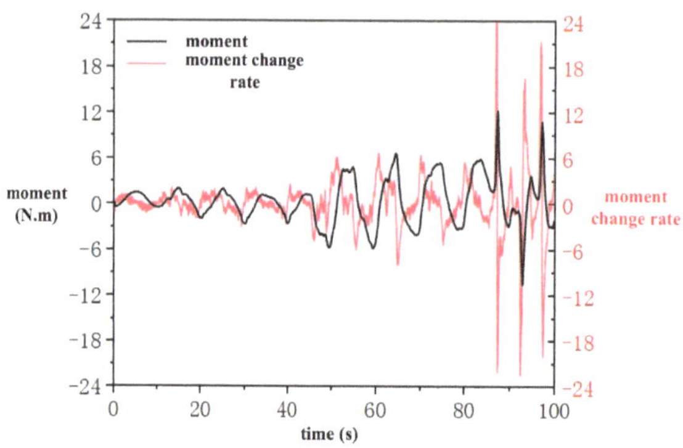

Figure 25. Experimental results of speed and speed change rate (C_t = 0.5, M_d = 5 N·m).

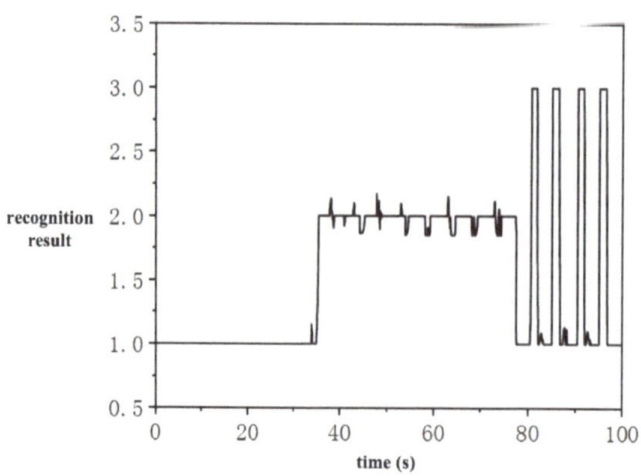

Figure 26. Experimental results of speed and speed change rate (C_t = 0.5, M_d = 5 N·m).

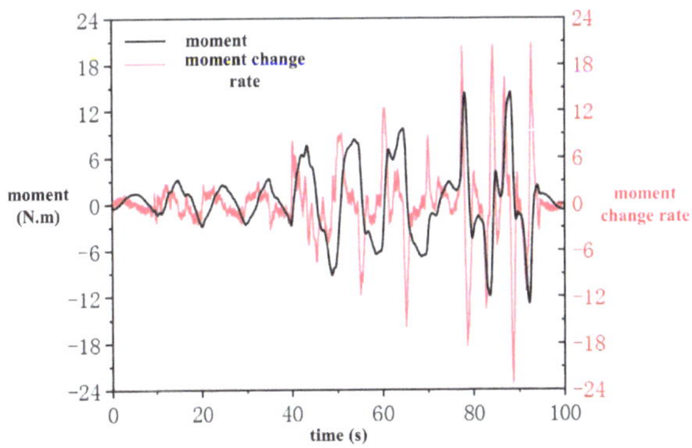

Figure 27. Experimental results of speed and speed change rate (C_t = 0.5, M_d = 5 N·m).

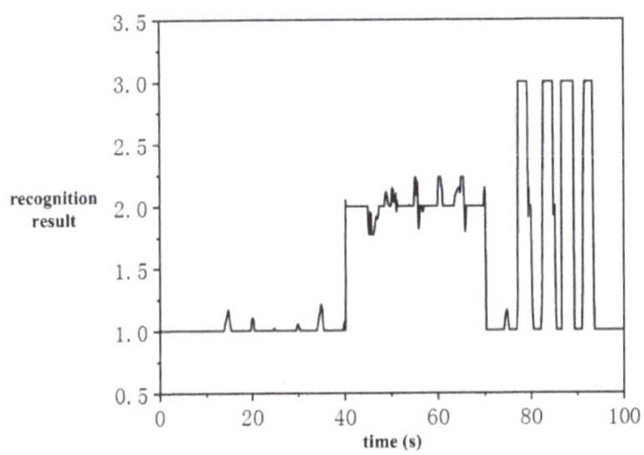

Figure 28. Experimental results of speed and speed change rate (C_t = 0.5, M_d = 5 N·m).

6. Conclusions and Outlook

This paper focuses on research on active–passive training control strategies for an upper limb rehabilitation robot. It establishes a man–machine interaction model for upper limbs and an overall control architecture for active and passive training and designs and optimizes the control algorithms. The rationality and effectiveness of the designed control algorithm are verified through experiments. The main conclusions are as follows:

(1) Based on the mechanism of upper limb rehabilitation movement, the upper limb training function was determined, and an upper limb man–machine integration model and upper limb man–machine integration dynamics model are constructed.

(2) According to the requirements of active–passive training, active and passive controllers based on feedforward compensation are designed. The moment output of the robot was also planned, and the output moment is correlated with the actual speed to ensure a smooth moment output. We design an active–passive switching controller that utilizes cosine function curves to plan and output the controlled moment and velocity, avoiding the shaking of the rehabilitation robot caused by sudden changes in control variables during training mode switching. A training intention recognizer based on an adaptive neural fuzzy algorithm is designed, and the collected rehabilitation training data are put into the neural network to obtain a fuzzy training intent recognizer that conforms to active and passive training, thereby improving the accuracy of judgment.

(3) By adopting the independently developed experimental platform for the upper limb rehabilitation robot, active–passive training, intention recognition, and training mode switching are achieved. The results show that the active and passive training processes are smooth, the training intention recognition is accurate, and the switching between active and passive training modes is steady. This verifies the feasibility and effectiveness of the

established mathematical model in upper limb rehabilitation training, thus enhancing the functionality, stability, and adaptability of rehabilitation robots to meet the functional needs of individuals in upper limb rehabilitation at different stages.

(4) In future research, we will consider discussing the topic of isokinetic muscle strength training control, which will not directly use torque control but adopt a torque × coefficient = speed scheme. This is safer to use for individuals in rehabilitation.

Funding: This research was funded by the Undergraduate Teaching Research and Reform Project of University of Shanghai for Science and Technology (JGXM24281, JGXM24263) and First Class Undergraduate Course Construction Project of University of Shanghai for Science and Technology (YLKC202424394).

Data Availability Statement: The original contributions presented in the study are included in the article, further inquiries can be directed to the author.

Conflicts of Interest: The author declares no conflicts of interest.

References

1. Gu, W.J. The effect of rehabilitation training on the motor function of stroke patients in the recovery period. *Guide China Med.* **2017**, *15*, 122–123. [CrossRef]
2. Yu, H.L. Rehabilitation Robots: Ten Future Prospects. *Chin. J. Rehabil. Med.* **2020**, *35*, 900–902.
3. Liang, F.Y.; Mo, L.F.; Sun, Y.Q.; Guo, C.; Gao, F.; Liao, W.H.; Cao, J.Y.; Li, B.B.; Song, Z.H.; Wang, D.; et al. Inter-limb and intra-limb synergy modeling for lower limb assistive devices: Modeling methods and feature selection. *Cyborg Bionic Syst.* **2024**, *5*, 0122. [CrossRef]
4. Yu, F.Y.; Liu, Y.; Wu, Z.X.; Tan, M.; Yu, J.Z. Adaptive gait training of a lower limb rehabilitation robot based on human–robot interaction force measurement. *Cyborg Bionic Syst.* **2024**, *5*, 0115. [CrossRef]
5. Hu, K.X.; Ma, Z.J.; Zou, S.L.; Li, J.; Ding, H.R. Impedance sliding-mode control based on stiffness scheduling for rehabilitation robot systems. *Cyborg Bionic Syst.* **2024**, *5*, 0099. [CrossRef]
6. Krebs, H.I.; Ferraro, M.; Buerger, S.P.; Miranda, J.N.; Antonio, M.; Michael, S.; Daniel, L.; Bruce, T.V.; Neville, H. Rehabilitation robotics: Pilot trial of a spatialextension for MIT-Manus. *J. Neuroeng. Rehabil.* **2004**, *1*, 5. [CrossRef]
7. Westlake, K.P.; Patten, C. Pilot study of lokomat versusmanual-assisted treadmill training for locomotor recoverypost-stroke. *J. Neuro Eng. Rehabil.* **2009**, *6*, 18. [CrossRef]
8. Shi, D.; Zhang, W.X.; Zhang, W.; Ding, X.L. A review on lower limb rehabilitation exoskeleton robots. *Chin. J. Mech. Eng.* **2019**, *32*, 74. [CrossRef]
9. Li, R.L. *Design of the Rehabilitation Robot with Human Upper and Lower Limbs'Active and Passive Motion*; Zhengzhou University: Zhengzhou, China, 2017.
10. Mason, M.T. Compliance and Force Control for Computer Controlled Manipula-tors. *IEEE Trans. Syst. Man Cybern.* **1981**, *11*, 418–432. [CrossRef]
11. Hogan, N. Impedance Control: An Approach to Manipulation: Part I—Theory. *J. Dyn. Syst. Meas. Control* **1985**, *107*, 1 7. [CrossRef]
12. Cao, J.H.; Xie, S.Q.; Das, R. MIMO Sliding Mode Controller for Gait Exoskeleton Drivenby Pneumatic Muscles. *IEEE Trans. Control. Syst. Technol.* **2018**, *26*, 274–281. [CrossRef]
13. Zhang, Y.K. *Design and Control of Position and Posture Maintaining Rehabilitation Training Robot for Human Upper Limbs*; Beijing Jiaotong University: Beijing, China, 2022.
14. Narayan, J.; Kalita, B.; Dwivedy, S.K. Development of robot-based upper limb devices for rehabilitation purposes: A systematic review. *Augment. Hum. Res.* **2021**, *6*, 4. [CrossRef]
15. Fu, R.R.; Liang, H.F.; Mi, R.F. Feature Identification of EEG Preparatory Response Signals Evoked by Motor Intention. *Acta Metrol. Sin.* **2023**, *44*, 1597–1601. [CrossRef]
16. Zhai, M.Q. *Research on Recognition Method of Human Lower Exremity Movement Intention Based on sEMG Signal*; Chongqing University: Chongqing, China, 2021.
17. Fan, Y.J. *Study on Lower Limb Exoskeleton for Rehabilitation Based on Multi-Source Information Fusion Including sEMG & Interactive Force and Its Clinical Trail*; Shanghai Jiaotong University: Shanghai, China, 2014.
18. Farina, D.; Jiang, N.; Rehbaum, H.; Holobar, A.; Graimann, B.; Dietl, H.; Aszmann, O.C. The extraction of neural information from the surface EMG for the control of upper-limb prostheses: Emergingavenues and challenge. *IEEE Trans. Neural Syst. Rehabil. Eng.* **2014**, *22*, 797–809. [CrossRef] [PubMed]
19. Liu, Z.J.; Lin, W.; Geng, Y.L.; Yang, P. Intent Pattern Recognition of Lower-limb Motion Based on Mechanical Sensors. *IEEE/CAA J. Autom. Sin.* **2017**, *4*, 651–660. [CrossRef]

20. Li, G.N.; Tao, L.; Meng, J.Y.; Ye, S.J.; Feng, G.; Zhao, D.Z.; Hu, Y.; Tang, M.; Song, T.; Fu, R.Z.; et al. Research on mode adjustment control strategy of upper limb rehabilitation robot based on fuzzy recognition of interaction force. *J. Biomed. Eng.* **2024**, *41*, 90–97. [CrossRef]
21. Tran, H.T.; Cheng, H.; Lin, X.C.; Duong, M.K.; Huang, R. The relationship between physical human-exoskeleton interaction and dynamic factors: Using a learning approach for control applications. *Sci. China Inf. Sci.* **2014**, *57*, 5–17. [CrossRef]
22. Wang, C.X.; Yang, X.J.; Xu, Q.; Ma, Z.C.; Sun, Y.N. Motion Recognition System for Upper Limbs Based on 3D Acceleration Sensors. *Chin. J. Sens. Actuators* **2010**, *23*, 816–819.
23. Li, Z.N.; Wang, T.; Wang, B.R.; Guo, Z.W.; Chen, D.J. Trajectory planning for manipulator in Cartesian space based on constrained S-curve velocity. *CAAI Trans. Intell. Syst.* **2019**, *14*, 655–661. [CrossRef]
24. Hong, H.C.; Chen, M.J.; Kang, J.X.; Xu, H.T.; Lin, H.J.; Xu, Z.Q.; Sun, H.J.; Zhou, X.L. Prediction of trihalomethane levels in tap water based on adaptive networkbased fuzzy inference system and simple water quality parameters. *Acta Sci. Circumstantiae* **2023**, *43*, 290–299. [CrossRef]
25. Mahmoodabadi, M.; Arshad, R.R. Long-term evaluation of water quality parameters of the Karoun River using a regression approach and the adaptive neuro fuzzy inference system. *Mar. Pollut. Bull.* **2018**, *126*, 372–380. [CrossRef] [PubMed]

MDPI AG
Grosspeteranlage 5
4052 Basel
Switzerland
Tel.: +41 61 683 77 34

Machines Editorial Office
E-mail: machines@mdpi.com
www.mdpi.com/journal/machines

www.ingramcontent.com/pod-product-compliance
Lightning Source LLC
Chambersburg PA
CBHW041406210926
51850CB00013B/3187